Transforming the Rural Asian Economy: The Unfinished Revolution

by

Mark W. Rosegrant
and
Peter B. R. Hazell

OXFORD
UNIVERSITY PRESS

Oxford University Press is a department of the University of Oxford.
It furthers the University's objective of excellence in research, scholarship,
and education by publishing worldwide in

Oxford New York

Athens Auckland Bangkok Bogotá Buenos Aires Calcutta
Cape Town Chennai Dar es Salaam Delhi Florence Hong Kong Istanbul
Karachi Kuala Lumpur Madrid Melbourne Mexico City Mumbai
Nairobi Paris São Paulo Singapore Taipei Tokyo Toronto Warsaw

with associated companies in Berlin Ibadan

Oxford is a registered trade mark of Oxford University Press

Published in the United States by Oxford University Press Inc., New York

© Asian Development Bank 2000

First published 2000
This impression (lowest digit)
1 3 5 7 9 10 8 6 4 2

All rights reserved. No part of this publication may be reproduced,
stored in a retrieval system, or transmitted, in any form or by any means,
without the prior permission in writing of the Asian Development Bank.
Exceptions are allowed in respect of any fair dealing for the purpose
of research or private study, or criticism or review.
Enquiries concerning reproduction should be sent to
the Asian Development Bank.

This book is sold subject to the condition that it shall not, by way
of trade or otherwise, be lent, re-sold, hired out or otherwise circulated
without the publisher's prior consent in any form of binding or cover
other than that in which it is published and without a similar condition
including this condition being imposed on the subsequent purchaser.

Published for the Asian Development Bank by
Oxford University Press

British Library Cataloguing in Publication Data
available

Library of Congress Cataloging-in-Publication Data
available

ISBN 0 19 592448 7 (Paperback)
ISBN 0 19 592447 9 (Hardback)

Printed in Hong Kong
Published by Oxford University Press (China) Ltd
18th Floor, Warwick House East, Taikoo Place, 979 King's Road, Quarry Bay
Hong Kong

Contents

Foreword xiii
Preface xvii

Introduction xix

Chapter I **Agricultural Growth and the
            Economic Transformation** 1

    Introduction 1
    Sources of Economic Growth in Asia 5
    Explaining Economic Growth 9
        Prudent Macroeconomic Policies 11
        Market-oriented Policies 11
        Open Trade Policies 12
        Industrial Policy 13
        Investment in Education 15
        Institutional Issues 16
    Importance of Agriculture in Promoting the
       Transformation 18
    Changing Role of the Agricultural Sector 26
    Conclusions 27

Chapter II **Growth and Policies for
            Poverty Reduction** 29

    Introduction 29
    Trends in Total Poverty 30
    Trends in Rural Poverty 33
    Growth and Poverty Reduction in Asia 36

Beyond Growth: Policies for Poverty Reduction 40
    Education 41
    Health and Nutrition 43
    Rural Infrastructure 44
    Land Reform 46
    Safety-Net Programs 48
Conclusions 53

Chapter III Agricultural Diversification and Commercialization 57

Introduction 57
Structural Change in Agriculture 57
Commercialization and Diversification 66
    Agricultural Research 66
    Economic Liberalization 67
    Rural Financial Markets 68
    Rural Infrastructure 72
    Property Rights 73
Agricultural Commercialization and Diversification
and the Poor 73
    Impact of Technological Change 74
    Impact of Commercialization 74
    Regional Disparities 77
Conclusion 78

Chapter IV The Rural Nonfarm Transformation 81

Introduction 81
Importance of Rural Nonfarm Activities 82
    Definitions 82
    Employment Shares 82
    Income Shares 85
    Importance to Women 87
    Importance to the Poor 89
Structural Transformation of the Rural
Nonfarm Economy 92

Overview 92
Changes in the Composition of the Rural
    Nonfarm Economy 93
Changes in the Spatial Structure of
    Rural Economies 96
Changes in Local Labor and
    Capital Markets 100
Impact on the Poor 103
Determinants of the Transformation 104
Agricultural Growth 106
Population Density 110
Rural Towns 111
Rural Infrastructure 112
Macro and Trade Policies 114
Rural Industrialization Policies 115
Conclusions 119

Chapter V Sources of Agricultural Growth 123

Introduction 123
Agricultural Output and Input Growth 124
Input Use Trends 125
Land and Labor Productivity Growth 134
Patterns of Public Spending on Agriculture 139
Size and Trends in Public Expenditure 140
Public Expenditures on Subsidies
    and Investments 144
Sources of Growth in Asian Agriculture 147
Conclusions 160

Chapter VI The Evolution of Cereal and Livestock Supply and Demand: Policies to Meet New Challenges 161

Introduction 161
Evolution of Cereal and Livestock Demand 162
Trends in Cereal Demand 163
Growth in Demand for Livestock Products 165

Evolution of Cereal Production: Trends, Challenges,
and Policies 167
    Challenges in Sustaining Cereal Production 169
    Policies for the Crop Sector 172
Evolution of Livestock Production : Trends, Challenges,
and Policies 178
    Challenges for Increasing Livestock
       Production 180
    Policies for the Livestock Sector 185
Conclusions 189

Chapter VII Impact of Trade, Macroeconomic, and Price Policy on Agriculture 191

Introduction 191
Historical Patterns of Agricultural Protection 193
Trade, Macroeconomic, and Price Policies in Asia 196
GATT and WTO 203
Regional Trade Agreements 209
Structural Changes in International
    Agricultural Trade 213
    Policy Reforms in Developed-Country
       Agriculture 214
    Changes in the Nature of
       Agricultural Trade 214
    International Price Variability 215
Conclusions 217

Chapter VIII Economic Reform in Asia Transition Economies 219

Introduction 219
Initial Conditions and Exogenous Shocks 222
Agricultural Reform 229
State Enterprise Reform 237
Trade, Macroeconomic, and Stabilization Policies 242
Provision of Social Protection During Transition 246
Environmental Protection During Transition 249

Lessons and Future Directions For Reform 250

Chapter IX The Financial and Economic Crisis in East and Southeast Asia 255

Introduction 255
Causes of the Crisis: Two Views 256
    Development of the Crisis 257
    Real Appreciation of Currencies 258
    Large Current-Account Deficits 259
    Large Short-term Capital Flows 259
    Overborrowing and Overlending in
        the Financial Sector 261
    The Onset of the Crisis 263
Impact of the Crisis 266
    Social Impact of the Crisis 266
    Declining Incomes 266
    Rising Absolute Poverty 268
    Increases in Malnutrition 269
    Increased Pressure on Rural Areas 269
    Declining Social Services—Threats to
        Education and Health 270
    Impact on the Agricultural Sector 272
Lessons for Policy 275
    Social Services and Safety Nets 275
    Good Governance 278
    Financial and Corporate Management 280
    International Capital Flows 281
    International Recovery Programs 283
Conclusions 287

Chapter X Environmental and Resource Challenges to Future Growth 289

Introduction 289
Cropland—Will the Base for Food Supply Shrink
    or Expand? 290
    Potential for Cropland Expansion 290
    Loss of Cropland to Urbanization 291

Net Expansion/Loss of Cropland 293
Water Resources—Does Scarcity Endanger
Future Food Supply? 294
Energy—Plentiful Enough to Support a
Growing Food Supply? 297
Land Degradation—What Extent and
Productivity Effects? 299
    Extent of Land Degradation 300
    Productivity Effects of Land Degradation 300
Improved Crop Productivity—The Key to a
Secure Food Supply? 305
    Fertilizer—Raising Yields or
        Damaging the Environment? 307
    Plant Genetic Resources—A
        Narrowed Base? 310
    Biotechnology—Higher Yield Ceilings? 312
Climate Change—Different Crop Productivity Effects
for Developing vs. Developed World............................ 315
Conclusions 319

Chapter XI Challenges for Less Favored Areas 323

Introduction 323
Extent of Poverty and Environmental Degradation in
Less Favored Areas 324
    Poverty and Productivity in the LFAs:
        Evidence of Links 326
    Environmental Degradation in the LFAs:
        Outcome of Low Productivity and
        Expansion of Cropped Area 327
Returns to Public Investments in the LFAs 328
    Returns on Investments in High-Potential
        Areas: New Realities Challenge
        Conventional Wisdom 329
    Productivity and Poverty Impact of
        Investment in LFAs: Emerging Evidence 330

Strategies for Developing LFAs:
    Recognizing Commonality Amid Diversity 333
        Promote Broad-Based Agricultural
            Development 335
        Improve Technologies and Farming
            Systems 335
        Ensure Equitable and Secure Access to
            Natural Resources 336
        Ensure That Risks Are Managed Effectively 339
        Invest in Rural Infrastructure and People 340
        Provide the Right Policy Environment 341
        Strengthen Public Institutions 342
Conclusions 343

Chapter XII Alternative Futures for Asian Agriculture and Food Security 345

Introduction 345
Global Trends—A Prominent Role for Asian
    Developing Economies 346
Alternative Scenarios for Agriculture
    in Developing Asia: Weak vs. Strong
    Investment/Policy Reform Strategies 348
        Low Investment and Weak Policy Reform 349
        High Investment and Strong Policy Reform 351
Alternative Scenarios: Impact on Rural
    Asian Food Markets and Food Security In 2010 353
        Scenario Effects on Supply, Demand, and
            Net Trade 353
        Scenario Effects on Food Security:
            Childhood Malnutrition 366
        Can Childhood Malnutrition in Asia Be
            Eliminated by 2010? 371
Conclusions 374

Chapter XIII Lessons Leaned 377

    Introduction 377
    Agricultural Growth: An Engine for Asia's
        Economic Development 378
    Making Growth Pro-Poor: A Key to Poverty
        Reduction in Asia 381
    Agricultural and Rural Economic Growth:
        Markets Enhanced by Public Policies 382
                Agricultural Research and Extension 383
                Investment in Rural Infrastructure 385
                Property Rights 386
                Economic Liberalization 389
                Rural Financial Markets 391
                Investment in Education, Health, and
                      Nutrition 393
    Completing the Transformation: Reversing
        Environmental Degradation 395
    Completing the Transformation: Reaching Less
        Favored Lands 400
    Extending the Transformation: A New
        Understanding of the Role of Governance 401
    Extending the Transformation: Managing a
        New Revolution in Agricultural Technology 403
    Extending the Transformation: Managing
        Globalization 407
    Alternative Futures: No Room for Complacency 409
    Conclusion 411

References 453

Appendix 453

    Asia In The Global Agricultural Economy 453
    IFPRI's IMPACT Global Food Model : Model
        Characteristics 454
        Projections for Cereals 456
    Projections for Livestock Products 472

Projections for World Prices and
  International Trade 479
Projections for Malnutrition Among
  Pre-School Children 483

Indexes

Author Index 486
Title Index 493

Foreword

An economic transformation has occurred in much of rural Asia since the Asian Development Bank (ADB) last undertook a survey of the region in 1976. The rural economy has become increasingly linked to a rapidly integrating world economy and rural society in Asia faces new opportunities and challenges.

The transformation of rural Asia has also been accompanied by some troubling developments. While large parts of the region have prospered, Asia remains home to the majority of the world's poor. Growing inequalities and rising expectations in many parts of rural Asia have increased the urgency of tackling the problems of rural poverty. The rapid exploitation of natural resources is threatening the sustainability of the drive for higher productivity and incomes in some parts of rural Asia and is, in general, affecting the quality of life in the entire region.

These developments have altered the concept of rural development to encompass concerns that go well beyond improvements in growth, income, and output. The concerns include an assessment of changes in the quality of life, broadly defined to include improvements in health and nutrition, education, environmentally safe living conditions, and reduction in gender and income inequalities. At the same time, the policy environment has changed dramatically. Thus, there has arisen a need to identify ways in which governments, the development community at large, and the ADB in particular, can offer more effective financial and policy support for Asian rural development in the new century.

Therefore, the ADB decided to undertake a study to examine the achievements and prospects of rural Asia and to provide a vision for the future of agriculture and rural development in Asia into the next century. The objective of the

Study was to identify, for the ADB's developing member countries in Asia, policy and investment priorities that will promote sustainable development and improve economic and social conditions in the rural sector.

The Study was designed as a team effort, using ADB Staff and international experts under the guidance of an ADB interdepartmental steering committee. To address the diverse issues satisfactorily and in a comprehensive manner, five thematic subject areas were identified to provide the analytical and empirical background on which the Study's recommendations would be based. Working groups comprising ADB staff were set up to define broadly the scope and coverage of each of the themes. The five working groups acted as counterparts to international experts recruited to prepare the background reports, providing guidance to the experts and reviewing their work to ensure high quality output.

A panel of external advisers from the international research community was constituted to review and comment on the approach and methodology of the study and the terms of reference for each of these background reports. The external advisers also reviewed the drafts of the reports. In addition, external reviewers, prominent members of academe and senior policymakers, were appointed to review each of the background reports and to provide expert guidance.

The preparation of the background reports included four workshops held at the ADB's headquarters in Manila: an inception workshop in May 1998; two interim workshops, in November 1998 and January 1999, respectively, to review progress; and a final workshop in March 1999, at which the background reports were presented by their authors to a large group of participants comprising senior policymakers from the ADB's developing member countries, international organizations, international and locally based nongovernment organizations, donor agencies, members of academe, and ADB staff.

The five background reports, of which this volume is one, have now been published by Oxford University Press. The titles and authors of the other volumes are:

The Growth and Sustainability of Agriculture in Asia
Mingsarn Santikarn Kaosa-ard and Benjavan Rerkasem, with contributions by Shelley Grasty, Apichart Kaosa-ard, Sunil S. Pednekar, Kanok Rerkasem, and Paul Auger

Rural Financial Markets in Asia: Policies, Paradigms, and Performance
Richard L. Meyer and Geetha Nagarajan

The Quality of Life in Rural Asia
David Bloom, Patricia Craig, and Pia Malaney

The Evolving Roles of State, Private, and Local Actors in Rural Asia
Ammar Siamwalla, with contributions by Alex Brillantes, Somsak Chunharas, Colin MacAndrews, Andrew MacIntyre, and Frederick Roche

The results and recommendations from the Study were presented at a seminar during the ADB 32nd Annual Meeting in Manila. These have since been published by the ADB as a book titled *Rural Asia: Beyond the Green Revolution*.

The findings from the Study will provide a basis for future discussion between the ADB and its developing member countries on ways to eradicate poverty and improve the quality of life in rural Asia. The volumes in this series should prove useful to all those concerned with improving the economic and social conditions of rural populations in Asia through sustainable development.

TADAO CHINO
President
Asian Development Bank

Preface

In addition to the generous financial support from the Asian Development Bank, the book would not have been possible without the guidance and assistance of many individuals. At the Asian Development Bank, Yang Weimin, Akira Seki, Hans-Juergen Springer and M. Tusneem provided overall guidance to the project. Bradford Philips and Shahid Zahid capably managed the project against a tight time frame. The ADB staff working group for our book provided innumerable constructive comments on several drafts of the manuscript. The working group was led initially by Sultan H. Rahman, and later by Shahid Zahid. Members of the group included David Edwards, Richard Vokes, Carl Amerling, Xianbin Yao, Hua Du, and Mandar Jayawant. Many other staff of the ADB also provided important inputs, through discussions, participation in a series of workshops, and provision of valuable papers and documents.

The book also greatly benefited from the insightful comments and intense interaction with the external reviewers, Walter P. Falcon and Saeed Ahmed Qureshi, and the Senior Advisory Team commissioned by ADB, in particular Klaus Lampe, Justin Lin, Obaidullah Khan, and Meryl Williams. The other members of the consultant team, Richard Meyer, Ammar Siamwalla, Mingsarn Santikarn Kaosa-ard, and David Bloom also provided important insights throughout the project. Finally, several colleagues, including Shenggen Fan, Francesco Goletti, Raisuddin Ahmed, and Randolph Barker provided helpful comments on parts of the manuscript.

Strong logistical and administrative support was provided by the Rural Asia support team of the ADB, including Elizabeth Tan (who also assisted in data collection and synthesis), Laura Britt, Lilibeth Perez, and Dang Nada. Sara Collins Medina did an excellent job in editing the book, saving us from many errors.

At IFPRI, Claudia Ringler and Julie Witcover provided invaluable research support, including literature review, data collection and analysis, drafting of background material, preparation of tables and figures, and technical editing. Beverly Abreu was responsible for final word processing of the first draft of the full manuscript, and Denise Dixon and Beryl Hackett-Perez handled additional word processing and editorial details.

Mark W. Rosegrant

and Peter B. R. Hazell

Introduction

Much of Asia was in a desperate situation in the early 1960s. Agriculture was the major source of income and employment for most of the population, but its productivity was low and stagnant. At the same time, populations were growing rapidly, leading to accelerating demand for food and a rapidly increasing number of rural workers to employ. Many countries faced critical and increasing food-security problems and the doubly difficult challenge of trying to absorb an increasing labor force while simultaneously increasing labor productivity. The food crisis in India during the mid-1960s was indicative of a worrying situation that existed more broadly all across Asia and of the increasing dependence of many Asian countries on food aid for their survival.

Instead of plunging into a Malthusian spiral of rapid population growth and famine, however, the Asian rural and general economy underwent a dramatic transformation over the past 30 years; the speed and level of agricultural, rural, and overall economic development achieved in most Asian countries far exceeded expectations. In South Asia, famine was averted and more: cereal production rose by 92 percent using only 4 percent more land during the almost three decades from 1969/71 to 1994/96, boosting total per capita food supply from 2,060 to 2,380 calories per day. Gains in East and Southeast Asia combined were likewise impressive. There, cereal production almost doubled between 1969/71 and 1994/96 while using 22 percent more land, pushing total per capita calorie availability up from 2,063 to 2,662 calories per day. Across the region, annual incomes per person rose several-fold from 1970 to 1995: in the People's Republic of China (PRC) from $91 to $473, in Indonesia from $207 to $706, and in India from $241 to $439 (all figures in 1987 real dollars). These strong upward trends in both incomes and agricultural production have contributed to a rapid reduction in poverty despite increasing population pressure: in 1975, six out of ten Asians lived in poverty, but by 1993, only two out of ten East Asians and four out of ten South Asians did.

Despite these remarkable successes, Asia continues to face major challenges. A devastating economic crisis struck East and Southeast Asia in 1997, raising doubts throughout Asia about the sustainability of rapid growth. Some negative environmental effects of past growth have emerged and have threatened to at least slow the pace of further growth. There have been growing disparities between countries and between regions within countries. Asian economies became more diverse, with rapid development in East and Southeast Asia (except for the Philippines), moderate growth in South Asia, and stagnation in many of the centrally planned (later transitional) economies prior to market reforms. In many cases there were also widening disparities in income and access to services and education across demographic groups. These disparities are particularly prevalent in South Asia and in the Philippines. And, while the green revolution that imported high-yield varieties of cereal crops and other high-technology inputs benefited irrigated areas throughout Asia, many rainfed areas were left behind.

This book describes the dramatic, but unfinished, transformation of the rural Asian economy, with an emphasis on public policies and strategies and their impact on agricultural and economic growth, poverty, and the environment. The book can be divided conceptually into three main sections. The first, comprised of Chapters 1–6, assesses in detail the processes of rural economic and agricultural growth. It examines the determinants of agricultural growth and economic transformation (Chapter 1), how this transformation has impacted upon Asia's rural poor (Chapter 2), the process of agricultural diversification and commercialization (Chapter 3), the rural nonfarm transformation (Chapter 4), sources of agricultural growth (Chapter 5), and the evolution of cereal and livestock supply and demand (Chapter 6).

In the second section, broader economy-wide issues and policies and how they affect agriculture and the rural economy are the focus. Chapter 7 examines the impacts of trade, macroeconomic policy, and price policy on agriculture. Chapter 8 assesses the process and impacts of economic reform in Asian economies undergoing the transition from central planning to market orientation. In Chapter 9, the financial and economic crisis in East and Southeast Asia that was ignited in 1997 is evaluated.

The third section focuses on priority challenges and alternative futures, including environmental and resource challenges to future growth (Chapter 10), the challenges facing less favored areas (Chapter 11), and alternative futures for agriculture and food security in Asia (Chapter 12). The concluding chapter summarizes lessons learned from the process of rural transformation.

I Agricultural Growth and the Economic Transformation

INTRODUCTION

Recent studies of the rapid economic transformation in East and Southeast Asia attribute the changes to pro-growth macroeconomic and trade policies, including low trade barriers, liberalized and competitive domestic markets, high domestic savings rates complemented by private capital inflows, and critical investments in infrastructure and education (see, for example, Collins and Bosworth 1997; Rohwer 1995; World Bank 1993a). These studies provide important insights into the sources of economic growth in Asia, but with few exceptions, the role of agriculture and the rural economy in the transformation is neglected. Thus, there is a missing link in the understanding of the preconditions necessary for successful economic development strategies. In this chapter, therefore, the economic growth performance in Asian countries in recent decades is summarized, the sources of economic growth are examined, and the crucial role of growth in agriculture and the rural economy as a prerequisite to the launch of broad-based economic growth is highlighted.

Many of the Asian developing economies had spectacular growth over the past three decades, and particularly after 1980, until the financial and economic crisis beginning in 1997 temporarily disrupted this growth (the causes and implications of the East and Southeast Asian crisis are explored in detail in Chapter IX). Figure I.1 and Table I.1 show the growth in gross domestic product (GDP) of the 13 largest Asian developing

Figure I.1: Growth in GDP, 1967-95 (percent per year)

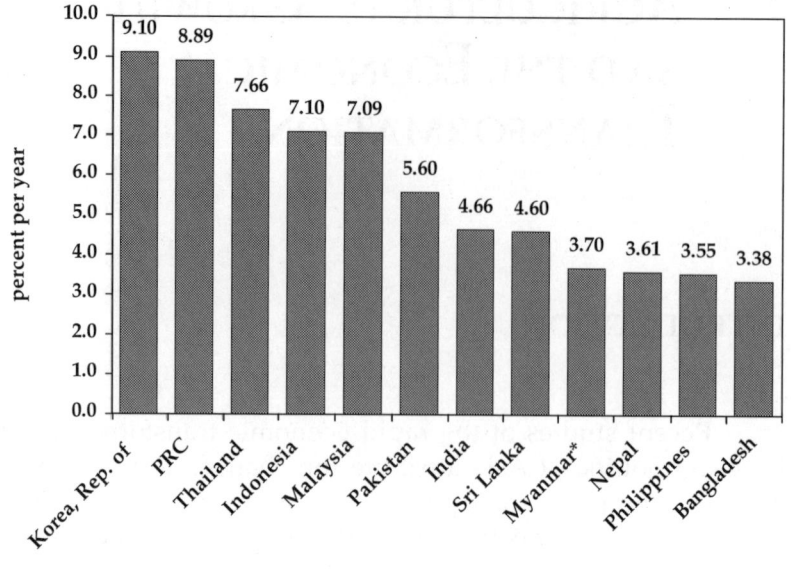

Country	Percent per year
Korea, Rep. of	9.10
PRC	8.89
Thailand	7.66
Indonesia	7.10
Malaysia	7.09
Pakistan	5.60
India	4.66
Sri Lanka	4.60
Myanmar*	3.70
Nepal	3.61
Philippines	3.55
Bangladesh	3.38

Note: Growth rates are 3-year centered moving averages.
Source: WDI 1998.

Table I.1: Total GDP in million US dollars (constant 1987)

Year	1970	1995	1967-80	1980-89	1989-95	1980-95
Bangladesh	10,784	24,165	2.30	4.32	4.33	4.32
PRC	75,472	577,486	7.35	9.84	10.87	10.25
India	133,961	407,880	3.58	5.84	5.26	5.61
Indonesia	24,851	139,321	7.42	6.08	7.95	6.83
Korea, Rep. of	30,870	254,426	10.04	8.76	7.59	8.29
Malaysia	10,865	62,655	7.40	5.53	8.80	6.82
Myanmar*	36,087	87,904	4.69	1.08	5.56	2.85
Nepal	1,784	4,463	2.41	4.35	5.13	4.66
Pakistan	13,527	49,810	5.35	6.54	4.75	5.82
Philippines	18,577	43,162	5.48	1.47	2.58	1.91
Sri Lanka	3,093	9,275	4.83	4.38	4.46	4.41
Thailand	17,420	107,921	7.24	7.48	8.86	8.03
Viet Nam	–	59,947	–	–	7.82	–

Note: For Myanmar, constant GDP values are only given in the local currency unit.
Source: WDI 1998.

economies (Viet Nam is not shown in Fig. I.1). Based on economic growth rates, the countries can be divided into three groups. The first is the group of very rapidly growing East and Southeast Asian countries, led by the People's Republic of China (PRC), with a growth rate of over 10 percent per year after 1980. Indonesia, the Republic of Korea, Malaysia, and Thailand grew at rates of 6.8–8.3 percent per year during 1980–95 and except for Korea, each of these countries grew significantly faster during the latter part of this period. Although a full time series on GDP is not available, Viet Nam also belongs to this group, at least for the 1989–95 period, with an annual rate of growth of 7.8 percent. The second group is the South Asian countries, which achieved moderate economic growth compared to East Asia, but still performed strongly compared to developing countries outside Asia. During 1980–95, Pakistan grew at 5.8 percent per year, India at 5.6 percent, and Bangladesh, Nepal, and Sri Lanka clustered at 4.3–4.7 percent per year. Finally, Myanmar and the Philippines had slow growth, 2.9 percent and 1.9 percent per year, respectively.

Similar trends can be observed with respect to annual growth of per capita GDP (Figure I.2, Table I.2). The PRC experienced the largest rate of growth, at 8.8 percent per year during 1980–95, even accelerating its annual rate of growth from the 1980s (8.3 percent) to the 1990s (9.6 percent). Only the Republic of Korea came close to this rate of growth in per capita GDP, at 7.1 percent per year during 1980–95, but its growth slowed down between the 1980s and the 1990s. Indonesia and Thailand, on the other hand, experienced accelerated growth, albeit at lower average levels of 4.9 percent and 6.5 percent annually, respectively. Malaysia achieved the most rapid improvement in its per capita growth rate, from 2.8 percent per year during 1980–89 to 6.2 percent annually during 1989–95, despite its rapid rate of population growth of 2.6 per year over the 1980–95 period. Viet Nam appears again at the lower end of the rapidly growing group, at 5.6 percent per year during 1989–95.

In the group of South Asian countries, the combination of higher population growth and lower GDP growth led to slower per capita GDP growth during 1980–95, ranging from a

4 Transforming the Rural Asian Economy

Figure I.2: GDP per capita, 1970 and 1995 (constant 1987 US$)

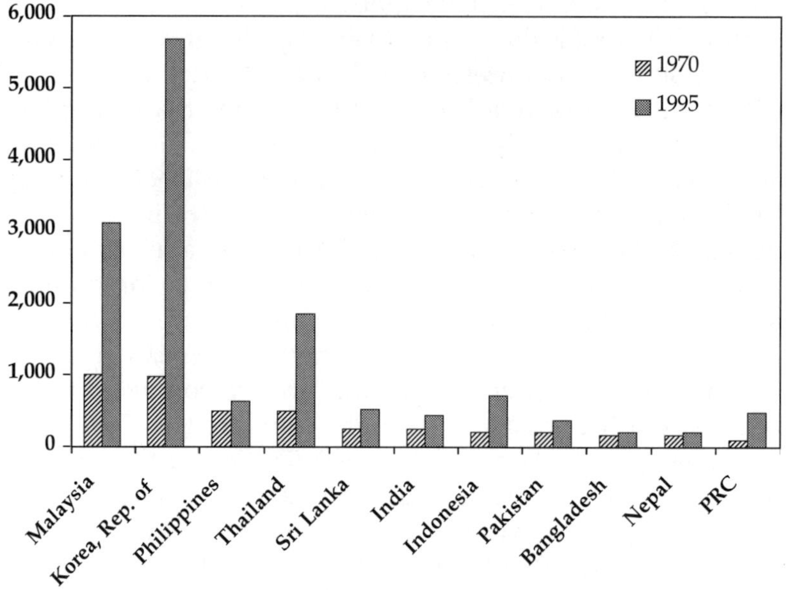

Source: WDI 1998. FAO FAOSTAT. 1998 (Population DataBase).

Table 1.2.: GDP per capita, (percent per year)

Year	1967-80	1980-89	1989-95	1980-95	1967-95
Bangladesh	-0.50	2.02	2.74	2.31	0.99
PRC	5.18	8.25	9.59	8.79	7.10
India	1.34	3.61	3.39	3.52	2.50
Indonesia	4.98	4.05	6.28	4.93	4.96
Korea, Rep. of	8.01	7.46	6.59	7.11	7.53
Malaysia	4.82	2.79	6.22	4.15	4.46
Myanmar*	2.36	-0.94	3.71	0.90	1.57
Nepal	-0.02	1.68	2.37	1.96	1.04
Pakistan	2.59	3.03	1.91	2.58	2.58
Philippines	2.70	-0.85	0.37	-0.37	1.05
Sri Lanka	2.95	2.89	3.41	3.09	3.03
Thailand	4.32	5.58	7.76	6.45	5.45
Viet Nam	–	–	5.64	–	–

Note: For Myanmar, GDP values are from the local currency unit. Growth rates are 3-year centered moving averages.

Source: WDI 1998. FAO FAOSTAT. 1998 (Population DataBase).

low of 2.0 percent annually in Nepal to a high of 3.5 percent per year in India. Growth in per capita GDP accelerated in Bangladesh, Nepal, and Sri Lanka, but decelerated in India (slightly) and in Pakistan (considerably). The annual rates of growth of per capita GDP in Myanmar and the Philippines improved from negative rates in the 1980s to positive growth in the 1990s. The rates of growth for the full period, however, were only 0.9 percent per year in Myanmar and negative 0.4 percent per year in the Philippines.

SOURCES OF ECONOMIC GROWTH IN ASIA

Considerable controversy remains about the relative contributions of technical progress and capital accumulation and the role of government policies in stimulating economic growth in Asia, but some areas of consensus are emerging. A substantial debate on the sources of economic growth in Asia was set off by Krugman (1994), who argued, drawing upon research by Young (1994, 1995) and Kim and Lau (1994), that the spectacular economic growth in East Asia was almost solely due to growth in capital and labor inputs, with virtually no contribution from productivity growth. A brief summary of this debate provides a useful base of comparison for the analysis of the contribution of productivity growth to agricultural output growth.

Analysis of total factor productivity (TFP) attempts to measure the amount of increase in total output that is not accounted for by increases in total inputs. Total factor productivity is computed as the ratio of an index of aggregate outputs to an index of aggregate inputs. Growth in TFP is therefore the growth in total output less the growth in total inputs (see Box I.1). Analyses of the contribution of TFP to general economic growth in Asia have produced a wide range of results, seemingly linked to the degree of adjustments made to account for input quality.

Young (1995) found, based on careful adjustments to input quality, that TFP growth accounted for about

> **Box I.1: Measuring TFP**
>
> Total factor productivity is the ratio between real output and real factor inputs. TFP can be estimated directly based on a production function, but more often it is estimated using a growth accounting framework derived from an underlying production function. TFP computed from growth accounting is the residual between output growth and input growth.
>
> Measurement of inputs can be difficult, with particular problems related to the measurement of capital inputs, including their aggregation, the valuation of capital, depreciation methods, and capacity utilization. Labor inputs can also be difficult to measure, because of the influence of education in improving the labor force quality and productivity. If capital and labor are not measured correctly, the quality changes in inputs will also be included in the growth accounting residual. If there are underadjustments in input quality improvement (as is often argued for developed countries), measured TFP will be higher than actual TFP. If there are overadjustments in quality (which Chen [1997] persuasively argues may often be the case for developing countries), measured TFP will underestimate actual TFP.
>
> Given these measurement uncertainties, it is useful to note that biases are likely to be relatively low in measuring TFP in developing-country agriculture because capital and labor are relatively undifferentiated and change slowly over time compared to the nonagricultural sector. Moreover, despite a range of methods used to adjust for input quality improvements, there is strong agreement across studies in the key conclusions regarding patterns of TFP growth in agriculture.
>
> (Chen 1997; Evenson, Pray, and Rosegrant 1999)

one third of general economic growth from 1966 to 1990 in Hong Kong, China; 28 percent in Taipei,China; 17 percent in Republic of Korea; and only 2 percent in Singapore. Other recent estimates for similar time periods establish a range for the contribution of TFP to economic growth of 32–46 percent for Hong Kong, China; 15–40 percent for Taipei,China; 9–37 percent for Republic of Korea; and –2 percent to 23 percent for Singapore (World Bank 1993a; Drysdale and Huang 1995; Kim and Lau 1994). For the PRC, alternative estimates indicate that TFP growth has contributed 24–28 percent of total economic growth; for Indonesia, 23–31 percent; for Malaysia, –8 percent to 24 percent; for the Philippines, –21 percent to 4 percent, and for Thailand 27–37 percent. There is less information on growth in TFP in South Asia (see Box I.2).

What can be learned from these results, aside from the difficulties inherent in the measurement of TFP? Based on the lower range of the estimates for the East Asian countries, Krugman (1994) argues that the results showed that the "East

---

Box I.2: Productivity Growth in South Asia

Compared to the explosion of studies on TFP growth in East and Southeast Asian economies in recent years, there has been relatively little work on South Asia. The available evidence, however, shows that economic growth has been slower in South Asia than in East Asia (2.7 percent increase in output per worker in South Asia, 1973-94, compared to 4.2 percent in East Asia, according to Collins and Bosworth [1997]). TFP growth of 1.3 percent per year has, however, been as high or higher over this period than in many East Asian countries, and has contributed a higher percentage to overall economic growth. Ali and Hamid (1996) find for Pakistan that the share of capital in the growth of the manufacturing sector was about 60 percent, the share of labor about 10 percent, and the share of technical change was a high 36 percent during 1973/74–1994/95. In the agriculture sector, on the other hand, technical efficiency only contributed 21 percent to growth during the same period. Sahota et al. (1991) find, based on a growth accounting framework, that TFP growth in the manufacturing sector in Bangladesh was actually declining between 1975/76 and 1984/85, and that the New Industrial Policy of 1982 had little impact on growth of total factor productivity.

Asian growth miracle" was in fact based on capital and labor accumulation similar to economic growth in the Soviet Union in the 1950s and 1960s, rather than on increases in productivity. Moreover, Krugman argues through analogy with the Soviet growth experience in the 1950s and 1960s that input-driven economic growth cannot be sustained. But Chen (1997) offers an alternative interpretation of the East Asian growth experience, noting that the relatively low TFP contribution reflects the importance of technological change embodied in inputs, such as quality improvements, together with economies of scale, increased capacity utilization, and structural transformation.

Modern growth theory shows that growth based on economies of scale and human capital accumulation can in fact be sustainable (see, for example, Lucas 1993). Moreover, a static view of sources of growth is inappropriate. The available evidence typically indicates that technological change is the major source of growth for industrialized countries and relatively unimportant in the early stages of growth in developing countries. The contribution of TFP growth should therefore increase over time as the East Asian economies continue to develop (Chen 1997). Rapid productivity growth through adoption and adaptation of technology from abroad is likely to depend on the stage of growth of an economy. Growth in the early stages of development may be primarily associated with physical and human capital accumulation. A significant potential for productivity growth through borrowing and use of technology may be possible only when a country has crossed a development threshold (Collins and Bosworth 1997). Grossman and Helpman (1994) argue that even if technological progress is the main driving force of long-term growth, factor accumulation will play an important role during a possibly prolonged transition phase.

In fact, studies that have disaggregated Asian TFP growth by subperiod confirm that the contribution of TFP to economic growth has increased over time. Chen et al. (1988) find that the contribution of TFP to industrial growth in the PRC increased from 14 percent in 1957–78 to 68 percent during 1978–85.

Borensztein and Ostry (1996) estimate negative TFP growth for the PRC during 1953–78, but positive growth of 3.8 percent per year from 1979 to 1993, accounting for more than 40 percent of GDP growth. Osaka (1997) finds that the Republic of Korea shifted from negative TFP growth before 1980 to a strong positive contribution from TFP thereafter, and that the Philippines and Thailand shifted from negative to positive TFP growth in the mid-1980s. Collins and Bosworth (1997) also show a remarkable increase in the contribution of TFP to overall growth in many East and Southeast Asian economies beginning in the mid-1980s. Comparing the periods 1973–84 and 1984–94, growth in TFP nearly doubled in Indonesia and Republic of Korea and more than tripled in Malaysia, Singapore, Thailand, and Taipei,China.

In summary, factor accumulation has been the primary source of growth in the East and Southeast Asian economies, while TFP growth has been a low to moderate contributor to overall economic growth. TFP growth is becoming an increasingly important source of economic growth as these economies continue to grow, however. Although it is difficult to generalize from the wide range of results, Indonesia, Malaysia, and Singapore seem to have relied relatively more on input-driven growth (with strong input quality improvements in the latter two countries), while Hong Kong, China; Republic of Korea; and Taipei,China have relied more heavily on technology-driven growth. Over the past ten to 20 years, many of the East and Southeast Asian countries have begun a shift of emphasis from capital accumulation to technology-based growth.

EXPLAINING ECONOMIC GROWTH

Explanations of productivity growth and overall economic growth in Asia have focused on a number of factors, including openness to trade, institutional quality, direct foreign investment, financial development, and macroeconomic policies. Collins and Bosworth (1997) note the difficulties in sorting out the impacts of the relevant policies and variables affecting economic growth. Much of the analysis of sources of growth implicitly or explicitly treats policies and underlying processes as affecting TFP growth, rather than factor accumulation; the latter, as shown above, has accounted for the lion's share of economic growth in East Asia. It may be more important to identify the policies that best promote high rates of savings and investment rather than policies that promote TFP growth (although there may also be considerable congruence between policies that promote TFP and those that induce high savings and investment). The available measures used to represent trade regimes, industrial targeting, and fiscal, monetary, and exchange-rate policy tend to be simple or highly aggregated and may capture differences poorly and be subject to measurement error. Moreover, it is difficult to separate the effects of policies that are implemented simultaneously.

Nevertheless, there is a fairly strong consensus in the literature on the importance to East Asian growth of conservative fiscal policies and sound exchange-rate policies as crucial components of macroeconomic stability, relatively low anti-export bias, and investment in education. The relative importance of these policies and the role of trade policy compared to investment incentives and the role of industrial policy remain open to debate. Collins and Bosworth (1997) find that fiscal discipline (low budget deficits) and exchange-rate stability are associated with higher economic growth. Krueger (1995) and World Bank (1993a) emphasize the importance of market-oriented policies. Amsden (1994) and Fishlow et al. (1994) argue that selective intervention and industrial policy have been important contributors to growth. Sachs and Warner (1995) find that export orientation and open trade policies are the most important government policies, while Rodrik (1995) argues that export promotion policies have not played a significant role. Other analyses emphasize the role of institutions in explaining economic growth in Asia. Rodrik (1997) stresses the primacy of institutional quality in explaining the differential growth performance among East Asian countries. In the following sections, we briefly review the evidence on the role of policies in economic growth in Asia.

Prudent Macroeconomic Policies

Although country experiences vary, most authors agree that all East Asian newly industrializing economies (NIEs), (the so-called "tiger" economies of Hong Kong, China; Singapore; Republic of Korea; and Taipei,China) and the Southeast Asian NIEs (Indonesia, Malaysia, and Thailand) had relatively small fiscal deficits and inflation rates were typically moderate, although not particularly low. Real interest rates have been quite stable, and black market exchange-rate premiums were comparatively low. The rapidly growing East Asian economies also achieved realistic real exchange rates. During 1978–86, the government deficit was 3.5–6.5 percent of GDP in Thailand and reached 15.5 percent in Malaysia during 1981–82. Inflation averaged 20 percent in Korea during 1974–81 and reached 40 percent in Indonesia in 1974. However, deficits and inflationary pressure were countered rather quickly in all the East Asian economies (Collins and Bosworth 1997; Rodrik 1997). Collins and Bosworth (1996) also show that the East Asian countries with smaller budget deficits and more stable exchange rates tended to grow faster, with smaller deficits supporting the accumulation of capital per worker and exchange-rate stability favoring productivity growth.

Market-oriented Policies

Market-friendly policies have played an important role in the growth experience of the East Asian NIEs. They can be considered as an intermediate approach, more proactive than laissez-faire policies and more cautious than active industrial policy. Market-oriented policies focus on the liberalization of trade, accompanied by decreased government regulation and intervention, and thus less rent seeking, all conducive to the functioning of a market economy. The East Asian exporters also faced fairly uniform incentives for exporting commodities across industries. Moreover, both imports and capital flow were gradually liberalized and the fiscal deficit was comparatively low. The share of government expenditures in GDP was relatively small.

The governments in all East Asian countries invested heavily in infrastructure and human capital to support rapid increases in manufactured output and exports. In addition, there was also little intervention in the labor market. Interventionist policies in the labor market, when they were adopted, appear to have been rather harmful for economic growth (Krueger 1995). World Bank (1993a) summarizes the market-oriented approach adopted by the NIEs as encompassing macroeconomic stability, human capital formation, openness to international trade, and an environment conducive to private investment and competition.

Open Trade Policies

Rapid growth in exports has been a central feature of the East Asian NIEs. Outward-oriented trade policies were accompanied by stable exchange-rate regimes. In the early 1960s trade policies focused on import substitution (with the exception of Hong Kong, China), followed in the mid- to late 1960s by export-promotion policies, in particular in Republic of Korea; Singapore; and Taipei,China (Collins and Bosworth 1997). Sachs and Warner (1995) conclude that open trade policies and related reforms represent the key to economic growth in developing economies. Utilizing a time-series regression analysis, the authors show that a high labor-to-land ratio (a given in most Asian economies) and high per capita income encourage early liberalization policies in developing countries. (However, some labor-intensive economies, like Bangladesh, India, Pakistan, and Sri Lanka, remained protectionist).

"Open" economies grew at 4.5 percent per year during the 1970s and 1980s, whereas closed economies grew only at 0.7 percent per year. Poor open economies grow faster than rich ones, a tendency that cannot be observed for closed economies. Finally, poor trade policies affect trade both directly and indirectly by limiting the rate of investment in physical capital.

Among the Asian economies, Hong Kong, China; Malaysia; Singapore; and Thailand can be considered as open economies throughout the past two to three decades; Taipei,China is considered as open after 1963, the Republic of Korea after 1968, and Indonesia after 1970. All other Asian economies only opened to trade in the late 1980s or early 1990s. Bangladesh was still considered a closed economy in 1994. More recently, the East Asian NIEs have started to shift away from extensive use of selected export-promotion policies.

Industrial Policy

Some analysts argue that industrial policies in the form of "soft" strategic interventions (as undertaken by the East Asian NIEs) can promote selected industries in order to create spillover benefits, enhance the overall competitiveness of the industry, and raise the future technological level of the economy (Wade 1994). Industrial policies include protectionist policies; export promotion; targeted credit and tax subsidies; and other industrial, technology, and education policies designed to support specific industries or sectors. Fishlow et al. (1994) attribute a very large role to industrial policies in the East Asian NIEs. Rodrik (1997) suggests that the comparatively poor performance of growth of investment as a share of GDP since the 1960s in Hong Kong can be attributed to its noninterventionist policies. But the experience with industrial policy in the East Asian NIEs is quite mixed, and even when industrial policies generate short-term benefits, they may be costly in the longer term. Whereas Hong Kong, China did not use any industrial policies, Thailand implemented few interventionist policies, Singapore relied on selected interventions; Taipei,China adopted a moderate number; and Japan and the Republic of Korea pursued highly interventionist policies. The Korean government focused on large conglomerates, whereas small, entrepreneurial firms were favored in Taipei,China (Rodrik 1994). Although foreign direct investment (FDI) is typically considered a primary way to transfer modern technology, the East Asian NIEs have also not adopted this strategy uniformly. Only Singapore actively encouraged FDI, Hong Kong adopted a laissez-faire approach, the Republic of Korea managed FDI carefully, and Japan practically prohibited FDI.

There is some evidence that industrial policy, in particular concerning directed credit and export promotion, and particularly in Japan, Republic of Korea, and Taipei,China, resulted in higher and more equal growth (World Bank 1993a). However, the positive effects were limited and it is difficult to transfer the experience to other developing countries. Lee (1992) for example, finds that trade restrictions and subsidized credit had adverse effects on growth in TFP in the Republic of Korea. There is also new evidence that, during the period 1960–90, Japanese industrial policies increasingly diverted resources away from high-growth sectors toward declining industries and did not have a positive impact on TFP growth within sectors. Similar results have been found for Taipei,China during the 1980s (Datt 1998).

Moreover, attempts to transfer the lessons of industrial policy to the rest of developing Asia have been largely unsuccessful. The extraordinary cost of mismanaged directed credit has been shown by the East Asian financial crisis (see Chapter IX). Direct picking of winners through industrial policy has also not fared well. In Malaysia, heavy industry was particularly promoted after 1981 when the Heavy Industries Corporation of Malaysia (HICOM) was created to target large-scale, capital-intensive iron and steel and other heavy industry companies. By 1988, nine companies had been involved in steel, cement, motor vehicles, and motorcycle engine manufacturing. But poor management, low profitability, and budgetary concerns rang in the end of the HICOM period at the end of the 1980s. Indonesia used public investments to promote industries designated as having high potential for technological learning. These investments were not linked up with private-sector efforts and were generally unsuccessful (World Bank 1993a).

Picking winners and providing protection to some industries also by definition penalizes other industries. As will be shown in Chapter IV, a clear example of this is the rural nonfarm sector, which has largely been ignored by policymakers and has suffered as a result of macroeconomic policies that discriminated against the agricultural sector. The few attempts to assist the rural nonfarm economy have generally favored manufacturing rather than service activities and large- rather than small-scale units of production, encouraging more capital-intensive patterns of development than is optimal. In summary, the industrial policy approach has not been successfully transferred from the early industrializing East Asian economies to the rest of developing Asia because of intrinsic difficulties. Industrial policies that have been implemented have not been based on a sound analysis of market failures they were intended to correct; they were not selective in addressing specific market failures; they ignored market signals in attempting to achieve efficiencies; they underestimated the informational requirements necessary for effective intervention; they overlooked the limited capacities and capabilities of government; and they overestimated the human and other resources available to build efficient industries (Jomo 1996).

Investment in Education

Human capital makes investment more productive, facilitates the adoption of modern technology, and enables the establishment of an efficient bureaucracy. All of the East Asian NIEs as well as Malaysia and Thailand had virtually universal primary schooling by 1960. Literacy rates were exceptionally high in all the countries except for Singapore (Rodrik 1994). According to Rodrik (1994), the initial conditions of high levels of primary-school enrollment and equality around 1960 can explain a large share of the growth of Republic of Korea, Malaysia, Taipei,China, and Thailand.

Most East Asian countries continued to invest heavily in education: in 1982, 85 percent of children in Taipei,China attended secondary school, 89 percent in Republic of Korea, and 92 percent in Japan, compared with 87 percent for all industrial market economies (Krueger 1995). Even Indonesia achieved a secondary enrollment rate of 46 percent by 1987. Only in Thailand was the secondary enrollment rate a low 28 percent, well below the 36 percent prediction based on its per capita income in 1987 (World Bank 1993a). Collins and Bosworth (1997) find that the contribution of improvements in education has been relatively large in East Asian countries compared to other regions in Asia, but remains a relatively small part of the growth story. Young (1995), on the other hand, discerns an important role for educational attainment in the exceptionally rapid income growth in the newly industrializing economies of Asia, in addition to large increases in labor-force participation and capital stock.

Institutional Issues

Good governance and strong institutional performance in the public sector are important for economic growth. The structure and orientation of the bureaucratic-institutional framework varies widely in Asia; bureaucracies have evolved from a series of different approaches and differ not only across countries but also across agencies within countries. Evans (1998) argues that successful growth has demanded a high level of bureaucratic capacity, close coordination between the public and private sectors, and the need to avoid rent-seeking behavior. Although East Asian bureaucracies have often been characterized as being relatively effective, all East and Southeast Asian bureaucracies, with the possible exception of Singapore, are characterized by considerable inefficiency (Evans 1998). As will be described in Chapter IX, the East Asian crisis has revealed fundamental weaknesses in governance in many Asian economies. Rodrik (1997) examines the impact of institutional quality measured by an index constructed from several factors:

- the quality of the bureaucracy, including autonomy from political pressure and expertise and efficiency in the provision of government services;

- rule of law, including soundness of political institutions, strong judicial system, and orderly succession of political power;
- risk of appropriation, with low values indicating high possibility of confiscation and nationalization; and
- repudiation of contracts by government, with low values indicating high risks (Knack and Keefer 1995, cited by Rodrik 1997).

Based on this index, Rodrik (1997) finds that Japan, Singapore, and Taipei,China are ranked high in the institutional quality scale, while Indonesia and the Philippines, both known for weak institutions, score particularly low. His analysis indicates that institutional quality, initial income, and initial education account for a large part of the differences in the growth experience of the East Asian countries. In the absence of superior institutions, Japan and Singapore would have been predicted to grow at rates below the regional average, whereas poor institutions are the primary culprits for low economic growth in Indonesia and the Philippines.

Overall, the evidence indicates that accumulation of physical and human capital was the most important factor in the East Asian growth experience, but that in most countries, productivity growth has also played a significant and increasing role in economic growth. Fiscal discipline, market-oriented policies, open trade policies, investment in education, and institutional quality have played a crucial role in economic growth, supporting both factor accumulation and productivity growth. Industrial policy has been less important to economic growth and may have negative long-term effects. In the next section, the role of growth in the rural and agricultural economy in establishing the preconditions for rapid economic growth is explored.

IMPORTANCE OF AGRICULTURE IN PROMOTING THE TRANSFORMATION

Most Asian countries (Hong Kong, China and Singapore excepted) began as predominantly agrarian economies, in which the agricultural sector accounted for the largest share of GDP, employment and export earnings (see Chapter III). Under colonial regimes, agriculture was also geared to the production of tropical export crops (oil palm, rubber, tea, etc.), which benefited from significant investment in research and infrastructure development, while the food-crop sector was neglected. After independence, most Asian countries depended heavily on their export crops and inherited a stagnant, low-productivity food-crop sector. Coping with low and highly unstable prices for their exports and a growing inability to feed themselves was the key challenge for the newly formed governments. Not surprisingly, national food self-sufficiency became a primary policy goal for most Asian countries. With uncertain foreign exchange earnings and limited ability to pay for food imports, increasing domestic food production to assure food security was a priority.

Until the food problem was solved, the development of the nonagricultural sector was necessarily constrained. Labor and capital could not easily be freed from a technologically stagnant agriculture and any significant growth of the nonagricultural sector would increase the demand for food. Food prices (and hence industrial wages) would then be driven up in the face of an inelastic aggregate agricultural supply. Not only would this reduce the competitiveness of the nonagricultural sector, but it would effectively transfer resources back to the agricultural sector where they would be subject to diminishing returns (Mellor 1973). Food imports, which might have provided a way out of this trap, were constrained by low and unstable export earnings.

A technologically driven transformation of the agricultural sector was a necessary condition for national economic growth. An agricultural revolution was needed not only to overcome the food constraint, but also to provide an engine of growth on the scale required to begin to transform the national economy. The nonagricultural economy was typically too small to play this role, even if it could have depended on cheap food imports. Even if the agricultural sector contains potential surpluses of capital, labor, and food output to fuel the rest of the economy, forced extraction of these resources from a stagnant agricultural sector would create widespread rural poverty and a drying up of these potential surpluses. Only a dynamic, rapidly growing agriculture can generate the sustained surpluses necessary to drive the economic transformation (Timmer 1988).

During the early stages of the transformation, agricultural growth contributed in several important ways. Foremost was its role in raising the living standards of the rural population, unleashing a massive increase in domestic demand for nonagricultural goods and services. This provided a nascent and growing market to spawn the growth of nonagricultural firms. The additional demand included input, service, and investment needs for a growing agriculture and, more importantly, it included growing consumer demands by millions of rural households as their incomes increased (Mellor 1976; Tomich et al. 1995).

Second, the agricultural sector provided relatively low-cost labor to the industrial and service sectors. Most of the labor required for the expansion of these sectors came initially from agriculture simply because there was almost no other source. Technological change in agriculture increases the output per unit of labor, allowing the release of labor to other sectors without causing stagnation in agriculture or driving up agricultural prices (Johnston and Mellor 1961; Timmer 1988).

Third, a growing agriculture was able to generate large amounts of capital to finance the nonagricultural sector. Part of this was through rural savings; savings rates have been exceptionally high in much of Asia. But part was extracted directly through taxes (particularly land and commodity taxes) and indirectly by turning the terms of trade against agriculture. These transfers have been highest during periods of rapid technological change in agriculture (Mellor 1973). They have also been very large in some cases. In Japan, for example, agricultural land taxes provided up to 70 percent of the central government's total revenue during the late 19th century (Bird 1974), and in Taipei,China between 1911 and 1960 annual net capital transfers from agriculture were generally in excess of 25 percent of the total value of agricultural production (Lee 1971).

Fourth, agricultural growth fostered the development of the agroindustrial sector, particularly firms that supplied key inputs (machinery, fertilizer, cement, etc.), and that processed agricultural output (food industry, textiles, jute, etc.). These activities were often the spawning grounds for the development of private firms that later entered export markets. Their growth also fostered the development of managerial skills and urban infrastructure that permitted later diversification into new nonagriculturally related activities. In Taipei,China and Republic of Korea, for example, the employment structure of rural manufacturing shifted markedly after the 1960s from processing primary products (food processing, beverages, tobacco, wood and bamboo products) toward much greater concentration in the production of chemicals, paper, printing, metals, and machinery (Otsuka and Reardon 1998).

Fifth, agricultural growth generated additional foreign exchange earnings, through both increased exports and reduced imports, which were key to effective industrialization in the early stages. Many Asian countries regulated the use of foreign exchange in the early years of the transformation, giving high priority to important capital goods and intermediate production goods.

Sixth, agricultural growth transformed rural regions, providing more diversified sources of income for rural households and increased livelihood opportunities without spurring massive migration to the cities. Even before the explosive economic growth of the last decade, the rural nonfarm economy accounted for one third or more of rural employment in many Asian countries, as will be shown in detail in Chapter IV. Small towns have grown rapidly with agricultural growth and now have economies that are dominated by service establishments and agroindustry firms.

Studies of the links between agricultural growth and the rural nonfarm economy in Asia have estimated regional income multipliers at between 1.5 and 2.0, (that is, the increase in value added in the regional nonfarm economy for each dollar increase in agricultural value added), and employment elasticities at about 1.0 (that is, the percentage increase in rural nonfarm employment for a 1 percent increase in agricultural gross output). See, for example, Bell, Hazell, and Slade (1982); Hazell and Ramasamy (1991); Hazell and Haggblade (1991); and Gibb (1974). These spatial patterns of development have helped foster the creation of dense patterns of rural infrastructure, which in turn has contributed to further rounds of growth and has helped to reduce widening interregional inequalities. It has also helped to contain the growth of large cities in many Asian countries.

The relationship between growth in the agricultural sector and the combined manufacturing and service sectors can be seen in Figures I.3 and I.4. Those countries with higher agricultural growth also experienced higher growth in nonagricultural activities during 1970–95. Indonesia and Malaysia, for example, fall almost directly on the fitted line showing this relationship, with annual per capita average growth in agriculture of 1.70 percent and 1.26 percent, respectively, fueling growth in the manufacturing and service sectors of 6.23 percent per year and 5.34 percent per year. Bangladesh and the PRC are at the opposite ends of the range of per capita agricultural growth rates, with Bangladesh experiencing negative growth of 0.67 percent annually during 1970–95 and the PRC excelling at 2.76 percent per year, on average. In the South Asian countries of India, Pakistan, and Sri Lanka, per capita agricultural growth translated into lower than expected growth in the manufacturing and service sectors for many of the reasons detailed elsewhere in this chapter.

The Philippines, and even more so the Republic of Korea, are outliers in the general trend. In the Philippines, per capita agricultural growth was slightly negative, on average, during 1970–95 (–0.06 percent per year) but growth in the manufacturing and service sectors was far below average for Asian developing countries—and was the lowest for the countries depicted in the graph—at 1.35 percent annually. Here, agricultural growth was highly inequitable and thus could not easily transform into nonagricultural growth; moreover, industrial policies themselves were not conducive to agricultural growth. In the Republic of Korea, on the other hand, growth in the manufacturing and service sectors took off rapidly, without supporting rapid growth in agriculture. This development was mainly due to the structure of the industrial sector in this country, with large industrial complexes concentrated in a few urban centers, demands for industrial products chiefly fueled by exports, and capital financed through large inflows (Mellor 1995). Thus, as Mellor (1995) points out, contrary to the evolution in most Asian developing countries, growing demands by the nonagricultural sector pulled agricultural production into faster growth. Per capita growth in agriculture accelerated during the 1980s, to 0.88 percent per year; but growth slowed down at the beginning of the 1990s, to –0.10 percent per year. When the outlier of Republic of Korea is taken out of the sample (Figure I.4) the estimated relationship indicates that about 88 percent of the growth one sector is associated with growth in the other sector. The estimated relationship suggests that for each 1 percent increase in per capita agricultural growth, growth in the manufacturing and service sectors accelerates by 2.1 percentage points.

Historically, all the rapidly growing economies in Asia (except island states like Hong Kong, China and Singapore) enjoyed successful agricultural revolutions prior to their industrialization. These agricultural revolutions derived from the intensification of key food crops (especially rice and wheat) through the spread of high-yielding varieties, fertilizers, pesticides, and massive investments in irrigation (see Chapters V and VI). Not all the countries that had successful agricultural revolutions went on to industrialize and grow rapidly, however. While agricultural growth was necessary during the early stages of the transformation, it was not sufficient to promote industrialization. Several other key factors are also needed to enable countries to successfully convert agricultural growth into national economic growth (Mellor 1976; Tomich, Kilby, and

Agricultural Growth and the Economic Transformation 23

Figure I.3: Per capita growth rate, nonagricultural vs. agricultural GDP (with Republic of Korea), 1970–95

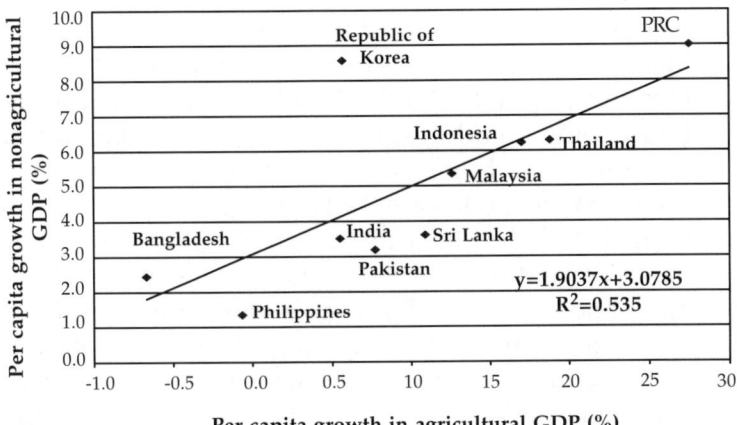

[Scatter plot showing data points: Bangladesh, Republic of Korea, Indonesia, Thailand, Malaysia, India, Sri Lanka, Pakistan, Philippines, PRC. Regression: $y=1.9037x+3.0785$, $R^2=0.535$. X-axis: Per capita growth in agricultural GDP (%), range -1.0 to 30. Y-axis: Per capita growth in nonagricultural GDP (%), range 0.0 to 10.0.]

Note: Growth rates are 3-year centered moving averages.
Source: WDI 1998 and FAO FAOSTAT 1998

Figure I.4: Per capita growth rate, nonagricultural vs. agricultural GDP (without Republic of Korea), 1970–95

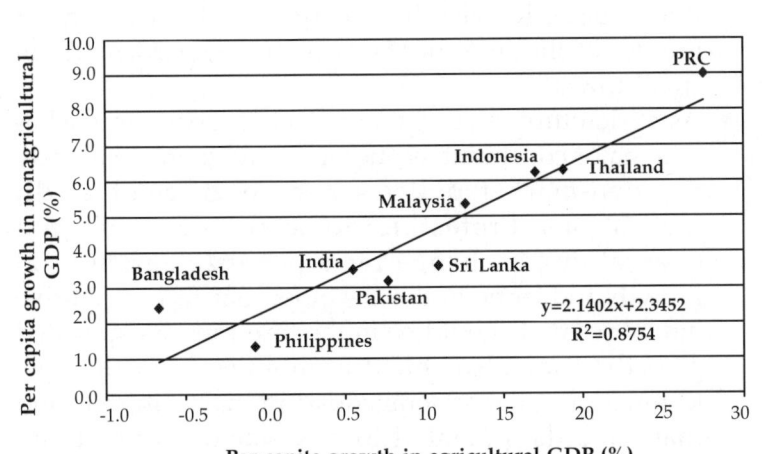

[Scatter plot showing data points: Bangladesh, Indonesia, Thailand, Malaysia, India, Sri Lanka, Pakistan, Philippines, PRC. Regression: $y=2.1402x+2.3452$, $R^2=0.8754$. X-axis: Per capita growth in agricultural GDP (%), range -1.0 to 30. Y-axis: Per capita growth in nonagricultural GDP (%), range 0.0 to 10.0.]

Note: Growth rates are 3-year centered moving averages.
Source: WDI 1998 and FAO FAOSTAT 1998

Johnson 1995; Ranis, Stewart and Angeles-Reyes 1990). These requirements can be clustered into three groups:

- Agricultural growth must be equitable, so that it puts increased purchasing power into the hands of the farming population. Small and medium-sized farm households spend much larger shares of incremental income on labor-intensive goods that are also nontradables; this leads to larger growth multipliers than if the additional income is concentrated in the hands of rural elites (Mellor 1995; Hazell and Roell 1983; King and Byerlee 1978). A prerequisite for equitable agricultural growth is an equitable distribution of land, with secure ownership and tenancy rights over land. Technologies should also be scale-neutral, and the required inputs should be available to all farmers regardless of size.
- A well-developed rural infrastructure is required to connect villages to local markets. This is needed for the efficient operation of agricultural input and product markets, and also to increase access to nonfarm goods and services to promote demand linkages (Hazell and Roell 1983). Rural infrastructure is also needed for growth of the rural nonfarm economy, particularly in small towns.
- As agriculture develops and food security diminishes as a major constraint, economies need to move quickly to open-market policies and to efficient credit institutions. Protecting domestic industries and overvaluing exchange rates penalize agricultural growth and impede the development of competitive industries that should be in the forefront of export-led growth. They also shield economies from the new technologies that are embedded in many imports and that have the potential to raise factor productivity significantly.

The earliest emerging economies in Asia were successful in meeting these requirements. Several (Japan; Taipei,China; and Republic of Korea) had major land reforms that led to equitable agricultural growth and they invested heavily in rural infrastructure and pursued pro-growth macro and trade policies. The PRC had a successful agricultural revolution from a technical perspective, but was only able to take full advantage of it once the organization of agricultural activity changed to a "household responsibility system" that released labor to the nonagricultural sector and effectively put additional purchasing power into the hands of the rural masses.

India and the Philippines have both had successful agricultural revolutions, but have been slow to convert their results into rapid economic growth. In India, agricultural growth has been relatively equitable and rural infrastructure is well developed, but the country has only recently begun to open its markets and to move to pro-growth macro and trade policies. The results are encouraging, with national income now growing at more than twice the rate of recent decades. In the Philippines, on the other hand, agricultural growth has been highly inequitable, with most of the gains siphoned off by the rich, and rural infrastructure remains weak. Even if fully implemented, pro-growth macro and trade policies on their own seem likely to lead to enclave patterns of industrialization and disappointing overall growth (see also Figures I.3 and I.4).

The importance of agricultural growth in supporting broad-based economic transformation in "late-blooming" countries such as India and the Philippines is brought out by Robinson, Roe, and Yeldan (1998). The authors developed a dynamic computable general equilibrium model of an archetype (or stylized) South Asian economy in order to simulate the likely growth paths resulting from alternative macro, trade, and investment policies over the 30 years 1990 to 2020. Results show the clear superiority of trade liberalization policies over "business-as-usual" patterns in South Asia. But they also show that trade liberalization needs to be accompanied by some fiscal restructuring in order to maintain government investment after the loss of revenue from trade tariffs. Finally, and especially important, the results show that government investment strategies do matter, even in a liberalized economy. In particular, in countries with a large agricultural base, there are sizable growth gains to be had from a pro-agriculture rather than a pro-industry public investment strategy. This result is consistent with similar findings by Adelman (1984), Yeldan (1989) and Adelman et al. (1989).

CHANGING ROLE OF THE AGRICULTURAL SECTOR

As the economic transformation of an economy proceeds, agriculture's share in GDP falls quite rapidly (see Chapter III) and its importance for economic growth diminishes. The nonagricultural sector becomes the primary engine of growth and is no longer dependent on resource flows from agriculture or on agriculture's demand linkages. The economic problem then is to absorb workers out of agriculture at a sufficiently rapid rate to stop their average productivity (and hence their incomes) from lagging too far behind the levels achieved in the nonagricultural sector. Typically, agriculture's share of total employment falls much more slowly than its share in national income, with the inevitable result that labor productivity (and hence per capita farm income) in agriculture lags behind the nonagricultural sector.

An important reason for this is that the employment elasticity in agriculture declines as commercialization proceeds, while the absolute size of the rural labor force continues to grow until quite late in the transformation process (Tomich, Kilby, and Johnston 1995). Improvements in labor productivity in agriculture require accelerated migration of workers to other sectors, either through rural-urban migration or economic diversification within rural areas, and labor productivity-enhancing investments in agricultural mechanization, larger farm sizes, and diversification into higher-value products. Many countries fail to manage this transformation at an adequate rate and are then confronted with an increasing political problem of relatively low incomes in the farm sector. The tendency then is to introduce income support policies for farmers, which as the experience of many countries in the Organization for Economic Cooperation and Development (OECD) shows, can lead to a continuing structural imbalance and an increasing burden on the public purse. The agricultural transformation in Japan led into this trap; the Republic of Korea and Taipei,China are already progressing down the same path. In Chapter VII the likelihood of other developing countries of Asia following the same trajectory of increasing agricultural protectionism is discussed.

CONCLUSIONS

With the exception of the city-states of Singapore and Hong Kong, China, the Asian countries that grew earliest and fastest experienced rapid agricultural growth in the early stages of growth. This agricultural growth was broad-based, benefiting small and medium-sized farms in particular, and hence was dependent on an equitable distribution of land. It was also characterized by cost-reducing technological change that led to significant improvements in total factor productivity, as will be shown in more detail in Chapter V. Rapid agricultural growth of this type proved to be a powerful engine for general economic growth, particularly in the regions in which it occurred, and helped to ensure that the benefits of growth were distributed widely across income groups in rural areas. The sheer size of the agricultural sector in GDP and employment (see Chapter II) not only made development of the agricultural sector essential, but meant that the productivity gains that were achieved had economy-wide significance.

Rapid growth in agriculture freed up labor and capital for the nonfarm economy, maintained a downward pressure on the prices of food and key primary inputs for agro-industry, contributed to foreign exchange earnings (through reduced food imports and increased agricultural exports), and provided a buoyant domestic demand for nonfarm goods and services. This led not only to rapid growth in the rural nonfarm economy, but also contributed importantly to the transformation of the urban economy.

The rapidly growing Asian economies began with a successful agricultural transformation, but as growth continued, matched this with stable macroeconomic policies, market-friendly policies, relatively open trade policies, and aggressive public investments in education and infrastructure. Countries that developed later or more slowly were less successful in adopting the green revolution, for example, because of unfavorable agroclimatic conditions, inappropriate farm-tenure structures, anti-agricultural policy biases, or failure to make the necessary public investments in research and rural infrastucture, and/or because their economies were not well positioned to exploit the potential growth linkage benefits emanating from agricultural growth. Economies with massive public intervention (for example, the centrally planned countries and India), or weak infrastructure, or that were inward-looking rather than export-oriented, were least successful in using the agricultural transformation revolution to stimulate a broader economic transformation.

The PRC is a good example of a country that enjoyed a widespread green revolution early on, but failed to capitalize on this for successful national economic growth until after the economy was liberalized. India and the Philippines also experienced limited national economic growth despite successful green revolutions, but their growth performance is improving as they reform and liberalize their economies. As economic growth proceeds and agriculture declines in relative size, economy-wide policies that support factor accumulation and productivity growth, including fiscal discipline, market-oriented policies, open trade policies, investment in education, and institutional quality are increasingly important in determining the pace of economic transformation.

II Growth and Policies for Poverty Reduction

INTRODUCTION

A sharp reduction in poverty has been one of the most remarkable achievements of the rapid agricultural and economic growth in developing Asia during the last three decades. Whereas in 1975, 60 percent of Asians lived in poverty, this ratio had fallen to less than one in three by 1995. The total number of poor declined from 1,149 million in 1975 to about 800 million in 1995 (a reduction of 25 percent), despite substantial growth in total population during this period (Table II.1).

These impressive gains mask considerable diversity of experience among Asian countries, however. The most striking divergence can be observed between the poverty reduction in East and Southeast Asia, on the one hand, and in the group of South Asian countries, on the other. Whereas some countries in Southeast Asia have roared ahead from poverty to middle-class prosperity within a mere two decades, others, chiefly in South Asia, have lagged behind. In this chapter, we review in more detail the patterns of—and differential rates of performance in—poverty reduction in Asia and examine the forces driving this reduction.

B. TRENDS IN TOTAL POVERTY

Table II.1 summarizes trends in total poverty in Asia. In most of developing Asia, rapid agricultural and economic growth has led to a dramatic reduction in both absolute and relative poverty over the past three decades. Poverty reduction in East Asia over the past two decades has been especially remarkable. Based on a head-count index defined as the share of the population below a constant (1985 prices) $1[1]-per-day per capita income poverty line, poverty in East and Southeast Asia (including the PRC, Indonesia, Malaysia, the Philippines, and Thailand) declined by two thirds between 1975 and 1995. In 1975, 50 percent of the population lived in absolute poverty, while in 1995 only 20 percent was considered poor. Because populations continued to grow, the percentage decline in poverty translates into a 40-percent decline in the absolute number of poor people, from 517 million to 310 million.

Table II.1: Poverty in Asia in the 1970s and 1990s

Country/Region	Poverty: persons below $1/day poverty line			
	(millions)		(percent)	
	1975	1990s	1975	1990s
SE Asia*	108.1	40.2	52.9	11.5
Indonesia	87.2	21.9	64.3	11.4
Malaysia	2.1	< 0.2	17.4	< 1.0
Philippines	15.4	17.6	35.7	25.5
Thailand	3.4	< 0.5	8.1	< 1.0
South Asia	472.2	514.7	59.1	43.1
PRC	568.9	269.3	59.5	22.2
Asia Developing*	1149.2	824.2	58.7	29.9

* Note that regional figures refer only to those countries listed in the table. The benchmark is the international poverty line, US$1 per day (Purchasing Power Parity, 1985 dollars).

SE Asia and PRC – from Ahuja et al. 1997 (PRC data from 1978 for rural PRC only); 1990s data are for 1995.

South Asia – 1990s data are for 1993 (World Bank 1996). 1975 figures are author's estimates based on synthesis of national poverty line computations adjusted to the international poverty line of US$1 per day.

---

[1] $ indicates US dollars throughout the text.

Moreover, the rate of decline accelerated in the more recent decade: whereas the total number of people in poverty fell by 27 percent in 1975–85, the decline in 1985–95 was closer to 34 percent. This pace of poverty reduction was faster than in any other region of the developing world and, as a result, the share of the world's poor living in East and Southeast Asia has been declining (Ahuja et al. 1997).

The two most populous countries in East and Southeast Asia, the PRC and Indonesia, recorded substantial declines in poverty between 1975 and 1995, with the head-count index falling by four fifths in Indonesia and by more than half in the PRC. In absolute terms the number of poor was reduced by more than one third in the PRC and fell by almost three fourths in Indonesia. Comparing poverty reduction in Indonesia and India illustrates the impressive performance in East and Southeast Asia relative to that of South Asia. Indonesia's headcount index was 63 percent in 1970—well above the 49 percent value for India at that time. But by the mid-1990s, the percentage of the Indonesian population estimated to be below the poverty line had fallen to 11 percent, which was less than one third of India's head-count index.

Other countries achieved even larger proportional reductions in poverty between 1975 and 1995. The proportion of the population below the poverty line in Malaysia fell by 95 percent (from 17.4 percent to less than 1 percent) and in Thailand by about 90 percent (from 8.1 percent to less than 1 percent). On the other hand, in the Philippines, which had much slower growth throughout most of this period, the headcount index fell much more slowly, from 35.7 percent in 1975 to 25.5 percent in 1995, and the number of poor actually increased from 15.4 million to 17.6 million. In addition to slower growth, policies in the Philippines were more heavily biased against agriculture during most of this period and income distribution was more unequal than in other Southeast Asian countries. However, there have been signs of progress against poverty in the early 1990s following economic reforms; furthermore, the slowdown in growth and setback in poverty reduction during the East and Southeast Asian economic crisis beginning in 1997 (see Chapter IX) were less severe than elsewhere in Southeast Asia (De Haan and Lipton 1998).

Progress on poverty reduction has also been achieved in South Asia, but it started at a later date and lagged behind the rate of poverty reduction achieved in East Asia. Using the World Bank $1-per-day head-count index, 59 percent of the South Asian population was below the poverty line in 1975 (Table II.1). The head-count index improved to 43 percent in 1993 (World Bank 1996c). However, despite this improvement, the number of poor in South Asia continued to increase, from 472 million in 1975 to 515 million in 1993, due to population growth (Table II.1).

The case of India is illustrative of broad patterns in poverty reduction in South Asia. Datt (1998) and Datt and Ravallion (1997) report that the period of the early 1950s up to the mid-1970s was characterized by fluctuations in poverty without a real trend in either direction. The head-count index averaged 53 percent in 1971–75 (based on a poverty line of 49 rupees per capita per month at October 1973–June 1974 all-India rural prices), nearly the same level as it had been in 1951–55. However, between 1969–70 and 1993–94, the national head-count index declined to 35 percent, and both the depth and severity of poverty decreased. It thus took more than 20 years for the poverty incidence to finally fall below the values of the early 1950s. In Bangladesh, progress in poverty reduction has varied greatly over time, as poverty declined from 1983/84 to 1985/86, then increased until 1991/92 and declined again until 1995/96 (Wodon 1999). The record of poverty reduction in Pakistan is relatively good, despite a weak record on literacy and health improvements, especially for women. National poverty indicators show a reduction in poverty from 54 percent in 1962 to 23 percent in 1984, with much of the progress attributable to growth in the service sector, trade, and noncrop agricultural sectors and to rural-urban migration (De Haan and Lipton 1998).

Historical data are unavailable for the Central Asian countries, but during the early stages of the economic transition a dramatic fall in average real income (see Chapter VIII) has been accompanied by sharply worsening income distribution. World Bank and United Nations Development Programme

(UNDP) country poverty assessments in Kazakhstan, the Kyrgyz Republic, and Mongolia indicate a significant worsening of poverty since 1990 (De Haan and Lipton 1998). UNDP (1998) cites estimates of poverty, based on a $100/month per household poverty line for Central Asian economies in the mid-1990s. Thirty percent of the population in Kazakhstan is below the poverty line, 80 percent in the Kyrgyz Republic, 20 percent in Turkmenistan, and 50 percent in Uzbekistan.

Although poverty declined overall in East, Southeast, and South Asia, regional differences in poverty indicators within Asian countries are large and are not in general converging. In India (and Pakistan), regional differences in poverty have remained or increased; poverty and regional variability fell in Indonesia and Malaysia; and declines in poverty in the PRC and Thailand have been accompanied by rising regional poverty differences. In Malaysia and Indonesia, the decline in regional differences is due in part to initial active targeting by the Malaysian Government of the poorer population, and by population redistribution policies of the Indonesian Government. In the PRC, most poverty is confined to resource-constrained, remote "backward regions" and is due to differentials of investments in human capital, infrastructure, fiscal decentralization, natural and geographical advantages, and policies that favored coastal areas. In Thailand, the concentration of economic growth in Bangkok increased regional poverty differences (De Haan and Lipton 1998).

TRENDS IN RURAL POVERTY

Poverty in Asia remains overwhelmingly a rural problem. Estimates of rural poverty in Asian economies are not directly comparable across countries, because they are based on national poverty lines; estimates of head-count ratios in a consistent numeraire such as PPP-adjusted dollars are not available. Even within individual countries, there are usually many alternative poverty estimates using different poverty lines and methodologies (see Tabatabai [1996] for a useful compendium of poverty estimates). However, there is enough consistency in the trends over time for most individual countries to get an overall picture of trends in rural poverty reduction. It appears from the available data that poverty has become more concentrated in rural areas over time and that urban poverty has declined more rapidly. Although rural poverty declined more in Southeast Asian countries and less in South Asian, rural poverty is predominant in both regions. In 1985, 79 percent of the Indian poor were located in rural areas; in Indonesia, 91 percent; in Malaysia and Thailand, 80 percent; and in the Philippines, 67 percent (World Bank 1990).

Who are the rural poor? The poor in rural areas tend to be illiterate and to depend on subsistence agriculture—often in resource-poor areas—and on (mostly low-skill) labor as their main asset. The rural poor also lack access to technology and credit; agricultural marketing costs are high because of distance to markets and poor rural infrastructure. Moreover, due to the small window of opportunity in rural areas, rural poverty has a large chronic component. A 1987/88 Bangladesh case study showed that the extremely poor owned less than half as much land, on average, as the nonpoor did. They also had less irrigated land and less land planted to high-yielding varieties; they depended on agriculture for a larger share of their income, compared to the nonpoor. Land ownership was a significant factor for determining absolute poverty. In the Philippines, the incidence of poverty was highest among farmers, and higher among self-employed households than among laborers. The rural poor households had a relatively larger family size, with more young and fewer well-educated household members. Poorer households had less access to use of modern agricultural technology, in part due to limited access to credit. Thus, there was only limited potential to increase productivity on the typically small land holdings. Moreover, access to social services, including health care and family planning, was constrained (Quibria and Srinivasan 1993).

Gender also influences poverty. Poor women typically have less income-earning employment and are more vulnerable to poverty, because they have fewer self-employment assets and marketable skills. They also experience disparities in wage and employment options that arise from the relatively higher labor intensity involved in domestic chores and subsistence tasks and from sex gaps in access to schooling, physical inputs, and credit (Bardhan 1993). Although in most Asian countries women and female-headed households are only slightly likelier to be poor, rural settings tend to increase the gender difference. In India, in 1983, it was estimated that rural women had a 12 percent higher probability than men of being poor—whereas men dominated urban poverty. In addition, female poverty tends to be more persistent, another characteristic caused and reinforced by lack of opportunities in the rural environment (De Haan and Lipton 1998).

Although the number of rural poor remains alarmingly high, many Asian countries were able to reduce the proportion of rural poor significantly during the 1980s, through a combination of rapid overall growth and more egalitarian policies (see below). Bangladesh has a very high incidence of rural poverty, but made considerable progress in poverty reduction during the 1980s: the incidence of poverty by headcount ratio dropped from very high levels of about 75 percent in the early 1980s to less than 50 percent in the last part of the decade (IFAD 1995). The incidence of rural poverty in India was higher than that of urban poverty throughout the 1960s and 1970s, but the figures began to converge by the end of the 1980s as rural poverty declined faster than urban poverty. However, despite the faster decline in rural poverty, poverty in India remains a predominantly rural phenomenon: in 1993–94, three out of every four poor people lived in rural areas (Datt 1998).

High economic growth rates in Indonesia in the 1970s and 1980s brought about a corresponding decline in the incidence of poverty both in absolute and in relative terms. The incidence of poverty in Indonesia as a whole declined from 39.8 percent in 1980 to 21.6 percent in 1987 and, in rural areas, from 44.6 percent in 1980 to 26.8 percent in 1987. Other East and Southeast Asian countries experienced significant declines in rural poverty as well, but despite the high growth rates in many economies, rural poverty persists. Accelerated growth in Thailand has been associated with a rapid decrease in the overall incidence of poverty, but growth has been uneven, with wide and growing rural-urban income disparities, and the incidence of poverty in rural Thailand declined at a much slower pace than that in urban areas (IFAD 1995). Poverty combined with poor education and health is also a near-universal condition in the mostly rural areas of the Mekong River Basin. In the Lao People's Democratic Republic (PDR), for example, where poverty levels are particularly high, 65 percent of the population is considered to be living in conditions of poverty, and two thirds of this number in "severe poverty," defined as households spending 80 percent or more of their consumption budget on food, according to recent consumption and expenditure surveys (Chagnon 1996).

Sri Lanka is an exception in that there was no reduction in rural poverty between 1970 and 1990, with the headcount ratio remaining virtually constant (Quibria and Srinivasan 1993). Its strong human-resource base and natural endowments would suggest that Sri Lanka could have achieved substantially higher growth rates and poverty reduction. However, the recent history of ethnic conflict, political unrest, and protectionist economic policies has slowed rural growth and poverty reduction. The Philippines also made slower progress in reducing rural poverty than the rest of Southeast Asia, with a decline in rural poverty from 57 percent in 1971 to 52 percent in 1991 (Balisacan 1994). Relatively slow growth, unequal income distribution, and policies that were heavily biased against agriculture contributed to the slow reduction in poverty.

GROWTH AND POVERTY REDUCTION IN ASIA

Rapid agricultural and economic growth, together with direct social spending, is the basis of poverty reduction in Asia. Econometric analysis indicates that economic growth explains between one third and one half of poverty reduction in Asia.

De Haan and Lipton (1998) report, based on a summary of various studies on developing Asia, that a 1-percent growth in per capita GDP is associated with a decline in the incidence of poverty of 0.82 percent. Moreover, contrary to the expectation of many observers, economic growth often improves both the relative and absolute incomes of the poor. Ravallion and Chen (1997) show that income distribution improves as often as it worsens in growing economies, and Deininger and Squire (1996) report that periods of economic growth were associated with increasing inequality in 43 cases and with decreases in inequality in 45 cases. Periods of economic decline were associated with increases in inequality in five out of seven cases (De Haan and Lipton 1998).

Since poverty is largely a rural phenomenon and since many of the poor depend, directly or indirectly, on the farm sector for their incomes, growth that raises agricultural productivity and the returns to farm labor is particularly important in reducing poverty. The contrast between Indonesia and India illustrates this point. Between 1970 and 1987 poverty in Indonesia declined by 41 percentage points; over the same period the purchasing power of agricultural value added rose by 2.6 percent annually per rural dweller. Between 1984 and 1987, a period of especially rapid declines in poverty, purchasing power grew by 5.0 percent per capita per year. In contrast, poverty in India decreased by 11 percentage points, and agricultural purchasing power grew by less than 0.4 percent a year. Most of the decline in poverty in India—7 percentage points between 1977 and 1983—took place at a time when agricultural purchasing power was growing at 1.5 percent a year (World Bank 1996c).

Econometric analysis confirms the link between agricultural growth and rural poverty in India. Datt and Ravallion (1997) found that differences in the growth rate of average agricultural output per unit of crop area were important in explaining cross-state differences in rural poverty reduction between 1958 and 1994. By contrast, differences in an Indian state's growth rate in nonagricultural output did not explain poverty reduction, reflecting the weak connections between urban economic growth and rural poverty reduction in India. Results showed that the urban poor also gained from rural growth, while the benefits to the urban poor of urban growth were partially dissipated by increasingly inequitable income distribution in urban areas. The initial endowments of physical infrastructure and human resources played a major role in explaining the trends in rural poverty reduction. Higher initial irrigation intensity, higher literacy rates, and lower initial infant mortality all contributed to higher long-term rates of poverty reduction in rural areas. A sizeable share of the variance in the trend rates of progress is attributable to differences in initial conditions of physical and human resource development—differences that reflect public-spending priorities (Datt and Ravallion 1997; Ravallion and Datt 1996).

In the PRC, the greatest progress in reducing poverty has occurred during periods of rapidly rising rural incomes. The most dramatic reduction in poverty occurred early in the 1980s, following reforms in agriculture. Between 1978 and 1985, the number of poor declined from 270 million to less than 100 million, according to the national poverty threshold of $0.60 per day (World Bank 1996c). Progress was also rapid relative to the international poverty line: a reduction from 569 million people in 1975 to 398 million in 1985 (Ahuja et al. 1997). Following this huge initial impact of agricultural reforms, progress in poverty reduction in the PRC has varied markedly, with the variations in the level of poverty directly related to the terms of trade for agriculture and to changes in government price policy for agriculture. The rate of poverty reduction slowed in the second half of the 1980s as rural incomes stagnated, due to declining gains in agricultural productivity combined with increased government intervention in agricultural markets, which reduced commodity prices. Relatively limited rural out-migration to fast-growing urban and coastal areas was a further constraint on poverty reduction. Thus, the number of poor declined relatively slowly from 1985 to 1993, from 398 million to 352 million (Ahuja et al. 1997). Large declines in poverty resumed in the 1990s, when a series of reforms in agricultural marketing policy (see also Chapters VII and VIII) improved the terms of trade for agriculture and boosted farm incomes (Rozelle et al. 1998). The poverty count fell significantly each year during 1991–96 (Park, Wang, and Wu 1998). Between 1993 and 1995 alone, the number of poor fell from 352 million to 269 million, nearly double the decline of the previous eight years (Ahuja et al. 1997).

Rapid agricultural and economic growth was the driving force behind the dramatic reduction in poverty in most of Asia. Agricultural growth that raises agricultural productivity and the returns to farm labor has been particularly important in reducing poverty because of the high concentration of poverty in rural areas and the dependence of many of the poor on the farm sector for their incomes. All agricultural growth is not equally beneficial to the poor. For agricultural growth to be pro-poor, it will ideally have a number of key attributes; these include

- a technology package that can be profitably adopted on farms of all sizes—such as the green revolution technology;
- a relatively equitable distribution of land with secure ownership or tenancy rights (see also below);
- efficient input, credit, and product markets so that farms of all sizes have access to needed modern farm inputs and receive similar prices for their products;
- a labor force that can migrate or diversify into the rural nonfarm economy; and
- policies that do not discriminate against agriculture in general, and small farms in particular (for example, no subsidies for mechanization).

The spatial configuration of agricultural growth is also important. If the growth occurs mostly in irrigated and other high-potential areas, then significant poverty problems may persist in less-favored areas even as the rest of a country forges ahead. Striking the right balance between public investments in different types of agroclimatic areas is also critical if agricultural growth is to have broad poverty-reducing impacts. These problems are addressed in more detail in Chapter XI.

The financial and economic crisis of 1997 raised the possibility of a reversal in progress in both urban and rural poverty reduction in the region. But recent spikes in poverty in countries like Thailand and Indonesia are unlikely to derail these countries from their long-term path of declines in poverty (see Chapter IX). The crisis should, however, be an important reminder for the governments of Asian developing economies of the need to increase their spending efforts on the mostly rural poor, and in particular, on their education, health, and nutrition.

BEYOND GROWTH: POLICIES FOR POVERTY REDUCTION

Although economic growth is the primary driver for poverty reduction in Asia and explains up to one half the decline in poverty, policies and investments in the fields of education, health, and infrastructure are also essential for sustained poverty reduction. Moreover, many of the important positive economic trends and beneficial policies described in Chapter I and elsewhere in this book may bypass the truly poor, or benefit them only gradually. Lipton and Sinha (1998) argue that, while globalization is changing the outlook for the rural poor by raising average incomes, it also tends to increase income variability both across regions (leaving some regions and countries behind) and across time, thus increasing the vulnerability of those who can least afford it. Moreover, macroeconomic and trade policy is being transformed by liberalization and globalization, producing large gains for many in both rural and urban areas (see Chapter VII), but relatively little for poor farmers and landless laborers, who often lack the skills, health, information, or assets needed to seize the new opportunities. The poor are thus increasingly concentrated in regions ill-equipped to gain from globalization/liberalization, e.g., in remote, backward areas of India and China.

Under these circumstances, growth alone will not solve the poverty problem. Policies must also reach out directly to the poor. Particularly important are investments in the human capital of the poor. Investments in health, nutrition and education not only directly address the worst consequences of poverty, but also attack some of the its most important causes. Even with rapid economic growth and active investment in social services, however, some of the poor will be reached slowly if at all. And even among those who do benefit to some extent, many will remain vulnerable to adverse events. These groups can be reached through income transfers, or through safety nets that help them through short-term stresses or disasters (World Bank 1990). In the remainder of this section, we will examine the role in alleviating poverty of investment in social services, rural infrastructure, land reform, and safety-net programs, including income transfers and income-generation schemes.

Education

As shown in Chapter I, education has a strong positive impact on economic growth. Education is also strongly linked to poverty reduction, through both direct and indirect influences. For agricultural areas, where most of the poor live, the direct impact of education works through the enhanced ability to adopt more advanced or complex technologies and crop-management techniques. Virtually all studies in this area confirm that better-educated farmers achieve higher rates of return on land (World Bank 1990). Education also encourages movement into more remunerative nonfarm work and induces migration to urban areas for industrial and service employment. The indirect effects of education are also significant: for example, educated mothers are more likely to ensure that their children receive an education and live healthily. Education of women, in fact, has powerful effects on nearly every dimension of development, from lowering fertility rates to raising productivity and improving environmental management (World Bank 1996c). These direct and indirect influences of education generate strong links between a region's or a country's educational attainment and its poverty reduction. This is especially true when primary education reaches women and when there is a high rate of completion of elementary school. Success in reducing poverty is usually enhanced by increasing the proportion of educational resources going to primary education and to the poorest groups or regions (Lipton 1998).

The cross-sectional evidence showing the strong education-growth-poverty reduction linkages, from household surveys and international data sets, is summarized in Gaiha (1994) and World Bank (1990). Evidence at the micro level also shows that education tends to reduce poverty. Wodon (1999) finds that, in Bangladesh, the educational level of both the household head and its spouse are important factors in reducing poverty. Jamison and Lau (1982) show that farmers and farm laborers improve their prospects of escaping poverty if they have some education (Lipton 1998). Singh and Hazell (1993) show that at the village level in India, there are strong synergistic interactions between better education, better health, higher earning power, and poverty reduction. They find that the combined effects of education plus land, or education plus bullock power, in reducing the incidence of poverty considerably exceed the sum of the individual effects, based on data for ten Indian villages over eight years (Lipton 1998).

Strategies to reduce the private cost of primary education for poor children in developing Asia include elimination of tuition and fees; introduction of vouchers that cover fees plus the costs related to school attendance; and the introduction of an augmented subsidy that covers fees, costs related to attendance, and opportunity costs forgone (i.e., students not working). The last measure has the best prospects to improve enrollment among the poorest. In Asia, an augmented subsidy was first introduced in Sri Lanka, and was later applied for poor girls in Bangladesh. Feeding programs at school have also enhanced the enrollment and retention rates among poor children in Bangladesh. Moreover, evidence of a strong association between female teachers and girls' enrollment suggests recruiting more female teachers, preferably from local communities (Hossain 1997).

Health and Nutrition

Investments in health and nutrition work together with education in poverty reduction. Such investments include development of safe drinking water, improved sewage disposal, and other sanitation measures, as well as immunization and public-health services. The effect of health and nutrition on productivity, and thereby on income enhancement and poverty reduction, is less well documented than the effect of education, but many studies now show positive effects of nutrition on agricultural productivity. Productivity impacts are most pronounced for activities in which most of the poor are engaged. For example, a study in India shows a significant link between wages and weight-for-height (a measure of short-term nutritional status) among casual agricultural laborers. In Sri Lanka, a significant positive effect of energy intake on real wages of laborers was found (World Bank 1990).

There is also a clear link between nutrition and education: better nutrition improves the learning ability and outcomes of children. Studies in the PRC, India, and other Asian countries show that protein-energy malnutrition contributes to lower test scores and poorer school performance. Micronutrient deficiency also weakens performance. In Indonesia, iodine deficiency reduced cognitive performance among children; in Thailand, provision of iodine supplements to schoolchildren has improved test scores. Investment in health and nutrition thus not only provides important health benefits, but also contributes to reducing poverty by boosting the productivity of the poor.

The private sector has a significant role to play in the delivery of health services, but there is a strong public-good rationale for interventions to improve health outcomes, particularly for the poor. Many essential public-health services are inexpensive and cost-effective, including immunization programs; provision of vitamins and micronutrient supplementation to school-age children; prevention programs for acquired immune-deficiency syndrome (AIDS), sexually transmitted diseases, and malaria; and community-based distribution of contraceptives. Moreover, basic health services, especially when administered through rural health centers, are particularly effective in reaching the poor and induce significantly greater reductions in poverty than higher-level services (Lipton 1998; Ahuja et al. 1996; World Bank 1990).

Rural Infrastructure

Rural infrastructure plays a key role in reaching the large mass of rural poor. When rural infrastructure has deteriorated or is nonexistent, the cost of marketing farm produce can be prohibitive for poor farmers. Poor rural infrastructure also limits the ability of traders to travel to and communicate with remote farming areas, limiting market access from these areas and eliminating competition for their produce. Construction of rural roads almost inevitably leads to increases in agricultural production and productivity by bringing new land into cultivation or by intensifying existing land use to take advantage of expanded market opportunities. In addition to facilitating agricultural commercialization and diversification, rural infrastructure, particularly roads, consolidates the links between agricultural and nonagricultural activities within rural areas and between rural and urban areas (IFAD 1995).

Binswanger, Deininger, and Feder (1993), in a study of 13 states in India, found that investments in rural infrastructure lowered transportation costs, increased farmers' access to markets, and led to substantial agricultural expansion. Better roads also lowered the transaction costs of credit services, resulting in increased lending to farmers, higher demands for agricultural inputs, and higher crop yields. Fan, Hazell, and Haque (1998) extend these results to show that rural infrastructure is not only an important driver for total factor productivity growth (TFP), but also directly contributes to a substantial reduction in rural poverty (see Box II.1).

Box II.1: Returns on Public Investments in Rural Areas in India

Few studies have directly assessed the relative effects of alternative public investments on agricultural productivity growth and equity. Fan, Hazell; and Haque (1998) address these issues for the case of India, based on an econometric model and state-level data for 1970–93. They find that all the productivity-enhancing investments considered offer a "win-win" strategy for reducing poverty while at the same time increasing agricultural productivity. There appear to be no tradeoffs between these two goals. There are sizable differences, however, in the productivity gains and poverty reductions obtained for incremental increases in each expenditure item.

According to the analysis, government expenditure on roads has by far the largest impact on poverty alleviation in rural areas, because it leads to new (non)agricultural employment opportunities, higher wages, and increases in productivity. If the government were to increase its investment in roads by 100 billion rupees (at 1993 constant prices), the incidence of rural poverty would be reduced by 0.87 percent. Moreover, the additional 100 billion rupees invested in roads would increase growth in TFP by 3.03 percent.

Second only to public investment in roads in alleviating rural poverty is investment in agricultural research and extension (R&E), which actually ranks first in its contribution to growth in TFP. Another 100 billion rupees of investment in R&E would increase growth in TFP by 6.98 percent and reduce the incidence of rural poverty by 0.48 percent. The effects of public investments in R&E on poverty alleviation are smaller than investments in roads because they only affect poverty through improved productivity and do not target the poor. If future investments in R&E were aimed more deliberately at the poor, they might well achieve a greater poverty impact (Fan, Hazell, and Haque 1998).

According to this analysis, other expenditures, including education, irrigation, and rural electrification, have lower but significant impacts on productivity and poverty reduction. Investments in irrigation and rural electrification have been essential in the past for sustaining agricultural growth. The levels of investment stocks achieved, however, may warrant maintenance of existing infrastructure rather than further increases.

Land Reform

Land reform can reduce poverty and enhance efficiency and growth though redistribution of land from large to small farms. However, not all policies that are labeled "land reform" increase productivity and reduce poverty; examples of policies that don't have the desired effects are enforced registration of individual titles, collectivization, state farming, prohibition of tenancy, and tenancy restriction in the absence of land distribution. The redistribution of land from big to small farms tends to sharply increase family labor use per hectare and, to a lesser extent, hired labor per hectare (Boyce 1987, Thiesenhusen and Melmed-Sanjak 1990). Lipton (1998), in a synthesis of the evidence, notes that the productivity and poverty reduction gains from genuine land reform are largest

- if land inequality is extremely large;
- if there are labor-intensive crop mixes and technical choices open to smallholders (and if choices of crop and method are considerable);
- if the rural person/land ratio is relatively high; and
- if complementary services (research, extension, credit, transport) are available (privately or publicly) to the new or enhanced smallholders.

A relatively equal land distribution in East and Southeast Asia, as compared to other developing regions, including most of South Asia, has contributed significantly to the strong record in poverty reduction in the region. Significant land reforms in the PRC, Viet Nam, Republic of Korea, and Taipei,China greatly improved the distribution of land in these countries, while Indonesia, Malaysia, and Thailand have historically had relatively small and widely-distributed land holdings (Ahuja et al. 1997).

There has also been some limited success with land reform in South Asia. India, where the incidence of poverty is highly correlated with lack of access to land, abolished the long-standing *zamindari* or "permanent settlement" system in the mid-1950s, ending one of the most iniquitous land ownership systems. Under the *zamindari* system, introduced at the end of the 18th century, feudal lords were declared proprietors of the land, peasants were transformed into tenant farmers, and rents were collected by a series of intermediaries who ruthlessly squeezed the farmers. This system was common in most of North India and covered around 57 percent of total area cultivated (Mearns 1999). The states of Kerala (known for its low income inequality and history of broad-based social programs) and West Bengal (where broad participation has been a key) have been particularly aggressive in transferring land to the people. Bangladesh and Pakistan undertook similar efforts in the 1950s to reform systems similar to the *zamindari*, but only limited efforts have been made to redistribute land since then.

In Central Asia, genuine land reform is probably the most crucial policy required to reverse the increase in poverty after 1990. Necessary reforms include the development of the legal framework for property rights, the establishment of private farms, and genuine privatization of state and collective farms. Progress on these reforms is described in Chapter VIII.

Land reform is politically difficult and further reform is highly uncertain. Yet recent analysis provides encouraging results that more modest and politically feasible tenancy reforms may provide significant poverty reduction benefits. Besley and Burgess (1998) find that, in India, both tenancy reforms that register tenancy and stipulate tenancy conditions and reforms that eliminate intermediaries between owner and tenant were linked to poverty reduction and increased agricultural wages that benefit landless laborers. Thus, although the effects on poverty are likely to have been greater had large-scale redistribution of land been achieved, partial, second-best reforms, which mainly affect production relations in agriculture, can play a significant role in reducing rural poverty. Mearns (1999), based on analysis of the Indian experience with land rights, proposes a series of guidelines for policy reforms to increase access to land for the rural poor; these include

- a selective deregulation of land-lease (rental) markets;
- a reduction in official and informal transaction costs by improving the management of land registration and land records;
- the promotion of women's independent land rights; and
- a strengthening of the institutions of civil society to provide checks and balances for a successful land reform.

Safety-Net Programs

Safety nets are programs that protect a person or household against two adverse outcomes: chronic incapacity to work and earn (chronic poverty) and a decline in this capacity because of a temporary situation that threatens survival in the face of limited reserves (transient poverty). Transient poverty is caused by events such as a sharp fall in aggregate demand, expenditure shocks (due to economic recession or transition, during unavoidable cutbacks in public spending, or as a result of a decline in production in sectors from which workers cannot migrate), or poor harvests (due to drought, flood, or pests, especially when they affect prices and production over a wide area). These shocks or disasters often cause the rural poor to lose their usual sources of protection offered by informal transfers (Subbarao, Braithwaite, and Jalan 1995).

Whereas in many Asian developing countries strong informal safety nets, including effective family-support systems and private income transfers, play a significant supporting role in cushioning the worst impacts of transient poverty, Central Asia under the Soviet system relied mainly on formal state-sponsored programs of social assistance. The collapse of the Soviet system has put considerable pressure on these systems (see Box II.2).

Safety-net programs usually provide either direct transfers of income to the poor or attempt to generate income. They include food subsidies, public-works programs, and credit programs aimed at the poor. Subbarao et al. (1997) and Lipton (1998) provide excellent syntheses of cross-country experience with safety-net programs; this section draws heavily upon these studies.

Food subsidies, including open general subsidies, quantity rationing, and food stamps, have been utilized for more than four decades in some Asian countries. Untargeted food—and other—transfer programs require large subsidies and have proven to be fiscally unsustainable, as they create explicit tradeoffs between spending on safety-net programs and investments for growth. Moreover, these programs are politically difficult to scale down or abandon. In Sri Lanka, for example, the universal ration program in operation prior to 1979 cost the government up to 5 percent of GDP. The unsustainably high cost drove the government to institute food-stamp programs, cutting costs to a still-substantial 1.3 percent of GDP. The Philippines and India have also cut back food-subsidy programs due to fiscal tightness and poor cost-effectiveness. India is slowly reforming its expensive and poorly targeted nationwide program, the Public Distribution System. But India also has some innovative and better-targeted programs, such as the Integrated Child Development Services and the Tamil Nadu Nutrition Program.

Targeted approaches to food subsidies are more cost-effective than universal schemes, but it is politically infeasible and administratively difficult to run finely targeted programs in countries where the incidence of poverty is high and households in dire need are difficult to separate from those that are not or are less so. Moreover, targeted systems such as food stamps often create disincentives to work and cause other kinds of consumption distortions. Approaches that impose an obligation on the part of the recipient (such as a labor or time requirement) are best in screening out the nonneedy. But such obligations should not be so onerous that they significantly increase transaction costs. In general, the poor should be supported with minimal distortions, local communities should be involved, and the approach should be demand-driven. A promising approach, although administration-intensive, is food transfers targeted to women and children, along with other services (such as immunization).

Unlike cash and in-kind transfers, such as food subsidies, income-generation programs oblige the recipient to exchange labor time for an income transfer. Two programs have been used

> **Box II.2: Safety Nets in Central Asia**
>
> Safety-net programs in Central Asia were built around comprehensive public social insurance programs, particularly universal pension systems and formal family assistance. The major goal of the family-assistance program was to maintain a level of per capita income as family size increased or to supplement wages. Many of these programs were typically administered by the state enterprises and were almost universal. However, some of the weakly targeted cash transfers had eroded in real terms over the years.
>
> When the Central Asian economies collapsed at the beginning of the 1990s, output contracted, real incomes declined sharply, and unemployment and inflation rose rapidly. Some jobs were artificially maintained, generating a new echelon of working poor. As a result, rising numbers of newly poor have spread throughout the society. Pensions and other benefits depreciated significantly. The elimination of widespread price subsidies, incompatible with the transition to market economies, the withdrawal of subsidies on items such as housing, heating, and food, and the reduction in the quality and availability of social services hurt people who often had low fixed incomes. Unemployment benefits often had to be newly introduced. Due to the precarious financial situations of these countries in transition, any assistance and transfer program that imposes a heavy fiscal burden has little chance of being sustained. As poverty is correlated with family size to some extent in all Central Asian republics, reformed family-assistance programs have been identified as likely contenders for targeting the larger
>
> (continued next page)

widely: labor-intensive public works and credit-based self-employment (livelihood) programs. Public-works programs provide mainly current benefits and in most countries offer only temporary employment during off-seasons, when agricultural work is limited. However, long-run benefits can be generated

> Box II.2 (continued)
>
> number of poor people. However, the line between redressing an old instrument to effectively target the poorest at a low administrative cost and simply maintaining an ongoing assistance program in the face of political backlash is difficult to draw. The Kyrgyz Republic, for example, has discontinued family assistance to children above age 18 and introduced means testing for assistance, which is crucial to minimize costs. However, means testing has become more difficult as informal employment and other means of increasing income have risen sharply following the contraction of the economies.
>
> Moreover, although family and other assistance might have increased as a share of GDP following the economic decline—as it did in the Kyrgyz Republic—the plunge in GDP and high inflation have prevented social assistance programs from adequately protecting the people in greatest need. In addition, when a large number of families is qualified for assistance—about 80 percent of families with children qualify in the Kyrgyz Republic—total available resources are thinly spread and the impact on poverty is minor. Thus, in the Kyrgyz Republic, slightly more than 40 percent of the transfers go to the nonpoor.
>
> The fiscal necessity of reducing entitlements like pensions still needs to be better balanced with the need to reach the newly poor. Adopting a more structured approach to identifying the newly poor requires an institutional capacity that is often not in place. Creating acceptance among the population of targeting resources to the poorest of the poor is proving similarly difficult, but effective policies are badly needed, with large numbers of people in Central Asia living in poverty with little or no social support.

if the public works themselves are designed to build up assets (savings, physical capital, skills, health, or infrastructure) owned by, or providing future employment income to, the poor. For example, India's Million Wells Scheme and its successors built irrigation wells on small and marginal farms, creating durable assets and enabling poor farmers to generate subsequent self-employment income. Some forms of village infrastructure, such as the drought management works favored by some public-works programs in India in the early 1980s and again recently, directly help the poor, as can human capital creation, for example employment schemes to build primary schoolrooms.

Cross-country experience indicates that public works can be a very effective means of consumption smoothing for poor households. Improved targeting can be achieved through various methods: self-targeting through type of work, form of payments, and level of wage; targeting of women through piece rates, location of work, or provision of daycare; and geographic targeting of regions adversely affected by shocks. Program wages should be close to the prevailing market wages for unskilled labor in order to minimize labor-market distortions. Careful attention should be given to the quality of assets created and their maintenance to improve program cost-effectiveness.

Credit programs for the poor have been undertaken both in response to transitional poverty resulting from economic shocks or natural disasters and to address chronic poverty in low-growth situations. Most credit programs aim to provide an income stream over the medium term, but, as will be described in more detail in Chapter III, public credit programs have usually had highly negative impacts on financial markets and have not been particularly effective in reaching target clients. Unlike other social assistance programs, credit programs require high, specialized levels of financial and managerial capacity. In addition, since credit is fungible, the funds may be used for purposes other than those for which they are intended.

The generally poor performance of massive, subsidized, state-directed credit in the 1970s and 1980s taught important lessons for improved performance of targeted credit:

- transaction costs for the poor should be reduced by simplifying loan-processing procedures to avoid delays and repeat visits;
- interest rates should not be subsidized;
- reliance on program-manager (bureaucrat) or income-

based identification of potential beneficiaries should be replaced by a reliance on communities, nongovernment organizations (NGOs), and other local groups;
- savings should be promoted as an integral part of the program; and
- covariant risks to borrowers and lenders should be mitigated through group lending, small loans, and repeat loans to those who promptly repay.

It should also be recognized, however, that even with appropriate reforms, the most successful credit programs have had only a modest record in reaching the ultrapoor. Therefore, caution should be exercised in using targeted credit interventions as a poverty-reducing tool. As will be shown in more detail in Chapter III, interventions in financial markets are usually a poor second-best approach for dealing with important social problems that require more direct policies.

CONCLUSIONS

Poverty has declined rapidly during the last three decades in developing Asia, but large pockets of poverty remain in virtually all developing countries in the region. Rural poverty has been particularly neglected in the past, because the rural poor are more dispersed, are more difficult and more expensive to target and reach out to, and are generally of less political interest. However, reaching out to the rural poor not only increases their well-being and future incomes, but also furthers overall agricultural and economic growth in the region.

Indonesia, Malaysia, and Thailand demonstrate the benefits of an appropriate balance between policies that spur growth and policies that enable the poor to participate in growth. All three countries achieved rapid and sustained growth in GDP that was relatively labor-intensive—with agriculture playing a leading role—which generated demand for labor, thereby benefiting the poor. These countries also provided for adequate social expenditures. As a result, they have achieved universal primary education and their infant-mortality rates are lower than those of many countries with similar incomes. The improvement in the skills and quality of the labor force enabled the poor to seize the opportunities provided by economic growth (World Bank 1990). In other countries the creation of opportunities for the poor and the development of their capacity to respond have not always been so well balanced. Pakistan also had rapid GDP growth, but reduction in poverty has been much smaller and slower than in East and Southeast Asia. Pakistan has one of the lowest rates of primary-school enrollment in the world, leading to a failure to improve the skills of the labor force that has limited the ability of the poor to benefit from growth.

Growth alone, then, will not be sufficient to meet the needs of South Asia's poor. Moreover, in Pakistan (as well as Bangladesh and India) social indicators are among the worst in the developing world; in many parts of the region, the economic growth of the 1980s was not accompanied by concomitant improvements in living standards. Out of every 12 children born in South Asia, at least one is expected to die before reaching the age of one; nearly half of all children do not finish primary school. The status of women is of particular concern: in many parts of South Asia, they are less well educated, fall ill more frequently, have a shorter life expectancy, and work far longer hours than men. Increasing women's access to services—particularly to basic education, health services, nutrition, water and sanitation, and family planning—and improving the quality of those services are essential (World Bank 1990, 1996c).

Thus, it is possible to have economic growth without adequate social progress. Conversely, at least some social indicators can be improved even in the absence of rapid economic growth. The experience of Sri Lanka shows that remarkable social progress can be achieved even at low levels of income. The benefits of Sri Lanka's long-standing support for social services can be seen in its under-five mortality rate, which was 66 per thousand in 1980, an impressive achievement for a low-income country. Infant mortality rates have fallen from

48 per 1,000 live births in 1970 to 16 per 1,000 in 1996, while life expectancy at birth has climbed from 67 to 75 years for women and 65 to 71 years for men over the past three decades (World Bank 1999). But slow economic growth makes it impossible to reduce poverty itself. This suggests that specific public policies can make the poor better off in some important aspects, such as reduced child mortality, but raising the incomes of the poor and thus lifting people above the poverty line require broad-based economic growth. The evidence shows that the countries that have been most successful in attacking poverty have achieved rapid agricultural growth and broader economic growth that makes efficient use of labor; they have also invested in the human capital of the poor. Rapid and relatively labor-intensive economic growth provides the poor with opportunities to use their most abundant asset, their labor. Investment in human capital improves their immediate well-being and increases their capacity to take advantage of the newly created possibilities (World Bank 1990).

III Agricultural Diversification and Commercialization

INTRODUCTION

The onset of rapid economic growth in Asia, stimulated in significant part by agricultural growth, in turn initiated a process of transformation of the structure of production across sectors. As the the share of manufacturing and services has expanded, the share of agriculture in the value of total output has declined. The transformation of the structure of output has been accompanied by a decline in the share of employment in agriculture. In addition, the decline in the relative size of agriculture in Asian economies has been accompanied by the commercialization and diversification of agriculture. As economies grow, there is a gradual movement from subsistence food-crop production to a diversified market-oriented production system. Many observers have expressed concerns about possible adverse effects of this process of diversification and commercialization on the poor, but for the most part, these concerns are not borne out by recent evidence.

STRUCTURAL CHANGE IN AGRICULTURE

The rapid economic development in most of Asia, which was described in Chapter I, fundamentally changed the relative position of agriculture in the region. A central and nearly universal characteristic of economic development is the structural change in the productive sectors, with a relative decline in the size of the agricultural sector in the economy. A number of factors contribute to this relative decline, including

- the effects of income and expenditure growth on food demand and food prices relative to other goods and services, in particular Engel's Law, which states that as real expenditures increase, the share of expenditures on food declines;
- differential rates of technical change, with more rapid growth in technical change in the nonagricultural sector leading to a decline in the share of agriculture in the economy; and
- capital accumulation and the resultant change in capital and labor endowments, leading to a decline in the share of total output of the relatively labor-intensive agricultural sector, countered by a relative increase in the output share of the capital-intensive nonagricultural sector.

The relative importance of these effects will depend on the structure and dynamics of development in each economy. Although Engel's Law has typically received most attention in the literature as a determinant of agriculture's decline, Martin and Warr (1992), for example, found that in Thailand, capital accumulation and technical change are far more important in explaining the decline of agriculture relative to the nonagricultural sector. In Indonesia, on the other hand, capital accumulation was by far the dominant cause for agriculture's relative decline, while changes in relative prices played only a small role; technical change was actually biased in favor of agriculture, slowing the decline in its relative size (Martin and Warr 1991).

The decline in the relative size of agriculture has proceeded rapidly in much of Asia, as can be seen in the share of agricultural value added in total GDP, in the share of agriculture in exports, and in the share of agricultural labor in total labor. The rate of relative decline in agriculture has been a direct function of the rate of GDP growth, with the rate of agriculture's relative decline faster in the rapidly growing countries. As can be seen in Figure III.1 and Table III.1, the share of agriculture in GDP in the Republic of Korea, the most rapidly industrializing of the countries shown here, declined from 34 percent in 1966 to 15 percent in 1980 and to under 7 percent by 1995. In 30 years, the share of agriculture in GDP declined by two thirds in Indonesia and Thailand, to 17 percent and 11 percent respectively, while in the PRC and Malaysia, the share was more or less cut in half, to 20 percent and 13 percent respectively. Viet Nam's share of agriculture in GDP significantly declined as economic growth boomed during the 1990s, to 28 percent in 1995, after variable increases and declines in the late 1980s.

The South Asian countries also experienced a substantial decline in the share of agriculture, although from a higher initial level and not as rapidly as the East and Southeast Asian countries. The share of agriculture in GDP in Bangladesh was variable, with no trend decrease, in the range of 50–60 percent during the stagnant growth period of 1966–78, before commencing a steady decline to 31 percent in 1995 as economic growth accelerated. The decline in India's share of agriculture in GDP began in the early 1970s; the value added to GDP by the agriculture sector declined from about 45 percent to 28 percent in 1995. Both Pakistan and Sri Lanka experienced relatively slow declines in the share of agriculture in GDP, from 37 percent to 26 percent, and from 28 percent to 23 percent, respectively, over the 30-year period. Nepal's structural change was more rapid, with a shift in the contribution to GDP by agriculture from 71 percent in the mid-1960s to 42 percent by 1995.

In the Philippines, structural shifts in the economy had already taken place by the mid-1960s; consequently, the decline in the value added to GDP by agriculture was rather modest during the last 30 years, from 26 percent in 1966 to 22 percent in 1995. Myanmar is the only exception in this group of Asian developing countries, with very slow growth in the nonagricultural sector leading to an increase in the share of agriculture in GDP from 34 percent in 1966 to 61 percent in 1995.

The declining share of agricultural exports in total exports in the region is directly linked with the declining contribution of agriculture to total GDP. Thus, despite the sometimes very rapid annual rates of growth in agricultural exports of some Asian developing economies over the 1967–95 period (for example, Viet Nam, 15 percent; Republic of Korea, 14 percent; Indonesia, 10 percent), the share of agricultural exports in total exports declined significantly in each Asian developing country shown in Table III.2. India experienced the slowest decline, at 3.3 percent per year, followed by the rest of the South Asian country group.

In several of the rapidly growing East and Southeast Asian economies, growth in the agriculture sector was the a major source of overall economic growth until the late 1980s. Consequently, the more rapid decline in the share of agricultural exports in total exports in these countries only began in the late 1980s or early 1990s, with agriculture's share in the PRC, Indonesia, Malaysia, Pakistan, Thailand, and Viet Nam declining to between 10 and 13 percent by 1995. The share of agricultural exports in total exports in Bangladesh declined substantially over the 1982–89 period, followed by an even more rapid decline at the beginning of the 1990s to only 4 percent of total exports in 1995. In the Republic of Korea, on the other hand, the agriculture sector already contributed only 10 percent to total exports by 1966 and 6 percent by 1970; the share dropped even further to 1 percent by 1995.

Parallelling the trend of a declining relative importance of agriculture in the economic structure is the share of agriculture in the total labor force, although these shifts are less dramatic. By far the largest structural shift in the labor-force participation over the last 30 years occurred in the Republic of Korea, with labor in agriculture declining almost fourfold, from 54 percent in 1966 to 14 percent in 1995 (Table III.3 and Figure III.2). Only Malaysia experienced a similar decline in the share of the economically active population in agriculture, from 58 percent to 23 percent. In the other East and Southeast Asian economies, the share of the labor force in agriculture has so far remained at or above 50 percent (the PRC, 70 percent; Indonesia, 52 percent;

Agricultural Diversification and Commercialization 61

Figure III.1: Agriculture, value added (% of GDP)

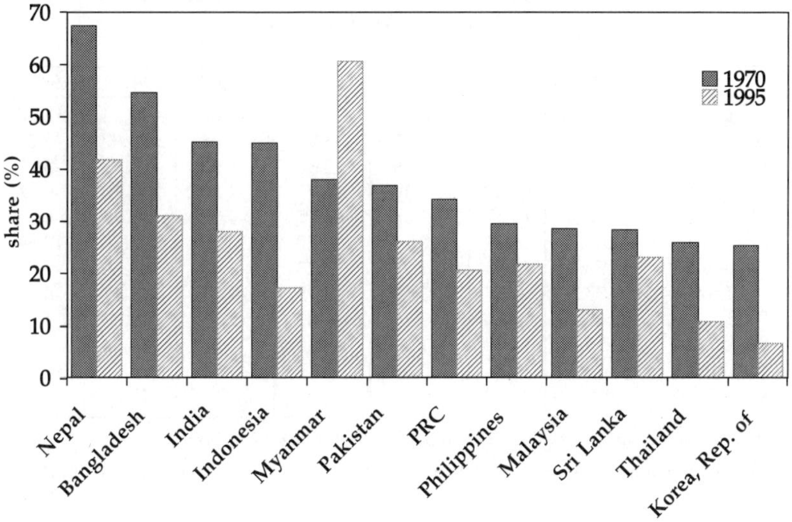

Source: WDI 1998.

Table III.1: Agriculture, Value Added (% of GDP)

Year	1966	1970	1975	1980	1985	1990	1995
Bangladesh	53.95	54.56	61.95	49.64	41.77	36.85	30.88
India	44.94	45.17	40.50	38.10	33.03	30.97	27.87
Indonesia	50.81	44.94	30.18	23.97	23.24	19.41	17.16
Korea, Rep. of	34.12	25.37	24.13	14.53	12.49	8.68	6.54
Malaysia	28.32	28.52	27.98	21.91	19.30	18.72	13.02
Myanmar	34.40	38.00	47.07	46.54	48.20	57.26	60.63
Nepal	70.51	67.29	71.76	61.77	51.71	51.63	41.77
Pakistan	37.07	36.83	32.05	29.52	28.53	25.98	26.02
Philippines	25.69	29.52	30.34	25.12	24.58	21.90	21.62
PRC	36.06	34.13	31.97	30.09	28.35	27.05	20.59
Sri Lanka	28.30	28.30	30.35	27.55	27.69	26.32	23.01
Thailand	33.40	25.92	26.87	23.24	15.81	12.74	10.84
Viet Nam	–	–	–	–	–	37.47	27.55

Source: WDI 1998.

Table III.2: Agricultural Exports as Share of Total Exports (%)
and Annual Growth in Agricultural Exports as Share of
Total Exports (%/year)

Year	1970	1995	1967-82	1982-95	1982-89	1989-95	1967-95
Bangladesh	39.57	3.51	-5.99	-11.73	-9.06	-14.74	-8.70
India	30.97	13.85	-2.52	-4.14	-7.01	-0.67	-3.28
Indonesia	36.9	10.33	-12.19	2.36	6.92	-2.71	-5.71
Korea, Rep. of	6.31	1.09	-8.10	-4.77	-4.83	-4.70	-6.57
Malaysia	42.26	10.1	-3.54	-6.75	-4.38	-9.45	-5.05
Nepal*	105.73	6.47	-12.17	-9.27	-6.19	-12.73	-10.84
Pakistan	26.2	10.23	-1.78	-7.23	-3.03	-11.89	-4.35
Philippines	33.53	6.98	-3.58	-8.95	-9.37	-8.47	-6.11
PRC	45.43	9.83	-3.16	-6.83	-2.13	-12.03	-4.88
Sri Lanka	55.07	14.49	-1.31	-8.80	-4.63	-13.44	-4.86
Thailand	46.49	12.85	-0.94	-9.19	-9.70	-8.59	-4.86
Viet Nam	–	21.10	–	–	–	-13.09	–

Note: Percentages above 100 percent in Nepal are due to data inconsistencies.

Source: (of agricultural exports) FAO FAOSTAT *1998*. Agricultural Trade; total exports: WDI 1998.

Thailand, 60 percent). In Viet Nam, the share has also declined only slowly, from 79 percent to 69 percent over the 30-year period. Both Pakistan and Sri Lanka experienced rates of decline in the share of labor in agriculture similar to the rapidly developing East and Southeast Asian countries, with shares of agricultural labor in 1995 of 49 percent and 47 percent, respectively.

The share of the labor force in agriculture declined more slowly in Bangladesh and India, to slightly more than 60 percent by 1995. In Nepal, almost the entire active population, 93 percent, is still employed in the agriculture sector, with virtually no change during the 1966–95 period. Although agriculture only contributed between 20 and 25 percent to GDP in the Philippines during the last thirty years, the sector still employed 43 percent of the labor force in 1995, down from 60 percent in 1966. In Myanmar, the increasing contribution of agriculture to GDP is not reflected in the employment structure. On the contrary, the share of labor in agriculture declined over the 30-year period, albeit not markedly, from 79 to 72 percent.

Table III.3: Share of Agricultural Labor in Total Labor Force (%)

Year	1966	1970	1975	1980	1985	1990	1995
Bangladesh	85.16	83.50	78.01	72.60	68.90	65.23	60.56
India	73.73	72.64	71.08	69.53	66.74	64.02	61.87
Indonesia	69.70	66.30	62.08	57.82	56.46	55.18	51.78
Korea, Rep. of	53.90	49.14	43.17	37.12	27.64	18.11	13.53
Malaysia	57.57	53.75	47.30	40.79	34.15	27.33	22.68
Myanmar	79.42	78.39	77.09	75.81	74.55	73.27	71.77
Nepal	94.65	94.41	94.09	93.78	93.67	93.65	93.37
Pakistan	65.03	64.55	63.92	63.16	57.50	51.80	49.40
Philippines	60.20	57.91	55.14	52.30	49.03	45.76	42.63
PRC	80.25	78.34	76.26	74.24	73.24	72.24	69.54
Sri Lanka	55.81	55.28	53.57	51.88	50.26	48.50	47.03
Thailand	81.37	79.82	75.37	70.91	67.48	64.06	60.32
Viet Nam	79.13	77.49	75.34	73.20	72.25	71.30	69.36

Source: FAO FAOSTAT 1998. *http://faostat.fao.org*. Population DataBase.

Figure III.2: Share of Agricultural Labor in Total Labor Force (%)

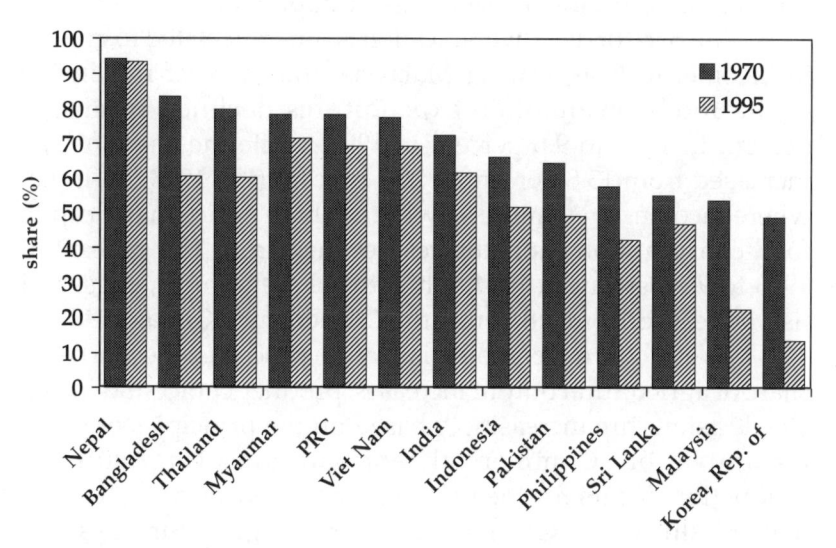

Source: FAO FAOSTAT 1998. Population DataBase.

COMMERCIALIZATION AND DIVERSIFICATION

The process of general economic growth with concomitant decline of the relative size of agriculture in the economy has been accompanied by the commercialization and diversification of Asian agriculture. As economies grow, there is a gradual movement out of subsistence food-crop production (mostly of basic staple crops) to a diversified market-oriented production system. The process of diversification out of staple-food production is triggered by rapid technological change in agricultural production, by improved rural infrastructure, and by diversification in food-demand patterns. The slowdown in income-induced demand growth for staple foods is accompanied by a shift of diets to higher-value foods such as meats, fish, fruits, and vegetables. These dietary transitions are induced by declining income elasticities of demand for staples as per capita incomes rise and by the rapid migration of population to urban areas (see also Chapter VI).

The pace of diet diversification is directly related to growth in income. In fast-growing Malaysia, the share of rice in per capita food-consumption expenditures declined from 24.1 percent in 1973 to 9.6 percent in 1993, while the share of fish increased from 13.8 percent to 20.2 percent (Lin 1998). In India, where incomes grew more slowly, the share of rice in per capita food-consumption expenditures declined from 15.1 percent in 1973 to 9.7 percent in 1994, while the share of meat, eggs, and fish increased from 1.8 percent to 2.1 percent (Kumar 1998).

As agricultural commercialization proceeds, the marketed share of agricultural output increases; product-choice and input-use decisions are increasingly based on the principles of profit maximization. Commercial reorientation of agricultural production occurs for the primary staple cereals as well as for higher-value crops. Commercialization of agricultural systems leads to greater market orientation of farm production; progressive substitution out of nontraded inputs in favor of purchased inputs; and the gradual decline of integrated farming systems and their replacement by specialized enterprises for crop, livestock, poultry, and aquaculture products. The farm-level determinants of increasing commercialization are the rising opportunity costs of family labor and increased market demand for food and other agricultural products. Family labor costs rise because of increasing off-farm employment opportunities, while positive shifts in market demand are triggered by urbanization and/or trade liberalization (Pingali and Rosegrant 1995).

This diversification in production in Asia can be seen at the aggregate level by the pattern of production growth of staple cereals compared to other, higher-value crops. For all Asian countries, livestock production grew at 6.4 percent per year and production of vegetables, fruits, and treenuts grew at 4.4 percent per year, whereas cereal production only increased at 2.7 percent annually during 1973–96 (Delgado and Ammar 1997). The process of commercialization has been accelerating in recent years, with the East Asian countries at the high end of this agricultural transformation process, the Southeast Asian countries moving rapidly towards commercialization, and the South Asian countries advancing more slowly. In East Asia, livestock production grew at 8.5 percent per year during 1989–95, while cereal production grew at 2.4 percent per year. In Southeast Asia, livestock production increased at 5.8 percent per year and cereal production grew at 2.8 percent per year during the same period. In South Asia, finally, livestock production grew at 4.3 percent per year and cereal production increased at 2.6 percent per year during 1989–95.

The rapid pace of production diversification is even more apparent in the pattern of change in value of output. In Asia as a whole, the value of vegetable production as a proportion of the value of cereal production increased from 17 percent in 1980 to 30 percent in 1993. Moreover, diversification is more rapid in the fast-growing East Asian economies than in South Asia. In East Asia, the value of vegetable production as a percentage of value of cereal production increased from 22 percent in 1980 to almost 40 percent in 1993, while in South Asia, the percentages changed from 10 percent to 15 percent over the same period (Kumar 1998).

Initially, diversification implies the addition of other crops and other enterprises at the farm-household level. As commercial orientation increases, however, diversification at the agricultural-sector level is created by household-level specialization, as households shift away from traditional self-sufficiency goals and towards profit- and income-oriented decision making, with increased responsiveness of farm outputs to market needs. Mixed farming systems give way to specialized production units that can respond more quickly to market price and product quality signals. The returns on intensive subsistence production systems that require high levels of family labor generally decline relative to the returns on market-oriented production. In addition, family labor is increasingly replaced with hired labor, as family members find more lucrative nonagricultural employment opportunities. The increasing opportunity costs of family labor due to the increase in off-farm employment opportunities lead to a substitution of traded inputs for the more labor-intensive nontraded inputs.

Thus, mechanical power substitutes for human and animal power, chemical fertilizer for manure and other organic fertilizers, and commercial feed for farm-produced fodder (Pingali and Rosegrant 1995; McIntire, Bourzat, and Pingali 1992; Han 1992). These household-level changes in the organization of production are facilitated

- by agricultural research that generates new technologies to increase productivity and farmer incomes;
- by economic liberalization that opens up new opportunities;
- by the development of rural financial markets that provide the credit for the expansion of commercial agriculture;
- by public investments in rural infrastructure; and
- by establishment of secure property rights.

Agricultural Research

Agricultural research can provide farmers with increased flexibility to make crop choice decisions and to move relatively freely between crops. Both substantial crop-specific research and system-level research efforts contribute to flexibility of crop choice. Crop-specific research includes increases in yield potential, shorter-duration cultivars, improved quality characteristics, and greater tolerance to pest stresses. System-level research includes land-management and tillage systems that allow for shifts of cropping patterns in response to changing incentives and farm-level water-management systems that can accommodate a variety of crops within a season.

Economic Liberalization

Liberalization of domestic markets is often a key step in starting or accelerating the process of commercialization. The main components of economic liberalization are a reduction in trade restrictions (elimination of quantitative restrictions, uniformization and reduction of import tariffs); realignment of macro policies (reduction of fiscal deficits, elimination of multiple exchange rates, easing of exchange controls); and a liberalization of markets in general, including financial markets and asset markets. This reform process opens up international trade opportunities and provides price signals to guide producer decisions, thus encouraging a diversification of the economy. The opening of domestic markets, however, may expose producers to increased risk due to the greater volatility of world prices, although the evidence is mixed (see Chapter VII). Governments have historically intervened heavily in domestic markets to protect and stabilize the prices of agricultural commodities; when effective, stabilization policies have reduced domestic producer price variability relative to international price variability (Islam and Thomas 1996).

The relationship between diversification and risk is thus crucial in the context of trade and macroeconomic reforms designed to align domestic prices more closely with international prices. Diversification can be an efficient mechanism for reducing the impact of risk on producers' welfare. Diversification of the crop mix at the household level is unlikely to greatly reduce the price risk, however; the prices of agricultural commodities are highly correlated, because of their common reaction patterns to aggregate, worldwide, and macroeconomic shocks. Furthermore, the process of diversification is itself likely to increase the correlation of prices. On the consumption side, the increasing flexibility of diets (as discussed above) means that more substitution in consumption will occur, resulting in higher correlation of prices. On the supply side, to the extent that investments in infrastructure and increased market integration make a more diversified output mix possible, there will be increased substitution possibilities in production, resulting again in higher price correlation. A strategy aimed at household-level diversification is likely to be self-defeating: the more each unit of production is diversified, the more positive the correlation between prices and the lower the gains from diversification.

Rural Financial Markets

Liberalization and integration of rural and other, broader financial markets can reduce the costs of increased price variability through risk pooling on an economy-wide basis. Financial integration for risk spreading is critical at the rural household level as well. In order to exploit the income-enhancing potential of commercialization of agriculture, financial markets must accommodate the increased ability of households to save and build up productive asset bases and improve human resources. Rapid development of rural finance systems at the grass-roots level is an impetus to commercialization, particularly since commercialization of agriculture often leads to large, lumpy payments of cash a few times a year. The process of commercialization itself can provide the critical market size required for efficient, unsubsidized rural banking with low overhead costs.

In addition, technological changes often require complementary investments that increase demands for working and investment capital. Some of these demands will be self-financed, others will be serviced by informal sources, but still others will require longer-term loans provided by formal institutions. Supplying reasonably priced loans, therefore, can speed the adoption of technology, expand the production of food supplies, and increase farm incomes. Effective rural financial institutions can assist in spreading the benefits of commercialization more widely across the community and region.

The strong complementarity between commercialization (and rural development in general) and rural financial-market development has led governments to intervene heavily to influence the availability and cost of credit to farmers. Five main types of government interventions have been used:

- lending requirements imposed on banks;
- refinance schemes;
- loans at preferential interest rates;
- credit guarantees; and
- lending by government-operated development finance institutions.

In a comprehensive review of rural financial markets in Asia that constitutes a complementary volume in this Rural Asia series, Meyer and Nagarajan (1999) note that the benefits from directed and subsidized credit and other government interventions have been small. The BIMAS project in Indonesia and Masagana 99 in the Philippines were typical. These programs had a limited impact on adoption of new technology but seriously impaired the banks, cooperatives, and specialized agricultural development banks that tried to implement them. The rapid expansion of government credit to agriculture in India, Bangladesh, and the PRC in the 1970s and 1980s similarly appears to have had little impact on agricultural production.

There was a myriad of other and largely unintended consequences. Government interventions in rural financial markets failed to provide savings and other financial services demanded by farmers. Middlemen and banks often captured the subsidies intended for borrowers. Interest rates were low but borrower transaction costs were high, banks earned low returns on their capital, and credit allocation may have worsened income distribution if the credit was skewed in favor of larger farms.

The clearest impact has been found in the damage that directed credit inflicted on rural financial systems. Many subsidized credits became nonperforming loans because cheap interest encouraged unprofitable investments. In some cases borrowers intentionally defaulted because they believed that governments would waive the loans or not take action against those in the priority sectors. Financial discipline was damaged and intermediaries weakened. Repayment rates often deteriorated over time, as occurred in the BIMAS program in Indonesia. Loans were frequently forgiven, especially in India and Bangladesh. Many credit institutions were weakened and failed or required refinancing; refinance schemes discouraged savings mobilization, leading to lower financial intermediation. Clearly the relatively small economic benefits that were realized from directed and subsidized credit came at a high cost to financial systems (Meyer and Nagarajan 1999).

The general failure of directed and subsidized credit has caused a shift in approach that limits the role of financial markets to financial intermediation rather than as a tool to stimulate production, compensate for distortions in other markets, and alleviate poverty. First, the appropriate role for government is seen as creating an environment in which competitive financial institutions can emerge. Among other things, this means macroeconomic stability, reasonably low levels of inflation, procedures to enforce contracts, the protection of property rights, and a regulatory and supervisory system to ensure prudent financial operations.

Second, governments need to avoid the temptation of attempting to use financial institutions to carry out social policies, such as subsidizing particular economic activities or groups within society. Financial-market interventions are a poor second-best approach for dealing with important social problems that require more direct policies to encourage human capital formation and improve access to productive assets. The new approach is shown by the move of a number of Asian governments to financial-market liberalization, reduced targeting of loans, and the setting of interest rates at high enough levels to cover costs.

Due to the success of some microfinance organizations (MFOs), such as the Grameen Bank in Bangladesh, these have been held up as possible models for this new approach. But the applicability of microlending techniques to specialized farmers who have highly seasonal cash flows or long-term capital requirements may be limited. Most MFOs use group lending to reduce their transaction costs, but group lending may not be effective in sparsely populated areas where group members have less information about each other and monitoring is more costly. Financial markets for the poor are highly segmented, with each microlender serving a small market niche; high information and transaction costs discourage competition and constrain the MFOs from rapidly expanding to serve new clients and regions. In addition, many MFOs obtain their resources from subsidized sources and have little experience in mobilizing savings.

Lessons for agricultural lending might better be drawn from the experience of the unit desas of Bank Rakyat Indonesia (BRI-UD). Unit desas are village banking units that channel loans to farmers. BRI-UD was reorganized in 1983/84, following the collapse of BIMAS, with the objective of serving rural low- and middle-income households. BRI-UD makes only individual loans, has millions of clients, provides loans at high, market-determined interest rates, and has been very successful at mobilizing rural savings (Meyer and Nagarajan 1999).

Considerable progress has been made in some Asian countries to improve the economic environment for rural financial markets, but many policies still discourage rural finance. Even in the relatively successful case of Thailand's Bank for Agriculture and Agricultural Cooperatives (BAAC), the government continues to impose subsidized interest rates on BAAC farm lending and restricts its ability to serve nonfarm clients in rural areas. Likewise, the huge Integrated Rural Development Program in India uses massive public funds and discourages the emergence of unsubsidized institutions to serve poor clients.

Interest rates set at low levels discourage innovation and competition, so interest-rate liberalization is a necessary first step if countries want to create an environment in which financial markets can flourish. Portfolio restrictions must be carefully examined because they limit the lenders' ability to reduce risks through portfolio diversification (Meyer and Nagarajan 1999). Finally, the recent financial and economic crisis in East and Southeast Asia has revealed serious problems in regulation and oversight of financial institutions generally, which indicate the need for significant reforms in prudential oversight. These broader financial oversight issues are discussed in more detail in Chapter IX.

Rural Infrastructure

Infrastructure investments play a crucial role in inducing farmers to move toward a commercial agricultural system. Government investments in infrastructure should emphasize genuine public goods, such as improving general transport, communications, and market infrastructure, while allowing the private sector to invest in commodity-specific processing, storage, and marketing facilities. Governments should not preempt private-sector decisions by taking a "pick-the-winner" attitude towards diversification. As in the case of research, however, it is often impossible to avoid some degree of picking winners (or at least giving differential support) across commodities. Investment in rural infrastructure is commodity-group specific, and often commodity-specific, because of the three-way correlation among regions, agro-ecological zones, and agricultural production. A decision to build a road in a specific location is also a decision about the mix of products to be produced and marketed. The priority-setting process is further complicated because infrastructural investments take time to implement and demand careful attention to sequencing. Adjustment of investment strategies to major changes in relative prices must also be consistent with macroeconomic adjustment objectives. The importance of rural infrastructure to agricultural growth and equity is discussed in more detail in Chapter II, Chapter V and Chapter XI.

Property Rights

General economic liberalization provides opportunities for diversification and commercialization, but also places a premium on flexible farmer response in allocation of water, land and other resources in response to changing prices, comparative advantage, and economic opportunities (Rosegrant, Gazmuri Schleyer, and Yadav, 1995). If rights to the basic resources such as land and water are poorly secured and enforced, these resources can remain locked into inefficient uses. Efficient land markets and secure property rights are critical for the efficiency gains that spur commercialization and agricultural growth (Feder et al. 1988; Binswanger, Deininger, and Feder 1993). Secure rights to land create the incentives farmers need to invest in land improvements that conserve and increase the long-term productivity growth that can be induced by the start of commercialization. Secure land rights are complementary to policies that aim at liberalizing and integrating capital markets, because secure rights increase the probability that farmers can recoup the benefits from long-term investments, thereby increasing their willingness to make them. Because they can serve as collateral for loans, secure land rights also increase lender willingness to offer credit, leading to easier financing of purchased inputs and land improvements (Feder et al. 1988).

AGRICULTURAL COMMERCIALIZATION AND DIVERSIFICATION AND THE POOR

Concerns have been raised over possible adverse consequences of agricultural commercialization and diversification on the welfare of the poor. Critics have focused on three areas of concern: first, that the uptake of modern technologies associated with commercialization may be an inequitable process that at least worsens rural inequality and more likely increases absolute poverty; second, that in the shift to cash cropping, small-scale farmers might sacrifice their own food crops and expose their families to greater food insecurity; and third, that commercialization might worsen regional inequities because it favors areas that have greater agricultural-production potential. These critiques are considered in turn in the following sections.

Impact of Technological Change

Concerns about the adverse impact of modern agricultural technologies on the poor reached their zenith in the 1970s, when critics debated the negative impacts of the green revolution. Critics argued that, because of their better access to irrigation water, fertilizers, seeds, and credit, large farmers were the main adopters of the new technology; smaller farmers were either left unaffected or were made worse off because the green revolution resulted in lower prices, higher input prices, and efforts by larger farmers to increase rents or force tenants off the land. It was also argued that the green revolution encouraged unnecessary mechanization, with a resulting reduction in rural wages and employment. The net result, some critics argued, was an increase in the inequality of income and land distribution, an increase in landlessness, and a worsening of absolute poverty in areas affected by the green revolution (see, for example, Griffin 1972, 1979; Frankel 1976; Farmer 1977; ILO 1977; Pearse 1980).

Although a number of village- and household-based studies conducted soon after the green-revolution technologies were released lent some support to the critics (e.g., Farmer 1986), the conclusions have not proved valid when subjected to more recent scrutiny (Barker and Herdt 1978; Blyn 1983; Pinstrup-Andersen and Hazell 1985; Hazell and Ramasamy 1991; Lipton 1998). Small farmers may have lagged behind large farmers in adopting the green revolution technologies, but most of them eventually did adopt them and benefited from increased production as well as from greater employment opportunities and higher wages in the agricultural and nonfarm sectors (This very important carryover from agricultural growth to the rural nonfarm economy is discussed in Chapter IV).

Nor did the distribution of land worsen in most cases, as is also shown in Chapter IV. Large numbers of other poor people also benefited from the green revolution through increased employment and business earnings in the farm and nonfarm sectors, and from lower food prices (Pinstrup-Andersen and Hazell 1985). This is not to say that the green revolution was equitable everywhere, but the conditions under which it and other yield-enhancing technologies are likely to be equitable are now reasonably well understood. These conditions, which also promote poverty reduction, are described in Chapter II.

Impact of Commercialization

Critics of commercialization also feared that small farms would be left out of the commercialization process and would be unable to compete in the market as competition increased and prices fell. Moreover, concerns were also raised that if small farm households should forgo some or all of their traditional food crops in order to grow more cash crops for the market, this would (a) increase their dependence on purchased foods, exposing the household to greater food-security risk because of volatile market prices and uncertain income from cash crops; and (b) lead to a reallocation of income within the household in favor of men (who typically grow cash crops), with possibly adverse nutritional consequences for women and children.

A recent study summarized by Von Braun (1995) and Von Braun and Kennedy (1994) largely refutes these critics of commercialization. The study examines a series of comparative studies of selected sites (including study sites in five Asian countries) where farm households had recently switched from semisubsistence staple-food production with low levels of external inputs to production of more crops for sale in the market or to production with more purchased inputs. These studies found that, with few exceptions, commercialization of agriculture benefits the poor by directly generating employment and increasing agricultural labor productivity. Both the households that were commercializing their production and the hired laborers received direct income benefits. Furthermore, at all but one study site, the increased household income generated by commercialization was associated with an improvement in nutritional status for children in the household.

Commercialization is less likely to lead to adverse outcomes for the poor if farmers have secure property rights over their resources. Some of the worst apparent failures of commercialization cited in the literature, such as evictions of farmer-tenants, can be traced mainly to poorly defined land rights, rather than to the process of commercialization itself (Von Braun 1995). In addition to secure property rights, the provision of social-support services increases the benefits and reduces the probability of adverse consequences from the process (see also Chapter II). Foremost among these social-support policies are health and nutritional services (Von Braun 1995). Nutritional improvements are determined by both health and food consumption. Negative health effects from poor household and community health and sanitation can overcome potential positive effects of income growth from commercialization. Increased income and food consumption help to reduce hunger but cannot solve the problem of preschool children's malnutrition, which results from a complex interaction of lack of food and morbidity. Health and sanitation in rural areas can be promoted through improvement of community-level health services to exploit fully the welfare effects of agricultural commercialization.

Regional Disparities

Finally, it has been argued that agricultural intensification and commercialization that proceeds in certain regions but not in others can worsen regional disparities, with lagging regions falling farther behind as commodity prices drop in the wake of increasing productivity in the rapidly growing regions. The widening productivity gap between commercializing regions and slower-growing, subsistence-oriented regions could not only accentuate relative income differences but even cause an increase in absolute poverty in the lagging regions. In the study sites examined in Von Braun (1995), however, indirect-income benefits were generated through the increased demand for goods and services by the direct-income beneficiaries as well as by increased demand for inputs for commercialized agriculture. The wage rate and other employment benefits from commercialization spread to other regions when labor migrated from other regions into scheme areas. The more mobile the labor force, the more the benefits from commercialization spread across the economy and other regions.

Similar results have been found for the spread of modern rice technology in Asia (a classic process of commercialization). In a comprehensive cross-country comparative study, David and Otsuka (1994) found that the differential impact of new rice technology across regions did not worsen income distribution, thanks to the significant indirect effects that worked through labor, land, and product markets. Interregional labor migration from unfavorable to favorable regions tended to equalize wages across regions, allowing landless labor and small farmers in unfavorable areas to benefit too. Landowners in lagging regions were sometimes worse off but also partially protected their incomes through diversification out of rice.

While smoothly functioning product and factor markets help to equalize wages and incomes across regions, they are not always sufficient. In India, for example, poverty levels in many low-potential rainfall areas have improved little, even while irrigated and high-potential rainfall areas have progressed (Fan and Hazell 1999). Regional inequalities have also worsened in the PRC in recent years (Knight and Song 1986). Worsening regional disparities seem most likely to occur when agriculture is still the predominant source of national employment but the nonfarm economy is growing at only moderate rates. In these circumstances, the opportunities for out-migration from, and rural income diversification in, backward areas are likely to be smaller than needed. Where regional disparities worsen, there is need for increased public investment in backward areas, particularly in roads, agricultural research and development, and education (Fan and Hazell 1999).

CONCLUSION

Rapid economic growth in Asia initiated a period of significant decline in the share of agriculture in the economy. Economic growth also triggered the commercialization and diversification of agricultural systems. While the rate at which this agricultural transformation occurs varies by country, the direction of change is the same across Asia. Structural adjustment and trade liberalization policies that are currently being implemented in much of Asia can be expected to increase the speed at which the commercialization process occurs.

The process of commercialization requires a paradigm shift in agricultural policy formulation and research priority setting. With economic growth, the paradigm of staple-food self-sufficiency, which was the cornerstone of agricultural policy in many developing Asian countries, has become increasingly obsolete. A more relevant development paradigm is one of food self-reliance, where countries import a part of their food requirements in exchange for diverting resources out of subsistence production. A future emphasis of agricultural policy ought to be on maximizing farm household incomes rather than generating food surpluses.

Agricultural commercialization should not be expected to be a frictionless process; significant equity and environmental consequences should be anticipated, at least in the short to medium term and particularly when inappropriate policies are followed. The absorption of the rural poor into the industrial and service sectors has significant costs in terms of learning new skills and of family dislocations. Accompanying investments in education will thus be crucial.

Commercial systems could also face higher environmental health costs, especially in terms of higher chemical input use. Higher opportunity costs of labor could increase farmer reliance on herbicides for weed control for rice and other staple-food crops that are currently managed through hand weeding. Insecticide and fungicide use for high-value crops, such as vegetables and fruits, is substantially higher than for staples; improper use can increase the incidence of pesticide-related illnesses. Also, where property rights are not clearly established, high-value crop production in upland environments could lead to higher risks of soil erosion and land degradation.

Appropriate government policies can alleviate many of the potentially adverse transitional consequences arising from the process of commercialization and diversification. Important long-term strategies to facilitate a smooth transition to commercialization include

- investments in rural markets, transportation, and communications infrastructure to facilitate the integration of markets;
- investments in education to facilitate labor movement across sectors;
- investments in crop improvement research to increase productivity and crop management and extension to increase farmer flexibility and reduce possible environmental problems from high input use;
- establishment of secure rights to land and water to reduce risks to farmers and provide the incentives for investment in productivity- and conservation-enhancing technology;
- development and liberalization of rural financial markets to provide liquidity and to spread the risks as commercialization proceeds; and
- provision of support services, including health, sanitation and nutrition, to transform the income benefits from commercialization into broader human welfare benefits.

IV THE RURAL NONFARM TRANSFORMATION

INTRODUCTION

During the economic transformation, the emergence and rapid expansion of the nonfarm economy in rural areas and the towns that serve them becomes a major source of growth in incomes and employment. From a relatively minor sector, often largely part-time and subsistence-oriented in the early stages of development, the rural nonfarm economy develops to become a major motor of economic growth in its own right, not only for the countryside but for the economy as a whole. Its growth also has important implications for the welfare of women and poor households, sometimes helping to offset inequities that can arise within the agricultural sector.

While the expansion of the rural nonfarm economy is one of the most important steps in the process by which agricultural growth generates economic growth, it is also one of the least recognized and reported. Both the level and type of growth in the rural nonfarm economy are sensitive to government policies, yet these policies tend to be piecemeal or the offshoot of policies targeted at other parts of the national economy; there is little understanding on the part of policymakers of the rural nonfarm economy as a sector with its own internal consistencies and interests (there are, for example, no ministries of rural nonfarm economy). The result is typically a sector that gets buffeted by general macro, trade, labor, agricultural, and industrial-sector policies, set at the national level, and by regional, usually urban-biased, development plans of local governments.

IMPORTANCE OF RURAL NONFARM ACTIVITIES

Definitions

Definitions of the term "rural" and "nonfarm" vary across countries, and are usually based on settlement or locality sizes (e.g., India defines as "rural" all settlements of fewer than 5,000 people). In this context, however, it is preferable to think of rural as depending on function rather than on the size of a locality. As in Gibb (1974) and Anderson and Leiserson (1980), rural can be defined as any locality that exists primarily to serve an agricultural hinterland. In contrast, urban economies are driven by manufacturing, government or some other economic base independent of agriculture. Given this view, rural areas include all the rural settlements, central market places and towns that are linked together through economic transactions related to the agricultural economy.

"Nonfarm" activities are defined by most countries to include all rural economic activity other than agriculture, forestry, and fishing. They therefore include agricultural processing and trade (conventionally classified as part of the manufacturing and commerce sectors, respectively), as well as construction, mining, transport, and financial and personal services.

Employment Shares

The most readily available indicator of the relative importance of the rural nonfarm economy is its employment share, and these shares are reported in Table IV.1 for a number of South and Southeast Asian countries. Comparisons across countries are complicated by differences in definitions of rural areas, of the total work force, and of different nonfarm sectors. "Rural" is defined in these census data on the base of settlement size, with country differences in the size criterion used. In all cases, the work force is defined on the basis of primary occupation, although some variation in definitions exists.

Table IV.1—Employment Shares by Activity in Rural & Urban Areas, Selected Countries (percent)

Economy	Rural Population (as % total) 1960	Rural Population (as % total) 1994	Total Employment Agr.	Total Employment Nonfarm	Mft.	Transport	Nonfarm Employment Trade	Nonfarm Employment Services	Nonfarm Employment Finance	Construction	Other
Bangladesh (1991)	95	82									
Rural			66.1	39.9	6.8	4.0	....	35.4	....	3.3	50.2
Urban			15.1	84.9	8.1	6.5	....	31.8	....	3.4	50.1
Total			54.6	45.4	7.3	5.1	....	33.8	....	3.3	50.4
Sri Lanka (1981)	82	78									
Rural			55.7	44.3	19.8	8.3	16.5	25.2	1.5	6.6	22.1
Urban			7.3	96.7	16.0	9.7	23.9	28.7	2.8	3.7	15.0
Total			45.2	54.8	18.5	8.8	19.2	26.5	2.0	5.5	19.5
Pakistan (1992/93)	78	66									
Rural			63.8	36.2	19.0	10.4	21.9	26.8	0.7	19.4	1.8
Urban			5.8	94.2	22.6	10.6	28.9	26.0	2.4	7.1	2.4
Total			47.6	52.4	20.8	10.5	25.4	26.4	2.6	13.2	2.1
India (1993/94)	82	73									
Rural			76.9	23.1	30.7	6.9	19.4	....	26.8	11.6	4.6
Urban			17.7	82.3	22.2	12.7	25.9	....	38.3	3.1	2.3
Total			61.5	38.5	28.5	8.4	21.1	....	29.8	9.4	2.8
Philippines (1980)	70	47									
Rural			74.0	26.0	20.9	11.9	13.2	32.1	3.0	11.5	7.4
Urban			18.3	81.7	19.4	11.3	14.9	35.9	7.1	8.1	3.3
Total			51.4	48.6	19.9	11.5	14.3	34.7	5.8	9.2	4.6
Indonesia (1995)	85	66									
Rural			63.1	36.9	23.8	8.2	31.7	24.2	0.5	9.4	2.2
Urban			9.4	90.6	20.0	8.0	30.1	31.1	2.4	6.8	1.6
Total			45.9	54.1	21.8	8.1	30.9	27.9	1.5	8.0	1.8
Thailand (1996)	87	80									
Nonmunicipal			49.9	50.1	30.3	5.1	22.1	....	19.7	21.5	1.3
Municipal			1.9	98.1	22.6	7.0	29.9	....	28.8	9.7	2.0
Total			39.7	60.3	27.6	5.8	24.8	....	22.8	17.4	1.6

Because part-time and temporary employment are important in many agricultural and nonfarm activities, the employment data tend to underestimate the importance of some activities, though the bias is likely to be small when expressed in share rather than in absolute terms. Nonfarm sectors are not always defined in the same way; the biggest differences tend to arise in the definitions of the "service" and "other" sectors (Bangladesh departs most from other country definitions).

Despite these differences in definitions, the data in Table IV.1 show a remarkably consistent story across countries. The nonfarm economy accounts for 40 to 60 percent of total national employment and the rural nonfarm economy accounts for 20 to 50 percent of total rural employment. Differences between South and Southeast Asian economies are also surprisingly small. While rapidly developing (until recently) economies like Indonesia and Thailand now have very little agricultural employment in their urban areas—only 9.4 and 1.9 percent, respectively, of total urban employment—the nonfarm share of rural employment is not much different from those of other countries. This no doubt reflects the fact that as rural settlements grow and diversify, they soon become classified as urban rather than rural areas in the census data. Even so, the share of nonfarm employment in total national employment was not much higher in Indonesia and Thailand in the mid-1990s than in Sri Lanka in 1981 or Pakistan in 1992/93.

The composition of the nonfarm economy also shows remarkable similarities across countries. Service activities dominate the nonfarm economy in both rural and urban areas, followed by manufacturing and trade. Service activities (including much of the "other" activities in Bangladesh and Sri Lanka) are more dominant in the lower-income South Asian countries, while trade and manufacturing are about equal in importance to services in the Southeast Asian countries.

Hazell and Haggblade (1993) calculated a breakout of rural nonfarm employment shares by rural and urban towns for 14 Asian countries. Their results show that, as suggested above, rural towns can be expected to have an employment structure that reflects their economic links to agriculture, while urban towns typically have a more independent economic base. On average, the nonfarm employment share for rural areas increases from 26 percent to 36 percent when rural towns are added to the definition of rural areas. Moreover, the nonfarm employment share increases quickly with size of locality and is 81 percent even for rural towns.

In India, nonfarm employment share also increases sharply with locality size. In rural towns (defined as having populations between 5,000 and 100,000), 76.4 percent of the work force was employed in nonfarm activities in 1971 (Table IV.2). In rural areas, services and household manufacturing activities are the most important sources of employment in rural areas, whereas employment in rural towns is more nearly dominated by trade and services. In urban towns, trade and services are also important sources of employment, but manufacturing dominates. Unlike rural areas and towns, manufacturing employment in urban towns is nearly all in formal nonhousehold activity; household manufacturing accounts for a mere 3.9 percent of total employment.

Income Shares

The employment shares discussed above may underestimate the relative importance of some nonfarm activities that have larger than average shares of part-time and seasonal employment. Daily earnings in various sectors also differ and this can affect their relative economic importance. Income shares provide a more reliable guide to the relative importance of different nonfarm sectors, but this kind of data is rarely available at the country level. Indeed, many countries do not even collect comprehensive production or income data on the output of informal (often household or cottage) activities that prevail in many nonfarm sectors (particularly the service and manufacturing sectors). It is therefore necessary to rely on household- and farm-level surveys, which are spotty in their regional and temporal coverage. Table IV.3 provides some relevant evidence on income shares.

Table IV.2—Employment Shares by Activity and Size of Locality, India (percent)

Economy	Total Employment		Manufacturing		Nonfarm Employment				
	Agric	Nonfarm	Household	Non-Household	Transport	Trade	Services	Construction	Other
India (1971)									
Rural	84.9	15.1	21.6	15.7	5.9	15.7	35.3	3.9	2.0
Rural Towns	23.6	76.4	8.6	19.5	10.5	25.4	25.4	4.5	1.8
Urban Towns	4.7	95.3	3.9	30.3	12.0	21.5	27.8	3.5	0.7

Source: Hazell and Haggblade (1991, 518).

Note: Rural towns are all urban areas under 100,000 in population; urban towns are urban areas with more than 100,000 population.

Table IV.3—Income Shares from Agricultural and
Nonfarm Activities, Rural Areas (percent)

	Agriculture	Nonfarm
India[a]		
1967/68	74.5	25.5
1975/76	69.5	30.5
1981/82	65.2-69.8	34.8-30.2
Republic of Korea[b]		
1971	81.9	18.1
1981	67.2	32.8
1991	53.7	46.3
Thailand[c]		
1978/79	43.3	34.7

[a] Taken from Hazell and Haggblade (1991, 516).
[b] Census data for Korea; farm households only.
[c] Taken from Ho (1986b, p.7)

Nonfarm income shares are typically 5 to 10 percent larger than nonfarm employment shares in rural areas, a direct measure of the importance of seasonal and part-time nonfarm activity. In the latest years for which data are available, nonfarm income shares account for one third to one half of total rural household income. These shares have also increased over time; up from 18.1 to 46.3 percent between 1971 and 1991 in the Republic of Korea, and from one quarter to one third between 1967/68 and 1981/82 in India.

Table IV.4 shows the sector composition of nonfarm employment for male and female workers in the rural labor force. Again, these data only capture workers defined on the basis of major occupation, excluding part-time and seasonal work.

Importance to Women

Manufacturing, service and trade activities account for the largest shares of employment for both male and female workers in rural areas (it can be assumed that much of the activity classified as "other" in Bangladesh and Sri Lanka also really falls into these categories). Women are relatively more concentrated in these

Table IV.4—Distribution of rural workers across nonfarm sectors, Male and Female (percent)

Country	Total Nonfarm	Mft	Transport	Trade	Services	Finance	Construction	Other
Bangladesh (1991)								
Male	100	6.5	4.6		39.4		3.7	45.8
Female	100	8.4	0.4		12.2		0.8	78.2
Sri Lanka (1981)								
Male	100	18.8	9.8	18.1	20.6	1.6	7.7	23.4
Female	100	24.9	1.6	8.9	46.1	1.4	1.2	15.9
India (1993/94)								
Male	100	26.8	8.4	21.1	27.2		12.6	3.8
Female	100	48.7	0.6	14.2	25.9		7.1	3.2
Philippines (1980)								
Male	100	16.5	19.0	9.6	22.9	3.1	18.6	10.4
Female	100	27.9	0.6	19.0	46.8	2.8	0.4	2.5
Indonesia (1995)								
Male	100	19.9	12.5	23.7	26.0	0.7	2.8	
Female	100	30.9	0.3	46.2	21.0	0.2	1.0	
Thailand (1996)								
Male	100	25.3	8.0	17.8	17.3	14.5	29.9	1.7
Female	100	37.5	1.0	28.2	23.1	0.3	9.7	0.5

activities than men in most countries. Trade seems to be more important for women in Southeast Asian countries than in South Asian ones, but transport and construction activities are much less important for women than men in all countries.

Table IV.5 shows the relative importance of women in total employment by sector in rural and urban areas. There is a dramatic difference between the South and Southeast Asian countries; women account for much smaller shares of total employment in every sector in the South Asian countries reported, in both rural and urban areas. Even in the manufacturing and service sectors that are most important to them (Table IV.4), their shares are one third or less of total employment. In contrast, women account for one third to one half of sector employment in manufacturing, trade, and services in the Southeast Asian countries and their importance in the financial-services sector is also great. Table IV.5 also confirms that women are very minor participants in the transport and construction sectors in all the countries reported.

There are surprisingly few differences between the data for rural and urban areas. Cultural and other biases that lead to sector differences in the way men and women are employed seem consistent in both rural and urban areas.

Importance to the Poor

The rural nonfarm economy is especially important to the rural poor. Landless and near-landless households everywhere depend on nonfarm earnings; those with less than 0.5 hectare earn as much as 80 percent of their income from nonfarm sources (Table IV.6).

Nonfarm shares are strongly and negatively related to farm size. Low-investment manufacturing and services—including weaving, pottery, gathering, food preparation and processing, domestic and personal services, and unskilled nonfarm wage labor—typically account for a greater share of income for the rural poor than for wealthier rural residents (Hazell and Haggblade 1993). The reverse is true of transport,

Table IV.5—Women's Share in Total Nonfarm Employment by Sector, Rural and Urban (percent)

Country	Total Nonfarm	Mft	Transport	Trade	Services	Finance	Construction	Other
Bangladesh (1991)								
Rural	14.7	18.2	1.5	....	5.1	....	3.9	22.8
Urban	12.1	15.9	1.4	....	5.7	....	3.3	15.7
Sri Lanka (1981)								
Rural	17.9	22.3	3.4	9.6	32.8	16.1	3.3	12.9
Urban	18.5	24.3	5.5	7.6	33.7	22.9	5.8	11.4
India (1993/94)								
Rural	19.4	30.3	1.8	14.0	....	18.6	....	16.9
Philippines (1980)								
Rural	38.6	51.5	2.0	55.4	56.3	36.5	1.2	13.1
Urban	38.2	34.6	5.3	39.1	56.0	38.9	1.9	15.9
Indonesia (1995)								
Rural	35.6	46.2	1.1	51.9	30.8	16.9	1.3	15.0
Urban	33.5	35.7	3.0	44.8	37.0	29.0	3.7	11.9
Thailand (1996)								
Rural	41.3	51.0	8.1	52.7	....	48.5	....	17.7
Urban	45.4	46.1	14.3	48.4	....	56.7	....	40.0

commerce, and such manufacturing activities as milling and metal fabrication, which require sizable investments.

Nonfarm income is also important to the poor as a means to help stabilize household income in drought years (Reardon et al. 1998). In a study of several villages in the semi-arid tropics of India, for example, Walker and Ryan (1990) found that nonagricultural self-employment and labor-market earnings became increasingly important sources of income during the 1980s, increasing mean income and dampening household income variability.

Table IV.6—Share of nonfarm income/employment in total household income/employment by farm size groups, selected countries

Farm size (ha)	Nonfarm share (%) Employment		Income
India	1987/88		
Landless	46.1		
0.01-0.4	29.3		
0.41-1.0	19.0		
1.01-2.0	14.0		
2.01-4.0	11.8		
4.01+	9.0		
North Arcot, India		1973/76	1982/83
0-0.1		15	35
0.1-1.0		22	23
1.0+		7	20
Republic of Korea		1970	1996
0-0.5		49	80
0.5-1.0		26	64
1.0-1.5		7	49
1.5-2.0		16	42
2.0+		5	37
Taipei,China			1979
0-0.5			67
0.5-1.0			58
1.0-1.5			48
1.5-2.0			40
2.0+			33
Thailand (4 regions)			1980/81
0-4.1			88
4.2-10.2			72
10.3-41.0			56
41.0+			45

Source: India, Chadha (1993); North Arcot, Hazell, Ramasamy, and Rajagopalan (1991); Korea, Choi (1997); Taipei,China, Ho (1986a); Thailand, Liedholm (1988).

STRUCTURAL TRANSFORMATION OF THE RURAL NONFARM ECONOMY

Overview

The present structure of the rural nonfarm economy in Asia results from an economic transformation that has being going on for many generations and at varying speeds in different countries. The process begins with a countryside dominated largely by self-sufficient and primarily agricultural households producing for themselves most of whatever farm and nonfarm goods and services they need. There is little trade or commerce and the prevailing agricultural technologies require few if any external inputs. Gradually, as population densities and market access increase, new technologies and modern farm inputs become available, leading to increased agricultural surpluses in some commodities and increased opportunities for trade. Increasing agricultural productivity also raises incomes, which in turn increases the types and amounts of consumer goods and services that rural households wish to purchase. Households begin to specialize, taking greater advantage of their particular skills, resource endowments, and market opportunities. Some nonfarm activities that were initially undertaken by farm households for their own consumption expand and are spun off as separate full- or part-time businesses. There is greater trade among rural households and with small market centers and rural towns. The latter begin to grow more rapidly.

With increasing specialization, not only do more nonfarm businesses emerge, but there is growing spatial concentration in small towns and market centers, because they offer bigger markets, easier and cheaper access to inputs, and better infrastructure. On the other hand, some service activities prosper in rural areas, particularly in the larger villages and rural market centers where they can better capture local demand (e.g., retail establishments, tea and coffee shops, and agricultural machinery service and repair shops). Rural towns grow in importance and as the rural economy continues to grow, trade with larger urban centers also expands and more urban goods become available. These often displace many traditional rural products, forcing structural changes in the composition of the rural economy and its towns. This process receives further impetus from rising wages, which drive workers out of many traditional but low-productivity nonfarm activities (e.g., factory-made shoes replace the products of the village cobbler; urban rice mills replace the local miller). As towns grow, they attract more workers from the rural hinterland, and the agricultural work force (though not necessarily the rural population) begins to decline. Towns grow as major sources in their own right of demand for nonfarm goods and services, for both production and consumption purposes, and their nonfarm activities expand to serve these needs as well as to export to other urban and rural areas. Agriculture becomes increasingly less important as the economic motor for the regional economy, eventually becoming a relatively minor economic activity in many rural regions as well as in the national economy. The transformation process is not identical in all countries, and is shaped in part by such factors as a country's comparative advantage in the production of tradable products (especially agriculture), population density, infrastructure, location, and government policies.

Changes in the Composition of the Rural Nonfarm Economy

Agriculture's share of income and employment in rural areas declines steadily as the transformation proceeds. In India, for example, agriculture's share of total employment in rural areas declined from 85.5 percent in 1972/73 to 78.4 percent in 1993/94. In Indonesia, its share declined from 67.8 percent in 1990 to 63.0 percent in 1995. In the Republic of Korea, it declined from 87.0 percent of total farm household employment in 1965 to 82.7 percent in 1991, and from 12.7 percent to 2.0 percent of total nonfarm household employment. The Philippines provides a contrasting story; between 1970 and 1980 agricultural employment increased from 70.6 to 74.0 percent of total rural employment. But this was just prior to the economic crisis of the early 1980s, a period best viewed as the reverse of the normal transformation process.

The declining share in agricultural employment need not imply a decline in absolute employment in agricultural activities. In countries with abundant and rapidly growing rural labor forces, agricultural employment may continue to grow, albeit at a slower rate, than nonfarm employment. Many of the South Asian countries fall into this category, as illustrated by the case of India in Table IV.7. India's agricultural employment grew at an average yearly rate of 1.3 percent between 1977/78 and 1993/94 in rural and urban areas; yet because nonfarm employment grew faster (by about 3.5 percent per year), agriculture's share of total employment declined. In contrast, in labor-scarce economies, workers are pulled out of agriculture at a rate that exceeds the growth in the rural labor force and the absolute number of agricultural workers declines. This has been characteristic of some East and Southeast Asian countries in recent years, as illustrated by Indonesia and the Republic of Korea in Table IV.7. In both cases, agricultural employment has declined (negative growth rates) over the periods analyzed, and at a relatively high rate in the case of the Republic of Korea.

Agriculture also declines in importance in rural household income as the transformation proceeds. In India, for example, agriculture's share of rural income declined from 74.5 percent in 1967/68 to between 65 and 70 percent in 1993/94, depending on how remittances are calculated (Table IV.3 and Hazell and Haggblade 1991). Agriculture also declined as a share of income among the Republic of Korea's farm households; from 81.9 percent in 1971 to 53.7 percent in 1991(Table IV.3).

Table IV.7 shows the average annual rates of growth in employment for different nonfarm sectors in rural and urban areas for India (1977/78 to 1993/94), the Philippines (1970 to 1980), Indonesia (1990 to 1995) and the Republic of Korea (1972 to 1991). Unfortunately, the available data for Korea are only partial. Annual average rates of growth in total nonfarm employment (rural plus urban) ranged from 3.0 percent in the Philippines to 6.2 percent in Indonesia. Employment grew more

Table IV.7—Average Annual Growth Rates in Nonfarm Employment by Sector, Rural and Urban Areas, Selected Countries (percent)

Country	Agr.	Total nonfarm	Mft.	Transport	Trade	Services	Finance	Construction	Other
India (1977/78 to 1993/94)									
Rural	1.31	3.45	2.47	5.20	3.46	3.29		5.95	
Urban									
Total	1.34	3.51	2.40	4.08	3.51	3.74		6.32	
Philippines (1970 to 1980)									
Rural	2.1	0.38	-3.25	2.76	0.06	1.82	18.36	1.53	-0.54
Urban	7.3	4.57	3.53	5.90	4.77	3.25	17.06	5.70	3.61
Total	2.7	3.02	0.65	4.75	3.12	2.80	17.26	3.81	1.25
Indonesia (1990 to 1995)									
Rural	-0.4	3.92	3.28	6.28	3.88	3.94	-3.97	5.71	-1.15
Urban	5.6	6.95	7.82	6.14	8.03	6.26	1.19	6.54	3.63
Total	0.0	5.45	5.36	6.21	5.92	5.28	0.28	6.08	0.84
Republic of Korea (1972 to 1991)									
Farm households	-2.6	0.8	1.2	na	na	na	na	2.2	na
Nonfarm households	-2.5	6.6	7.5	na	na	na	na	8.4	na
Total	-2.6	6.2	7.0	na	na	na	na	7.8	na

slowly in rural than urban areas in all four countries, however, particularly in the Philippines (0.38 percent for rural areas compared to 4.57 percent for urban areas) and the Republic of Korea (0.8 percent amongst rural households compared to 6.6 percent for nonfarm households) (Korea Statiscal Yearbook 1996). In fact, rural employment in the Philippines grew hardly at all in any nonfarm sector except financial services (which are a small source of employment to begin with) during the 1970s, with nearly all the growth in employment occurring in urban areas. This helps to confirm Ranis, Stewart, and Angeles-Reyes (1990), who have articulated the extreme urban bias that characterizes the Philippines' economic development. In India and Indonesia, rural employment grew at 2.5 percent per year or better in nearly every nonfarm sector, with particularly strong growth in the transport and construction sectors.

Nonfarm employment growth has been more beneficial to men than women workers (Table IV.8). In the Philippines, women's employment grew fastest in the transport, financial services and construction sectors, but these are very minor sources of employment to begin with (Table IV.4). On the other hand, women lost employment in key sectors like manufacturing and trade (which together accounted for 46 percent of their total employment in 1980), with the net result that their total employment declined by 1.63 percent per year between 1970 and 1980. Rural women have fared better in India and Indonesia in recent years, gaining significant employment in the manufacturing, trade and service sectors, but not as much as men.

Changes in the Spatial Structure of Rural Economies

As the transformation proceeds, there is greater urbanization of rural regions through the rapid growth of centrally placed villages and small towns. Many areas that were previously classified as rural in national census data achieve sufficient population size so that they are reclassified as urban. This partly explains the very high rates of urbanization reported for many Asian countries in recent years; only part of it represents movement of rural people to large cities. The growing prosperity attracts many workers to move out of agriculture and to resettle in local towns.

This need not lead to any decline in the number of farm households, at least not until quite late in the transformation process. As Table IV.9 shows, the number of farm households has continued to increase in most Asian countries, despite the ongoing rural transformation, with a reduction in average farm size and an increase in the share of small farms. The Republic of Korea is one of the few exceptions, where farms have become fewer and larger in recent decades. This pattern of change is to be expected in South Asia, where agriculture is often still challenged to absorb surplus rural workers. But it is surprising in East Asia, where workers have been "pulled out" of farming by growing labor shortages and higher wages in the nonfarm economy. It parallels earlier developments in Japan, where farm households diversified their income sources (with workers often taking part- or full-time jobs in local factories) while continuing as small but nonpoor smallholder farmers. Dispersed towns and dense patterns of rural infrastructure make this kind of development possible in rural Asia, at least as long as government polices favor spatially dispersed patterns of rural industry (more on this in a later section).

Much of the shift in farm workers involves younger people who leave home and form new households in the towns. This raises the prospect of an aging farm population, with perhaps a more significant exodus out of farming in the future as older generations retire. In the Republic of Korea, for example, about half (52 percent in 1994) of the adult farm population (more than 15 years old) is at least 50 years old and 30 percent are over 60 (Korea Statistical Yearbook, 1996).

Table IV.8—Average Annual Growth Rates in Rural Nonfarm Employment by Sector, Male and Female Workers (percent)

Country	Total nonfarm	Mft.	Transport	Trade	Services	Finance	Construction	Other
Philippines (1970 – 1980)								
Males	1.91	-0.31	2.66	2.61	1.63	15.78	1.47	3.31
Females	-1.63	-5.33	10.69	-1.56	1.97	25.20	9.51	-11.16
India (1977/78 - 1993/94)								
Males	3.73	2.37	5.30	3.94	3.65		6.21	na
Females	2.39	2.73	1.41	1.13	1.91		4.32	na
Indonesia (1990 - 1995)								
Males	4.40	4.25	6.16	3.71	4.43	-4.28	5.85	-0.43
Females	3.10	2.22	22.54	4.04	2.90	-2.31	-2.40	-4.30

Table IV.9—Changes in Farm Population, Cultivated Area, and Size Structure (ha)

Country (year)	Number Farms	Average Size	Total Cultivated Area	\multicolumn{5}{c}{Cultivated Land Size (ha)}
				0-1	1-2	2-3	3-5	5-10	10+
	(thousands)	(ha)	(10³ ha)						
Indonesia									
1963	13,236	1.1	13,460	70.1	18.2	5.7	3.5	1.8	0.7
1993	19,714			70.8	16.8	7.4	3.7	1.2	0.2
Korea, Rep. of									
1968	2,578	0.90	2,319	65.1	26.0	5.2	1.5		
1981	2,029	1.08	2,188	66.4	26.8	4.1	1.2		
1991	1,702	1.23	2,091	58.4	30.0	7.1	2.4		
1995	1,499			57.6	27.8	8.2	4.7		
Philippines									
1971	2,355	3.61	8,494	13.6	47.5		23.7	10.4	4.9
1980	3,441	2.83	9,725	22.8	46.1		17.1	10.5	3.4
1991	4,770	2.09	9,775	38.0	42.0		11.0	6.8	2.3
India									
1953/54	48,890			49.6	17.5	28.1		4.7	
1961/62	63,950			54.8	17.1	24.7		3.2	
1971/72	70,810			58.6	17.1	21.9		2.3	
1982/83	83,200			62.4	16.6	19.5		1.6	
1992	103,300			68.0	15.1	15.6		1.0	

				\multicolumn{5}{c}{Cultivated Land Size (ha)}
				0-0.94	0.95-2.38	2.38-4.78	4.79-9.58	9.59+
Thailand								
1963	3,214	3.47	11,149	18.5	29.4	27.5	19.2	5.4
1978	4,018	3.72	14,955	15.9	27.4	29.0	21.4	6.3
1993	5,648	3.36	19,002	19.7	30.1	28.1	17.2	4.8

Source: National agricultural census data. The data for India were complied by Dr. S.K.Thorat

Changes in Local Labor and Capital Markets

Labor

Increases in nonfarm employment are not necessarily accompanied by increases in wages. The two will increase together when the rural nonfarm economy is growing as a result of increased demand and increasing labor productivity. This is often characterized as a "pull" situation, where the nonfarm economy is attracting, or pulling, workers out of farming to better-paying jobs. In this situation, average labor productivity increases over time and wages increase. Since the pull is from the nonfarm sector, one should expect nonfarm wages to be higher than agricultural wages.

An opposite situation can arise if there is a growing surplus of rural workers and the agricultural sector is unable to absorb them all at a subsistence wage. In this situation, workers are "pushed out" of agriculture into nonfarm activities that can contribute to their subsistence. The rural nonfarm economy then acts as a residual sector of employment and there is an expansion of low-productivity work with declining wages. Since the push is from agriculture, agricultural and nonfarm wages are likely to be similar.

Available wage data for several Asian countries show that farm wages have lagged behind nonfarm wages in rural areas in recent years, for both male and female workers. This is consistent with a "pull" scenario in which a growing nonfarm sector is attracting workers out of farming. Even in a labor-surplus country like India, wage rates in nonfarm activities are considerably higher than in agriculture (40 percent higher for men and 22 percent higher for women), indicating that much of the growth in nonfarm employment is not a distress phenomenon but rather the result of the pull of more productive and higher-paying jobs (Bhalla 1997). This is confirmed by the observation that there is also less poverty among nonagricultural workers (Bhalla 1991).

These findings are all the more impressive when one considers that the employment elasticity in agriculture seems to have declined in recent years. In India, for example, the elasticity has fallen from about 0.75 in the 1970s (Tyagi 1981) to perhaps as low as 0.3 today (Bhalla and Singh 1998). For a pull situation to arise at a time when the rural labor force is still growing quite rapidly suggests that not only is the growth of the rural nonfarm economy strong, but the growth is also labor-intensive. This is consistent with the high employment shares noted earlier for service and commerce activities, which are not only labor-intensive on average but also have high employment elasticities (often close to 1.0).

In a pull situation, rural labor markets play a key role in shifting the composition of rural nonfarm activity. Increases in real wages raise the opportunity cost of labor, thereby making low-return nonfarm activities uneconomic. This leads to the demise of many traditional and low-paying craft and service activities and to the growth of new types of employment in trade, commerce and manufacturing. Hossain (1988) provides evidence from the green-revolution experience in Bangladesh. In villages with a majority of rice cropped in high-yield varieties, he identifies higher agricultural incomes, higher agricultural wages, and higher nonfarm income per capita compared to villages still dependent on traditional varieties. The higher nonfarm income in prosperous villages reflects a greater concentration of high-return nonfarm activity (transport and services) and less low-wage cottage industry, construction and earth hauling.

Capital

Agricultural growth has led to rapidly rising savings rates among farm households, reaching 25–35 percent of total household income in many Asian countries (Meyer and Nagarajan 1999). With rising incomes and savings rates, farm households can generate huge amounts of surplus capital. Part of this capital flows into the rural nonfarm economy, especially into businesses started by farm households or their relatives. But part also flows through kin networks, traders, money-lenders and financial institutions to urban areas.

Evidence from regional Social Accounting Matrices (SAMs) for the Muda river region in Malaysia (Bell, Hazell, and Slade 1982) and the North Arcot district in Tamil Nadu, India (Hazell, Ramasamy, and Rajagopalan, 1991) suggests that annual net capital outflows from rural regions that have benefited from significant growth in agricultural productivity can be huge. In the Muda region, the capital outflow in 1972 was equal to 8 percent of regional value added and 56 percent of total household savings. In North Arcot, the capital outflow in 1982/83 was equal to 4 percent of total regional value added and to 18 percent of total household savings; and this was during a severe drought year when total regional value added was down about 30 percent from normal (Hazell, Ramasamy, and Rajagopalan 1991).

Why do such large capital outflows occur from rural regions, and why is more of it not plowed back into the development of the local economy? An important reason is the opportunity for earning higher returns elsewhere, particularly in urban real estate and financial markets. But imperfectly developed local financial markets may also be partly to blame. Rural financial institutions have focused more on lending to agriculture and rural industries in the past and failed to recognize that rural households needed access to deposit or savings accounts that give reasonable returns. They have also failed to provide needed financial services to many small-scale and part-time nonfarm businesses, especially in the service sector, as witnessed by the recent explosion in microfinance provided by NGOs in rural areas. As financial services improve in rural areas, it is possible that larger shares of rural savings will be captured in rural areas (including local towns), and that this will further facilitate the growth of the rural nonfarm economy.

The financial needs of agriculture are changing with the transformation. Farms are getting smaller on average (see Table IV.9), but more cash-oriented and more productive per hectare. The vast majority of farm households are also reducing their dependence on agriculture by diversifying into nonfarm sources of income. This helps raise total household income, leading to higher savings, and gives them greater access to cash income that is likely to be less seasonal in nature than agricultural receipts. Taken together, these changes improve seasonal and annual cash flow for most farmers, thereby reducing their need for conventional forms of agricultural credit.

The financial needs of such farmers are becoming more complex and diverse: they include access to deposit and savings accounts and sometimes investment loans for nonfarm business activity (often undertaken by women rather than men) as well as for agriculture. More flexible and customer-oriented financial services are required to meet these needs. At the same time, a minority of large farms, which still account for significant shares of the total cultivated area (for example, in Thailand and Pakistan, the largest 7 percent of the farms control 32 and 40 percent of the total cultivated land, respectively) remain highly specialized in farming and are likely to need access to more conventional forms of agricultural credit.

Nonfarm businesses in rural towns also have growing financial needs and especially require improved access to working capital for purchasing inputs (Haggblade and Mead 1998). Many nonfarm businesses are initially capitalized by their owners using family (own plus relatives') capital and then grow by plowing profits back into them. Better access to long-term investment capital might facilitate growth and perhaps the entrance of additional firms.

Impact on the Poor

Nonfarm income shares have increased among the landless and among the smallest farm sizes in many Asian countries (Table IV.6) and this has undoubtedly been beneficial to many of the rural poor. Poor people have also benefited from increased employment opportunities in local towns, and many have migrated to urban areas. As shown in Chapter II, rural poverty declines quite quickly during the economic transformation of rural areas and the nonfarm economy plays an important part in that process.

The expectation that the poor would gain from agriculture-led regional development is confirmed by detailed analysis of changes in household incomes in the Muda river irrigation region in Malaysia and in North Arcot district, Tamil Nadu, India. Results of detailed semi-input-output models of these two regions track the changes in per capita income, by household type, induced by large increases in agricultural output (Table IV.10). In the Muda region the change in output was induced by a large irrigation project (Bell, Hazell, and Slade 1982). In North Arcot the change stemmed from a decade of agricultural growth during the era of the green revolution (Hazell, Ramasamy, and Rajagopalan 1991).

In both regions landless agricultural workers are the poorest household group, but these workers gained proportionally more income than any other group as a result of the growth in agriculture. The largest share of their income gain came from an increase in agricultural wage earnings, but 10 to 15 percent of their total gain was derived from nonagricultural sources. Small farmers also had significant increases in nonagricultural income; it accounted for 29 percent of the increase in total farm income in North Arcot. Large farms gained relatively little from nonagricultural sources in both regions. The real gains from the growth in the nonfarm sector accrued to the specialized, nonagricultural households, especially those residing in the local towns.

These results demonstrate that the rural poor do gain in absolute terms from agriculturally induced growth in the rural nonfarm economy. But it is also clear that, because the nonpoor gain even more, the distribution of income could nevertheless worsen.

DETERMINANTS OF THE TRANSFORMATION

Why does rural nonfarm activity vary so much over time and across countries? Resource endowments, location, towns, ethnicity, historical happenstance, and government policies all

The Rural Nonfarm Transformation 105

Table IV.10—Project-induced Changes in Per Capita Income in Two Regions, by Household Type

Household group	Preproject Income Per Capita[a] Agriculture	Preproject Income Per Capita[a] Nonfarm	Preproject Income Per Capita[a] Total	Ratio of post- to pre-project income	Percentage of Income Increase Agricultural Sources	Percentage of Income Increase Nonfarm Sources
Muda irrigation region, Malaysia, 1972 *(Malaysian dollars)*						
Landless paddy workers	65	68	133	1.71	89.4	10.6
Small paddy farms	138	71	209	1.59	88.7	11.2
Large paddy farms	250	92	342	1.64	91.3	8.7
Nonproject farms	116	269	385	1.06	50.0	50.0
Nonfarm households	17	896	913	1.14	17.5	82.5
North Arcot, India, 1982 *(Rupees)*						
Rural Villages[b]						
Landless laborers	295	118	413	1.33	85.5	14.5
Nonfarm households	49	592	641	1.20	15.0	85.0
Small farms	457	309	766	1.32	70.9	29.1
Large farms	1,246	347	1,593	1.32	89.1	10.9
Urban villages[b]						
Agriculturally dependent	648	384	1,032	1.28	82.2	17.8
Employed nonfarm	40	1,282	1,322	1.15	8.0	92.0
Self-employed nonfarm	55	2,684	2,739	1.18	4.5	95.5
Towns[c]						
Agriculturally dependent	616	556	1,172	1.30	64.3	35.7
Employed nonfarm	22	1,366	1,388	1.18	3.6	96.4
Self-employed nonfarm	4	4,191	4,191	1.27	0.1	99.9

[a] For the Muda regions, the "project" is the irrigation scheme; for North Arcot, it is a decade of growth induced by the green revolution.
[b] The classification of rural villages, urban villages, and towns is based on census definitions. Rural villages have populations of fewer than 5,000 people.

Source: Hazell and Haggblade (1993), based on Muda: Bell, Hazell, and Slade 1982; and North Arcot: Hazell, Ramasamy, and Rajagopalan 1991.

play a role. But agriculture, because of its size and initial dominance, also plays a key role in developing a growing market for the output of the rural nonfarm sector.

Agricultural Growth

Agriculture can influence nonfarm activity in at least three ways: through production, consumption, and labor market linkages. On the production side, a growing agriculture requires inputs (for example, fertilizer, seeds, pesticides, pumps, sprayers, machinery repair services), either produced or distributed by nonfarm firms. Moreover, increased agricultural output stimulates forward production linkages by providing raw materials that require milling, processing and distribution by nonfarm firms. Consumption linkages arise when growing farm incomes boost demand for a range of consumer goods and services. Demand increases as rising per capita incomes induce diversification of consumption into nonfood goods and services, many of which are provided by local firms. Consumption diversification into nonfoods proceeds rapidly as per capita incomes rise, with expenditure elasticities for home improvements, durables, services, transport, and education often well in excess of 1.0 (see, for example, Hazell and Roell 1983; Evans 1990).

The strength of the consumption linkages depends not only on the level of per capita farm income, but also on how that income is distributed. Studies by King and Byerlee (1978), Hazell and Roell (1983) and Deb and Hossain (1984) suggest that larger-sized farms and higher-income groups generate the greatest consumption linkages with the rural nonfarm economy, because they allocate larger shares of incremental income to locally produced nonfood goods and services. But since the largest farm sizes in these studies vary between 5 and 15 hectares, it cannot be concluded that really large, wealthy farms and estates should receive priority treatment. On the contrary, there are good reasons to expect that large farm households have much weaker consumption linkages to the local economy and that they spend much larger shares of incremental income in cities and abroad.

Demand for production inputs varies across agricultural zones and with technological change. Irrigated agriculture demands considerably more inputs than rain-fed agriculture, while mechanized and animal traction systems require more tools, equipment and repair services than do hand-hoe cropping systems. Technological change that increases the demand for modern inputs (for example, improved seeds, fertilizer and pesticides) also enhances the strength of agriculture's linkages to the nonfarm sector.

Agriculture also influences the supply side of the rural nonfarm economy, primarily through the labor market. Wages in agriculture set the opportunity cost of labor directed to nonfarm activities, while seasonality of labor demand in agriculture affects the supply of labor available for nonfarm activities. The type and volume of agriculture's output also influences the kind of agricultural marketing, processing, and transport systems that can prosper.

Cross-country plots of the relationship between agricultural income (measured as agricultural income per capita of the rural population) and the nonfarm share of total rural employment show a positive relationship between the two (Hazell and Haggblade 1993). The relationship is particularly strong when rural areas are defined to include rural towns as well as rural areas. Figures IV.1 and IV.2 plot more recently available data for selected Asian countries and for individual states in India. In both cases, a positive relationship is evident, even though the data do not include rural towns. The time series observations for India and Indonesia in Figure IV.1 are particularly interesting, showing sharp rises in nonfarm employment shares over time as per capita agricultural incomes have increased.

To infer causality, more formal analysis of the relationship between agricultural growth and the nonfarm economy is required. Most studies of Asia using semi-input-output models (which are more defensible than input-output models for this purpose; see the review in Haggblade, Hammer, and Hazell

108 Transforming the Rural Asian Economy

Figure 4.1: Rural nonfarm employment as a function of
agricultural income - by country

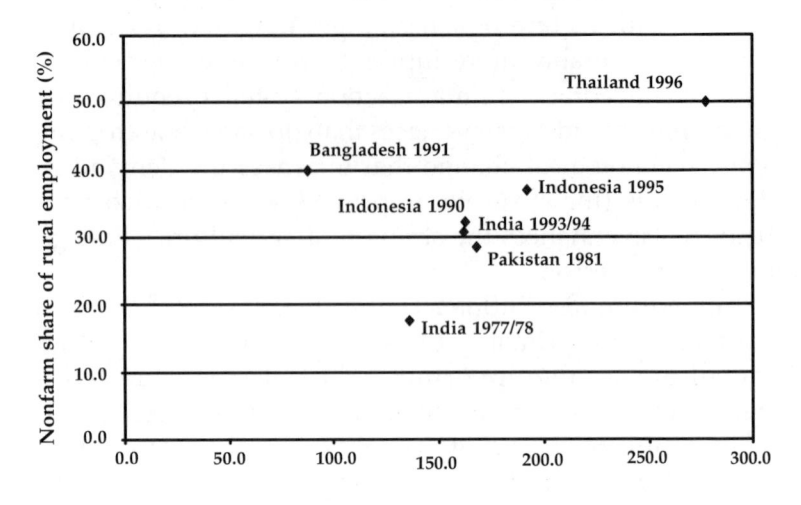

[Scatter plot with y-axis "Nonfarm share of rural employment (%)" from 0.0 to 60.0, x-axis "Agricultural income/rural population" from 0.0 to 300.0. Data points: Thailand 1996 (~280, 50); Bangladesh 1991 (~110, 40); Indonesia 1995 (~210, 37); Indonesia 1990 (~140, 30); India 1993/94 (~180, 30); Pakistan 1981 (~190, 27); India 1977/78 (~160, 17)]

Agricultural income/rural population

1991), estimate regional income multipliers at between 1.5 and 2.0. That is, each dollar increase in agricultural value added leads to an additional $0.5 to $1.0 of value added in the rural nonfarm economy (which is defined to include local towns). See, for example, studies by Bell, Hazell and Slade (1982) of the Muda river region in Malaysia; by Hazell, Ramasamy, and Rajagopalan (1991) of North Arcot district, Tamil Nadu, India; and more generally, Hazell and Haggblade (1991). These studies also attributed the largest shares of the multiplier to household consumption linkages (between two thirds and 80 percent), showing that the production, or inter-industry linkages, are relatively less important. This confirms expectations by Mellor and Lele (1972) and Mellor (1976), as well as the careful descriptive analysis undertaken by Gibb (1974) in Nueva Ecija province, the Philippines. Gibb documented the changes in the number and composition of firms in the region's towns over a decade of rapid agricultural growth and found the most rapid expansion to be among firms that served consumer-related

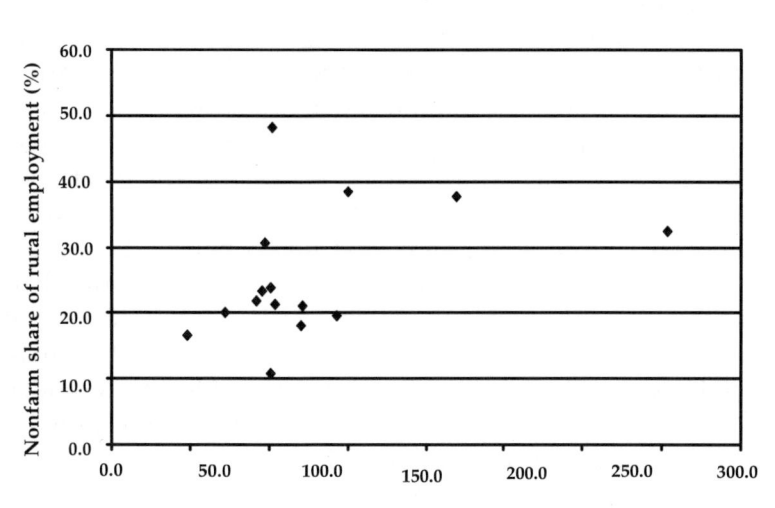

Figure 4.2: Rural nonfarm employment as a function of
agricultural income - by country
Indian states, 1993/94

[Scatter plot: x-axis "Agricultural income/rural population" from 0.0 to 300.0; y-axis "Nonfarm share of rural employment (%)" from 0.0 to 60.0]

needs. Gibb also documented the rapid gains in employment among nonfarm firms and found that each 1 percent of growth in agricultural output led to a 1-percent increase in nonfarm employment.

Agriculture's growth linkages to the nonfarm economy have been most studied in India. The mere fact that the nonfarm share of total national employment did not change for over a century until the full force of the green revolution was underway in the 1970s provides strong circumstantial evidence of the importance of agricultural growth as a motor for the nonfarm economy. Although some initial studies of the relationship were negative (for example, Vyas and Mathai 1978), subsequent time series evidence from fast-growing states like Punjab and Haryana was more convincing (Chadha 1986; Bhalla et al. 1990; Bhalla 1981). Moreover, based on an analysis of an India-wide sample of districts, Hazell and Haggblade (1991) estimated that each 100-rupee increase in agricultural income generated between 37 and 54 rupees in additional rural nonfarm income, the difference depending on whether or not the feedback effects of rural nonfarm growth on agricultural income are included in the analysis.

They also found that the income multipliers from agricultural growth were stronger in areas with better infrastructure, higher population density, and higher per capita agricultural incomes. The multipliers were particularly large in Punjab and Haryana, which score highly on these characteristics, and small in Bihar and Madhya Pradesh, which score poorly. Hazell, Ramasamy, and Rajagopalan (1991) used a semi-input-output model to estimate the income multiplier arising from the spread of high-yield rice varieties in North Arcot district in Tamil Nadu during the early 1980s, and found that each dollar of additional agricultural income led to an additional $0.80 of income in the region's nonfarm economy as a result of growth linkages. In an econometric study of the national economy, Rangarajan (1982) found that a 1-percent addition to the agricultural growth rate stimulated a 0.5 percent addition to the growth rate of industrial output, and a 0.7 percent addition to the growth rate of national income.

Population Density

Higher population density can impact on the rural nonfarm economy in a number of ways. By increasing the need for agricultural intensification, it leads to greater private investment in new technologies and land improvements and hence to increased production and investment demands for nonagricultural goods and services (Boserup 1965). Higher population densities also make possible more rapid attainment of minimum efficient scales of nonfarm production and service delivery. More densely populated areas are also more likely to be able to afford and maintain a denser network of roads and other rural infrastructure and to populate the rapidly growing towns, both of which activities foster development of nonfarm activity. Finally, higher population densities help keep wages low, at least in the early stages of the transformation when population growth is more likely to outstrip growth in agricultural employment; this is important for the nonfarm economy because of the labor intensity of much of its activity (Mellor 1976; Gibb 1974; Johnston and Kilby 1975).

Cross-country plots show a positive relationship between rural population density and the share of the rural labor force engaged in nonfarm activity (Haggblade, Hazell, and Brown 1989). Population density also emerged as a significant explanatory variable in the cross-state regression analysis undertaken by Hazell and Haggblade (1991) to explain differences in rural nonfarm employment in India.

Rural Towns

The development of rural towns has positive effects on rural nonfarm economic growth, because they offer large enough markets to capture economies of scale and scope for many types of nonfarm firms and because their higher levels of infrastructure development help reduce costs and facilitate communications and market reach. As the economic transformation proceeds, towns also become important centers of demand for their own production and consumption needs; this in turn creates new market opportunities for agriculture and rural nonfarm activity. Recent examples in Asia have been the explosion in urban demand for higher-value agricultural products (especially milk, meat, vegetables, flowers, and fruits), and the subcontracting of many lower-level manufacturing processes to rural nonfarm enterprises, both of which have boosted income and employment opportunities in surrounding rural areas (Otsuka 1998). These reverse linkages to the rural hinterland take on particular importance as rural towns become better integrated into the urban economy and develop manufacturing and service activities that serve urban and export demands in addition to rural needs. Such towns can become important growth poles for their surrounding regions and, as in the Japanese and Taipei,Chinese experiences, can lead to considerable income diversification even amongst farm households (Ho 1986a). Rapid growth of the urban economy in India in recent years has stimulated corridors of development along major highways and transport routes (Bhalla, 1997).

Rural towns can also stimulate additional agricultural production by improving the range and quality of available farm inputs, financial services, and agricultural marketing and processing services. A common example in Asia has been the benefit to agriculture of the development of local manufacturing of farm machines (Johnston and Kilby 1975). Often beginning as an outgrowth of traditional blacksmithing, local entrepreneurs respond to increasing demands for simple tillage, pumping, and threshing machines and provide products that are much better adapted to local conditions than machines purchased from outside. In the Indian Punjab, the initial spread of tractors spurred many blacksmiths to diversify into service and repair activities as their traditional work declined. Some of these "firms" subsequently expanded into the manufacture of locally designed tillage instruments for tractors, and then later into threshing machines and combine harvesters.

Rural Infrastructure

The development of rural infrastructure has powerful growth effects on the rural nonfarm economy. In the first place, it stimulates agricultural growth by lowering transport costs and increasing access to markets; this in turn leads to additional demands from the nonfarm sector. It also increases the access of rural people to nearby towns and rural market centers, enabling them to diversify and expand their consumption of nonfood goods and services.

Infrastructure development also impacts on the supply side of the rural nonfarm economy. Electrification, for example, is especially beneficial to small manufacturing and processing enterprises, shops, and service establishments, giving them a more reliable and cheaper source of power. Rural roads facilitate the movement of raw materials to rural towns and villages and of final products to their main markets—and at lower cost; they also enable firms to increase market size by giving them improved access to larger geographic areas and to increase rural labor mobility so that more village-based workers can take advantage of nonfarm employment opportunities in nearby towns. Telecommunications are increasingly important in linking rural firms to their customers and to the larger economy, enabling them to provide better and more timely service.

Improved infrastructure also facilitates the most economical location for different types of nonfarm activity. While many manufacturing and wholesale trading activities tend to concentrate in rural towns, many small-scale manufacturing activities (e.g., cottage industry and milling) and service activities (e.g., retail shops, coffee and tea shops, and personal services) expand in villages and rural market centers as infrastructure and agricultural development proceeds (Wanmali 1983). Ahmed and Hossain (1990) document how villages with better infrastructure increased their share of nonfarm income over similar villages with poorer infrastructure.

Infrastructure development also opens up the rural economy to greater competition from outside. This may take the form of cheaper products from lower-cost sources of supply or new or improved products that may displace some locally produced items. Improved infrastructure increases the exposure of rural people to urban tastes and products and this leads to changes in consumption behavior. The availability of electricity in a village, for example, creates demand for electrical goods (like radios, televisions and refrigerators) that are imported or produced in urban areas. Better roads and transport also lead rural people to travel to town more often and, once there, to purchase goods and services that they could not easily obtain before or that cost more at home. Some traditional rural and cottage industries lose their markets, but other types of activities expand and prosper (Jayaraman and Lanjouw 1998). The resulting changes in the composition of rural nonfarm activity have already been described in an earlier section.

Macro and Trade Policies

Apart from some manufacturing activities, the rural nonfarm sector until recently was largely ignored by policymakers. Because it depended heavily on agriculture either directly or indirectly for much of its demand, it suffered as a result of policies that discriminated against the agricultural sector. Recent macroeconomic policy reforms that have benefited the agricultural sector (e.g., currency devaluations and trade liberalization) should, therefore, have led to positive growth-multiplier benefits for the rural nonfarm economy. The policy reforms have also favored tradable goods production in general and this should have been directly beneficial to much rural industry.

The policy reforms can be a two-edged sword for the rural nonfarm economy, however. On the one hand, they create more opportunities for rural regions to produce and sell nonfarm tradables beyond their boundaries. They also reduce the price of nontradable goods relative to tradable goods, and this should help expand demand for many rural nonfarm services. On the other hand, the policy reforms also increase competition within rural areas between locally produced goods and cheaper, often higher-quality goods produced in urban areas or abroad, leading to the displacement of many traditional nonfarm activities.

The rapid growth in rural nonfarm activity in many Southeast Asian countries in recent years (e.g., Indonesia and Republic of Korea; see Table IV.6) suggests that the net impact of the policy reforms on the rural nonfarm economy has been very positive. While some changes in the composition of nonfarm activity have been necessary, the sector has been able to make these adjustments quickly.

A negative impact of the policy reforms in some countries has been a cutback in public expenditure in rural areas. Reductions in government investments in rural infrastructure can be expected to slow agricultural growth and the development of the rural nonfarm economy. In some countries, too, government expenditure in rural areas has been an important direct source of nonfarm employment, both for government employees residing in rural towns (civil servants, military personnel, etc) and for poor people participating in government employment schemes. In India, for example, government expenditure on rural-development and drought-relief employment schemes was a major and growing source of employment for the rural poor; the cutbacks in public spending on these schemes experienced during the policy reforms of the early 1990s contributed to a surge in rural poverty (Sen 1996).

Rural Industrialization Policies

Recognizing that the rural nonfarm economy can play an important role in creating employment and containing the growth of large cities, many governments have intervened with policies to promote the growth of their rural nonfarm economies. Typical interventions include

- creation of industrial estates in rural areas and smaller towns where firms receive privileged treatment in terms of infrastructure support and technical assistance;
- subsidies, tax breaks, foreign exchange licenses, etc., that give targeted firms a competitive edge in the market;
- technical-assistance and subsidized-credit programs that are targeted to certain types of nonfarm firms; and
- macro and trade policies that promote exports and protect domestic firms against cheap imports, at least during the early stages of an industry's development. Such polices have been used in a number of East Asian countries (beginning with Japan) to promote manufacturing more widely, but they also impact on rural industry as well.

Some countries, for example, the PRC and India, have also developed their own unique approaches to rural industrialization. The PRC's successful experience with "township-village enterprises" represents a rational response to the unique economic conditions created in rural areas by the absence of efficient factor and product markets (Otsuka 1998). Since workers and capital have not been free to migrate to larger cities, many small and medium-sized industries have grown up in rural towns and villages that might otherwise have been expected to locate in larger towns and industrial regions. This approach may well have relevance to other transforming economies in Asia. India developed a policy of protecting certain reserve industries as the unique domain of small-scale firms and cottage industry, prohibiting imports or the emergence of large-scale competitors. This approach helped create a great deal of employment, but retarded technological advance and growth in factor productivity (Chadha 1993; Singh 1990).

Apart from India's attempt to protect selected small-scale industries, policies to assist the rural nonfarm economy have generally favored manufacturing rather than service activities and large- rather than small-scale units of production. In many cases, small firms have effectively been placed at a competitive disadvantage against their larger-scale rivals (e.g., they do not receive the same subsidies and tax benefits) and this has encouraged more capital-intensive patterns of development than is optimal.

Even with this limitation, the success of rural industrialization programs has varied widely across countries. Rural nonfarm activity has flourished in Japan and Taipei,China in the post-World War II period, while in the Republic of Korea it has not, despite the promotion of rural industry in all three countries. In 1980, farm households in Japan and Taipei,China earned 80 percent and 65 percent, respectively, of their income from off-farm sources, three fourths of it in high-paying wage employment in rural towns and urban areas. Yet Republic of Korea farmers earned only 33 percent of their total household income from nonfarm sources (15 percent if remittances are excluded), with less than half in wage employment (Ho 1986a; Oshima 1986; Park 1986).

In explaining this disparity, many analysts point first to differences in agricultural performance (see Ho 1979, 1982, 1986a; Kada 1986; Oshima 1986; Park 1986; Saith 1987). They identify lower initial agricultural productivity in the Republic of Korea and a relative neglect of agriculture and its consequently lower growth, particularly since 1970. Weaker agricultural growth diminished rural consumption linkages and at later stages restricted the prospects for labor release from agriculture to high-paying, full-time, off-farm employment. In addition to more rapidly growing agricultural incomes, Japan and Taipei,China invested more heavily in rural roads, railroads, and electricity and adopted a policy environment supportive of dispersed manufacturing, commercial, and service activity. By the early 1960s, Japan and Taiwan boasted paved road and rural electrical networks with densities more than five times those in the Republic of Korea (Saith 1987), which chose instead to concentrate its industrial infrastructure in Seoul and Pusan.

Like the Republic of Korea, the Philippines has followed macro, trade and sector policies that have favored capital-intensive industry and urban areas. There has been rapid growth of urban areas, especially cities. In 1980, nearly half the total work force was employed in the nonfarm sector (48.5 percent), and 40 percent lived in urban areas (up from 32.7 percent in 1970). The Philippines now has one of most concentrated industrial sectors in the world for its level of per capita income; two percent of the firms produce 85 percent of industrial value added (Ranis, Stewart, and Angeles-Reyes 1990). There are few small and medium-sized enterprises but many cottage industries with low productivity. Urban concentration means that there are fewer opportunities for rurally based people to participate in the nonfarm economy and this has weakened the potential for farm-nonfarm linkages to create additional income and employment.

The experience with direct assistance programs for nonfarm firms has also been mixed and has led to generally disappointing results (see Haggblade and Mead [1998] for a recent review). Industrial estates for rural areas are widely viewed as expensive failures, while subsidized credit programs seem to have run into many of the same problems as agricultural credit programs (poor targeting, poor repayment, fungibility of capital, financial health of the lender, etc.). Technical assistance programs have had mixed results (Haggblade and Mead conclude that only about one third of the technical-assistance programs reviewed had a favorable cost/benefit ratio); the strongest performers were focused on high-leverage interventions in specific commodity subsectors (e.g., rattan production for export in Indonesia). Hyman (1998) also reports on some successful examples of NGOs developing and transferring "appropriate" technologies to small-scale firms.

Microfinance (or minimalist credit) programs to assist very small firms and part-time nonfarm activity (especially among women and the poor) are currently very popular and are assisting a segment of the rural nonfarm economy that has not been widely reached before. These programs may well lead to favorable poverty-reduction effects for the direct beneficiaries, but whether they lead to any growth in total nonfarm income and employment is a moot point. Because most of the service and cottage industry activities promoted by microfinance programs are regional nontradables that are constrained by local demand, any increase in their aggregate supply that might be induced by microfinance could simply depress prices and incomes for other producers, some of whom may also be poor.

It also needs to be recognized that many farm and landless households are too poorly educated or skilled to become successful nonfarm entrepreneurs. There are powerful self-selection processes at work in the marketplace and the less able are all too often left behind. In targeting assistance programs, including microfinance, it is necessary to recognize these limitations and to provide the necessary training and skills. Ravallion and Wodon (1997) find that in Bangladesh, the poor can realize significant nonfarm income despite their inherent disadvantages. Although the gains are lower in the poorest rural areas, they find that efforts to promote the nonfarm sector there can nonetheless be justified.

A promising development in recent years has been the emergence of subcontracting arrangements between large urban-based industrial firms and small-scale nonfarm units in rural areas (Hayami 1998; Otsuka 1998). These are especially popular for some metal-fabrication and textiles activities, where key components of the production process can be partitioned and the more labor-intensive components contracted out to small rural firms that have relatively cheap labor costs. Subcontracting requires good rural infrastructure and communications networks, effective laws and institutions to facilitate the enforcement of contracts, and a favorable growth environment for the industrial sector in general. The system played a key role in the industrialization of rural areas in Japan and has since spread to other East Asian economies.

CONCLUSIONS

This discussion of the rural nonfarm economy suggests several entry points where policy makers can help promote its growth. First and foremost is the need for agricultural growth. Because the rural nonfarm economy produces many regional nontradables, it depends primarily on agricultural production and farm incomes for its demand. This is especially true during the early stages of the rural transformation. Polices to promote agricultural growth are critical if the nonfarm economy is to grow. Moreover, agricultural growth should be broad-based, involving small and medium-sized farms, because these kinds of farms have the strongest demand linkages to the rural nonfarm economy. Supply-oriented policies in the face of stagnant demand are counterproductive to rural nonfarm producers.

But even if agriculture grows, the linkages to the nonfarm economy are constrained without good infrastructure. Villages need to be connected to local towns so that agricultural inputs and outputs can flow freely and so that people can go shopping. Local towns also need good infrastructure, especially roads, electricity, schools, sewers, water, and communications, in order to attract new firms and to grow.

A good legal and regulatory environment (e.g., to secure property rights and enforce contracts) and effective financial institutions are also required in order to promote trade, commerce, and manufacturing. Efficient rural financial markets that serve the full range of financial needs of farmers and nonfarm enterprises are more important than targeted credit programs. As the transformation proceeds, rural areas generate large capital surpluses; these need to be captured and managed more efficiently by the financial sector than seems to have happened in the past. Microfinance programs to help women and poor people develop nonfarm enterprises may contribute to poverty alleviation, but will probably contribute little to overall nonfarm economic growth. Microfinance also needs to be accompanied by appropriate training programs to give women and poor people the skills they need to compete in the market.

Rural people need adequate training if they are to have relevant technical, entrepreneurial, and management skills. The rural nonfarm economy provides much of its own training through apprenticeship schemes and on-the-job learning, but in an increasingly technical and communications-oriented world, specialized training schemes (e.g. computing, accounting) are needed, including programs for women, who dominate many service and trading activities.

Industrialization policies should foster the development of all kinds of rural nonfarm firms and not just manufacturing. It is an important fact that rural manufacturing (including agricultural processing and informal household manufacturing as well as formal manufacturing) only accounts for about 20 percent of total rural nonfarm employment in Asia. Most of the other 80 percent is to be found in the service, trade and construction sectors, which are dominated by small, labor-intensive, often part-time, and often women-led firms. Yet policymakers have been enamored by the manufacturing sector, and rural industrialization policies have showered manufacturing firms with all kinds of preferential tax, subsidy, licensing, and regulatory benefits, as well as with targeted and subsidized credit and technical-assistance programs. Moreover, these policies have typically favored larger capital-intensive manufacturing firms (the lure of the shiny rice mill or shoe factory) and neglected small labor-intensive firms and informal household manufacturing activities. Policymakers need to level the playing field and to revamp rural industrialization policies to a) be more inclusive—they should become rural "enterprise" rather than rural "industry" policies—and b) remove all unnecessary subsidies and protective policies that prevent rural firms from becoming competitive in the marketplace. It is no accident that for all the publicity about rural industries in China, the most rapid growth in recent years has been in small private firms specializing in service, trade, and construction activities.

Finally, as the nonfarm sector grows, there is need for increasingly open-trade, and pro-market policies to encourage greater efficiency and expansion into export markets.

The rural nonfarm sector has tremendous potential to create additional productive employment in rural Asia, which will be critical for those countries that still have rapidly growing labor forces. Rapid growth of the rural nonfarm economy also helps keep rural people living in rural areas, preventing excessive migration to the cities and keeping families intact. But this potential is too often constrained by poor rural infrastructure and inappropriate industrialization policies. The rural nonfarm economy deserves greater and more enlightened attention in the future.

V Sources of Agricultural Growth

INTRODUCTION

As discussed above, the rural transformation across Asia has resulted in the decline in agriculture's share of gross domestic product. Agricultural output has nevertheless grown strongly in much of the region and continues to grow, albeit at lower rates than before. And while rural population growth rates have dropped, there are nonetheless more people in rural Asia than ever before. Today it is harder to increase agricultural productivity beyond what has already been achieved: input levels are often already high and marginal increments in yields are diminishing, so new investments to produce productivity-driven, rather than input-driven, agricultural growth are needed.

What is happening instead is that with agriculture in a less predominant economic position, agricultural growth and its attendant problems no longer command the same attention in policy-making circles as they did 20 years ago. Public investment in agriculture and in the rural areas has been falling since the 1980s and problems arising as a consequence of agricultural growth, like environmental degradation, are not being addressed. New investment in agriculture should go toward increasing productivity rather than continuing to increase inputs: public agricultural research and extension are the main sources of agricultural productivity growth. New public investment in agriculture should be in research; even private investment in such research, which can also have public benefits, should be promoted.

AGRICULTURAL OUTPUT AND INPUT GROWTH

Asian agriculture grew rapidly during the past several decades. From 1967 to 1995, the rate of growth in agricultural output averaged 3.8 percent per year throughout the region, significantly more than the world average of 2.2 percent (Table V.1). This regional average masks a good deal of cross-country variation, however. The PRC had the highest rate of growth at 4.4 percent per year from 1967 to 1995, an impressive achievement considering the sheer size and limited natural resources of this country. Malaysia achieved the second highest rate of growth in overall agricultural output at 4.3 percent, due to the rapid diversification of its agricultural sector and a sharp increase in the export of tree crop products such as rubber, palm

Table V.1: Growth in Net Agricultural Production
(international 1989–91 dollars)

Period	1967–82	1982–95	1982–89	1989-95	1967–95
	(percent per year)				
Bangladesh	1.44	1.83	2.17	1.43	1.62
PRC	3.43	5.45	4.95	6.04	4.36
India	2.98	3.39	3.68	3.06	3.17
Indonesia	3.95	4.19	4.91	3.37	4.06
Korea, Rep. of	4.17	2.64	2.86	2.38	3.46
Malaysia	4.61	3.97	4.70	3.14	4.31
Myanmar	4.26	1.98	-0.04	4.39	3.20
Nepal	2.36	3.35	4.41	2.14	2.82
Pakistan	3.26	4.61	4.95	4.22	3.89
Philippines	3.79	2.20	1.41	3.12	3.05
Sri Lanka	2.12	0.85	0.03	1.82	1.53
Thailand	4.12	2.15	2.31	1.96	3.20
Viet Nam	3.27	4.61	4.18	5.12	3.89
AVERAGE	3.32	4.36	4.18	4.57	3.80
World	2.29	2.15	2.35	1.93	2.22

Note: Net Agricultural Production is gross production minus feed and seed. Growth rates are 3-year centered moving averages.

Source: FAO FAOSTAT. 1998 (Agricultural Production Indices).

oil, and, increasingly, cocoa, as well as livestock products like poultry and pork. Bangladesh and Sri Lanka, on the other hand, had the lowest annual rates of growth, at 1.6 percent and 1.5 percent, respectively. Growth in agricultural output in the other Asian developing economies ranged from 2.8 percent per year (Nepal) to 4.1 percent per year (Indonesia) during 1967–95.

The growth performance of Asian agriculture has improved over time. Growth increased from an annual average of 3.3 percent during 1967–82 to 4.4 percent in 1982–95. Moreover, compared to 1982–95, growth accelerated during 1989–95 at 4.6 percent per year—a period when global agricultural output growth was slowing down. The PRC, Myanmar, Philippines, Sri Lanka, and Viet Nam contributed to this remarkable recent agricultural growth performance.

Input Use Trends

Agricultural areas in the Asian developing economies span a wide range, from 1.9 million ha in Sri Lanka to 170 million ha in India (1995 data, see Table V.2). The expansion of agricultural land (defined as arable land and area under permanent crops) has contributed little to output growth in Asian developing countries, averaging 0.5 percent per year. Although annual growth in agricultural area actually increased between the two sub-periods, 1967–82 and 1982–95, from 0.3 percent to 0.8 percent, growth slowed considerably at the beginning of the 1990s, to 0.2 percent. The relatively high growth during the 1980s was due to increases in agricultural area in the PRC, Indonesia, and Malaysia. Over the 30-year period, agricultural area actually contracted in Bangladesh, Republic of Korea, and Myanmar, and growth in area was very small in India. Growth in agricultural land use was most significant in Malaysia, at 2.1 percent per year during 1967–95, due to a substantial increase in plantation production. Expansion of area was also relatively high in Nepal and Thailand, at 1.6 percent per year. These trends of negative or relatively low rates of growth in agricultural land area place the burden of output growth on improvements in land productivity.

Table V.2: Agricultural Land Use in Asia

Year/Period	1970	1995	1967–82	1982–95	1982–89	1989–95	1967–95
	(1000 hectares)			(percent per year)			
Bangladesh	9,097	8,800	0.05	-0.29	0.66	-1.39	-0.11
PRC	102,505	134,693	0.03	1.98	3.29	0.47	0.93
India	165,060	169,700	0.19	0.05	0.07	0.03	0.13
Indonesia	26,000	30,180	0.00	1.22	2.78	-0.57	0.57
Korea, Rep. of	2,298	1,985	-0.38	-0.70	-0.36	-1.10	-0.53
Malaysia	4,430	7,604	1.03	3.29	4.07	2.39	2.07
Myanmar	10,430	10,110	-0.21	0.03	-0.05	0.12	-0.10
Nepal	1,980	2,968	1.56	1.72	0.16	3.58	1.63
Pakistan	19,332	21,600	0.33	0.44	0.61	0.24	0.38
Philippines	6,952	9,520	1.72	0.53	0.40	0.68	1.17
Sri Lanka	1,894	1,886	-0.05	0.10	0.29	-0.13	0.02
Thailand	13,808	20,445	2.52	0.57	1.15	-0.11	1.61
Viet Nam	6,145	6,757	0.54	0.20	-0.44	0.96	0.38
TOTAL	369,931	426,248	0.28	0.80	1.34	0.18	0.52

Note: Land use includes the FAO categories 'arable area' and 'permanent crops'. Growth rates are 3-year centered moving averages.

Source: FAO FAOSTAT. 1998 (Land Use Domain).

Trends in irrigated area are shown in Table V.3. By 1995, Pakistan had a phenomenal 80 percent of its agricultural area covered by irrigation. Only the Republic of Korea has a similarly high share of area irrigated. Seven Asian developing countries cluster around 30 percent of agricultural area irrigated. Malaysia has the lowest proportion of irrigation coverage, mainly due to its tropical climate and the high proportion of unirrigated plantation agriculture. The average annual rate of growth in irrigated area in the major Asian developing countries was 1.8 percent during 1967–95. Growth varied substantially among the Asian countries. Whereas the South Asian countries of Bangladesh and Nepal achieved very rapid rates of growth, at 5.2 and 7.9 percent per year, respectively, growth was lowest in Indonesia, 0.6 percent annually, and the Republic of Korea, 0.1 percent per year. In the remaining countries, annual growth ranged from 1.2 percent in Pakistan to 3.6 percent in Thailand. With the exception of Bangladesh and Myanmar, the rate of growth in irrigated area slowed down significantly between the periods of 1967–82 and 1982–95, from 2.1 to 1.6 percent per year, on average. However, in six of the Asian developing countries, namely PRC, India, Indonesia, Malaysia, Myanmar and Viet Nam, there is some recovery in growth after 1989 compared to 1982–89.

The agricultural labor force increased more rapidly than agricultural area in all Asian developing countries except Malaysia (Table V.4). As a consequence, land-to-labor ratios declined throughout the region from an average of 0.57 ha per agricultural worker in 1970 to 0.45 ha per worker in 1995. The economically active population in agriculture reached 955 million people in 1995, up from 617 million in 1967, a reflection of the high rate of growth in the rural population throughout the region. In 1995, 75 percent of all labor employed in agriculture was located in the Asian developing economies, up from 68 percent in 1967. Annual growth in the agricultural labor force slowed significantly throughout the 30-year period, from 1.8 percent during 1967–82 to 1.4 percent during 1982–95. Indonesia and Viet Nam, however, both showed accelerated growth in the second period and declines at the beginning of the 1990s. In the Republic of Korea

Table V.3: Irrigated Area as a Percentage of Agricultural Area and Growth in Irrigated Area

Year/Period	1970	1995	1967–82	1982–95	1982–89	1989–95	1967–95
	(percent)			(percent per year)			
Bangladesh	11.63	37.56	4.95	5.39	6.35	4.29	5.16
PRC	37.18	37.02	1.59	0.79	0.36	1.28	1.22
India	18.44	31.82	2.64	2.42	1.55	3.44	2.54
Indonesia	15.00	15.18	0.65	0.49	0.22	0.80	0.58
Korea, Rep. of	51.52	60.76	0.73	-0.65	0.43	-1.89	0.09
Malaysia	5.91	4.47	2.34	0.22	0.19	0.26	1.35
Myanmar	8.04	15.38	1.87	3.37	-0.50	8.08	2.56
Nepal	5.91	29.82	12.55	2.67	5.51	-0.54	7.85
Pakistan	66.99	79.63	1.39	0.89	0.97	0.79	1.16
Philippines	11.88	16.60	3.80	1.33	2.08	0.46	2.65
Sri Lanka	24.55	29.16	1.91	0.21	0.01	0.45	1.12
Thailand	14.19	22.70	4.23	2.78	3.31	2.17	3.55
Viet Nam	15.95	29.60	3.71	1.29	1.11	1.49	2.58
TOTAL	25.17	33.24	2.05	1.56	1.10	2.10	1.82

Note: Growth rates are 3-year centered moving averages.

Source: FAO FAOSTAT. 1998 (Land Use Domain).

Table V.4: Land to Labor Ratio and Growth in Economically Active Population in Agriculture

Year/Period	1970	1995	1967–82	1982–95	1982–89	1989–95	1967–95
				(percent per year)			
Bangladesh	0.33	0.25	1.07	1.03	0.87	1.22	1.05
PRC	0.31	0.27	1.92	1.39	1.99	0.70	1.67
India	0.94	0.68	1.59	1.21	1.03	1.42	1.41
Indonesia	0.86	0.64	1.41	2.01	2.52	1.42	1.69
Korea, Rep. of	0.41	0.68	-0.07	-4.64	-4.87	-4.37	-2.22
Malaysia	2.21	4.14	0.57	-1.13	-1.22	-1.03	-0.23
Myanmar	0.97	0.60	1.93	1.76	1.84	1.66	1.85
Nepal	0.36	0.32	1.82	2.34	2.16	2.56	2.06
Pakistan	1.32	0.88	2.41	1.72	1.46	2.02	2.09
Philippines	0.86	0.80	1.90	1.23	1.24	1.21	1.59
Sri Lanka	0.79	0.53	1.69	1.44	1.58	1.28	1.57
Thailand	1.00	0.99	2.17	1.10	1.61	0.51	1.67
Viet Nam	0.39	0.26	1.58	2.19	2.56	1.77	1.86
AVERAGE	0.57	0.45	1.76	1.36	1.66	1.02	1.57

Note: Growth rates are 3-year centered moving averages.
Source: FAO FAOSTAT. 1998 (Population Domain, Land Use Domain).

and Malaysia, the labor force in agriculture has actually contracted during the last decades, a trend that will probably be seen in all Asian developing countries in the longer term. The PRC's economic reforms, initiated in 1978, opened up new opportunities for farm workers to seek employment in rural nonfarm industries. Thus, despite a relatively rapid rate of population growth in rural areas, labor use in agriculture has increased only slowly, especially during the 1990s.

Labor use in South Asian agriculture has grown more rapidly than in East and Southeast Asia. Indeed, in Nepal, growth in agricultural labor has accelerated during the past three decades, with a growth rate of 2.1 percent per year, on average, during 1967–95. In Bangladesh, India, and Pakistan, growth accelerated at the beginning of the 1990s after a slowdown during the 1980s. Lack of rural nonfarm employment opportunities and a relatively rapid growth in the rural population contribute to this trend.

One of the most dramatic changes in Asian agriculture during the past several decades has been the sizable increase in the use of both chemical fertilizers and agricultural machinery (Figure V.1, Tables V.5 and V.6). In 1970, the average farm in an Asian developing country applied 24 kg of fertilizer per hectare of agricultural land, less than half the global average. By 1995, the application rate had increased to 171 kg per hectare, on average, more than two thirds above the global rate and virtually on par with the fertilizer application rate in the United States of 173 kg per hectare. The most rapid growth occurred during the onset and takeoff of the green revolution in the 1960s and 1970s (10.8 percent per year). Crop yields increased markedly in response to the adoption of new, fertilizer-responsive varieties released by national and internationally sponsored research programs. National policies designed to subsidize fertilizer use by farmers were also at their peak in most of Asia at this time. During 1982–95, growth slowed substantially, to 5.9 percent per year. It slightly recovered at the beginning of the 1990s, however, a time when global application rates actually contracted. This recovery was partly due to the significant slowdown of growth in agricultural area in the region.

Figure V.1: Fertilizer application per unit of land, 1970 and 1995

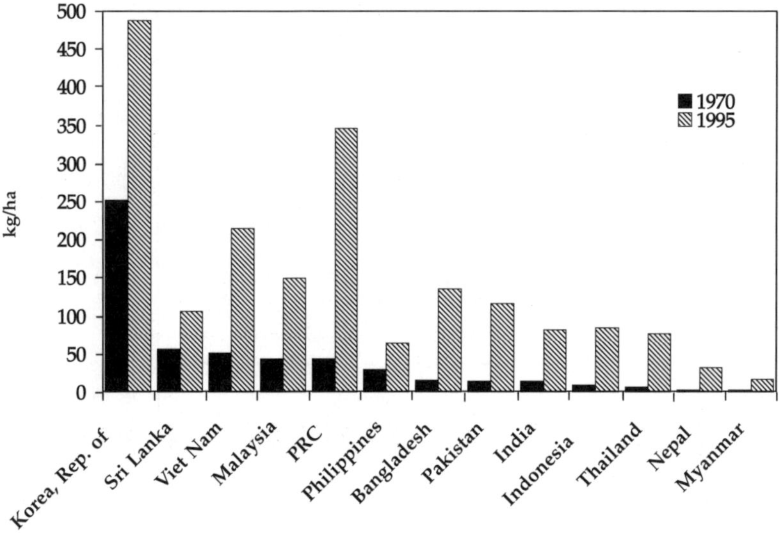

Note: Fertilizer includes Nitrogenous, Phosphate, and Potash fertilizers. Land is measured as arable land plus permanent crops.
Source: FAO FAOSTAT. 1998 (Means of Production Domain, Land Use Domain).

Fertilizer use has varied greatly throughout the region. The Republic of Korea stands out for its high fertilizer application rates. By 1970, it applied 2.5 times more fertilizer on a hectare of agricultural land than a typical developed country, and by 1995, it applied nearly five times more fertilizers. The PRC, a land-scarce and labor-abundant country, has a similarly high fertilizer application rate, at 346 kg per hectare in 1995. Viet Nam experienced the greatest growth in fertilizer application on a per hectare basis at the beginning of the 1990s, at 15.9 percent annually, resulting in an application rate of 214 kg per hectare in 1995, up from 51 kg per hectare in 1970. By 1995, Bangladesh, Malaysia, Pakistan, and Sri Lanka had also applied fertilizer in excess of 100 kg per hectare, whereas India, Indonesia, and Thailand had levels around 80 kg per hectare, and the Philippines close to 65 kg per hectare. Fertilizer application is quite low only in Myanmar and Nepal (Table V.5).

Table V.5: Fertilizer Application per Unit of Land, and Growth in Application

Year/Period	1970	1995	1967–82	1982–95	1982–89	1989–95	1967–95
	(kg/hectare)		(percent per year)				
Bangladesh	15.7	135.5	11.41	7.46	8.30	6.48	9.56
PRC	43	346.1	11.93	7.37	6.70	8.16	9.79
India	13.7	81.9	10.49	5.70	8.19	2.86	8.24
Indonesia	9.2	84.7	15.14	2.92	3.89	1.79	9.29
Korea, Rep. of	251.7	486.7	3.16	3.12	4.30	1.76	3.14
Malaysia	43.6	148.6	8.51	3.82	4.40	3.15	6.30
Myanmar	2.1	16.9	14.69	0.19	-7.59	10.09	7.71
Nepal	2.7	31.6	14.90	7.26	11.23	2.81	11.29
Pakistan	14.6	116.1	12.84	4.98	5.98	3.83	9.12
Philippines	28.9	63.4	5.17	4.10	6.40	1.47	4.67
Sri Lanka	55.5	106	2.32	2.01	3.01	0.86	2.18
Thailand	5.9	76.5	7.43	10.41	11.23	9.46	8.80
Viet Nam	50.7	214.3	7.28	12.95	10.52	15.85	9.87
AVERAGE	23.9	171.1	10.75	5.92	5.75	6.11	8.48
World	59.7	102.2	4.29	0.47	1.47	-0.69	2.50

Note: Fertilizer includes Nitrogenous, Phosphate, and Potash fertilizers. Land is measured as arable land plus permanent crops. Growth rates are 3-year centered moving averages.

Source: FAO FAOSTAT. 1998 (Means of Production Domain, Land Use Domain).

Table V.6: Tractors per Thousand Agricultural Labor and Growth in Use per Thousand Agricultural Labor

Year/Period	1970	1995	1967–82	1982–95	1982–89	1989–95	1967–95
	(tractors/000 workers)			(percent per year)			
Bangladesh	0.08	0.15	8.64	0.07	0.75	-0.73	4.57
PRC	0.38	1.36	11.13	-2.64	-1.38	-4.09	4.51
India	0.03	0.03	4.81	-0.52	4.75	-6.34	2.30
Indonesia	0.45	0.67	5.17	-0.38	0.12	-0.96	2.55
Korea, Rep. of	17.84	461.15	13.94	13.83	15.58	11.81	13.89
Malaysia	4.23	32.64	1.60	16.18	15.82	16.60	8.13
Myanmar	0.01	5.93	35.01	22.61	25.89	18.90	29.10
Nepal	0.78	4.65	5.31	10.66	13.22	7.75	7.76
Pakistan	0.05	0.19	10.24	2.49	5.77	-1.22	6.57
Philippines	2.59	25.61	14.22	5.90	9.02	2.37	10.28
Sri Lanka	2.83	3.25	2.18	-0.21	-2.44	2.46	1.06
Thailand	0.51	7.20	8.23	13.78	10.70	17.48	10.77
Viet Nam	0.18	3.74	14.76	7.81	-4.57	24.28	11.48
TOTAL	0.46	2.99	11.08	3.45	3.65	3.22	7.47
World	17.31	20.50	1.52	-0.13	0.64	-1.02	0.75

Note: Growth rates are 3-year centered moving averages.
Source: FAO FAOSTAT. 1998 (Means of Production Domain, Population Domain).

Although fertilizer application rates seem excessive in some Asian developing countries, in others there is a potential for considerable additional output increases from a judicious increase in fertilizer application.

At the beginning of the green revolution, few tractors were in use in Asian agriculture. By 1970, only 18 tractors were available for every 1,000 people employed in agriculture in the Republic of Korea, 4 in Malaysia, and not quite 3 each in the Philippines and Sri Lanka (Table V.6). At that time, 1,000 people working in agriculture had access, on average, to about 17 tractors globally, 167 tractors in developed countries, and 1,372 tractors in the United States. The Asian developing economies rapidly increased the number of tractors per agricultural worker during 1967–95, at 7.5 percent annually, with the most rapid growth, 11.1 percent, concentrated during the green revolution. Only in Malaysia, Nepal, and Thailand was growth faster during the second period. Moreover, growth accelerated during 1989–95 in Malaysia, Sri Lanka, Thailand, and Viet Nam, compared to 1982–89, although this is partially due to the substantial slowing of growth in agricultural employment: rapid economic growth raises the opportunity cost of farm labor, thereby inducing a substitution of machine power for labor in agriculture. The Republic of Korea experienced the most thorough growth, with rates above 10 percent for all periods. By 1995, tractor use in Korea surpassed the average developed-country usage, but was still less than a third of use in the United States. Usage in the average Asian developing economy, however, was still far below global use rates. Tractor use has remained particularly low in the South Asian countries of Bangladesh, India, and Pakistan, with less than one tractor per one thousand laborers even in the 1990s.

Land and Labor Productivity Growth

Figure V.2 shows the values for land productivity, measured as the value of aggregate agricultural output per hectare of agricultural land, in 1970 and 1995; Table V.7 captures in more detail the growth rates over the 1967–1995 period. Land productivity, measured as the value of aggregate agricultural output per hectare of agricultural land, increased by an average of 3.3 percent per year from 1967 to 1995 in Asian developing countries. Growth in land productivity was highest in the Republic of Korea (4 percent per year), followed by the PRC, Indonesia, Pakistan, and Viet Nam, all at or above 3.4 percent annually. In Nepal, on the other hand, growth was comparatively low, at 1.2 percent per year. In Indonesia, Republic of Korea, Malaysia, Myanmar, the Philippines, Sri Lanka, and Thailand, growth was highest during the green revolution. In the South Asian countries of Bangladesh, India, Nepal, and Pakistan, on the other hand, growth accelerated during 1982–95. In Viet Nam, growth was also higher during the 1980s, when the country entered its agricultural liberalization phase.

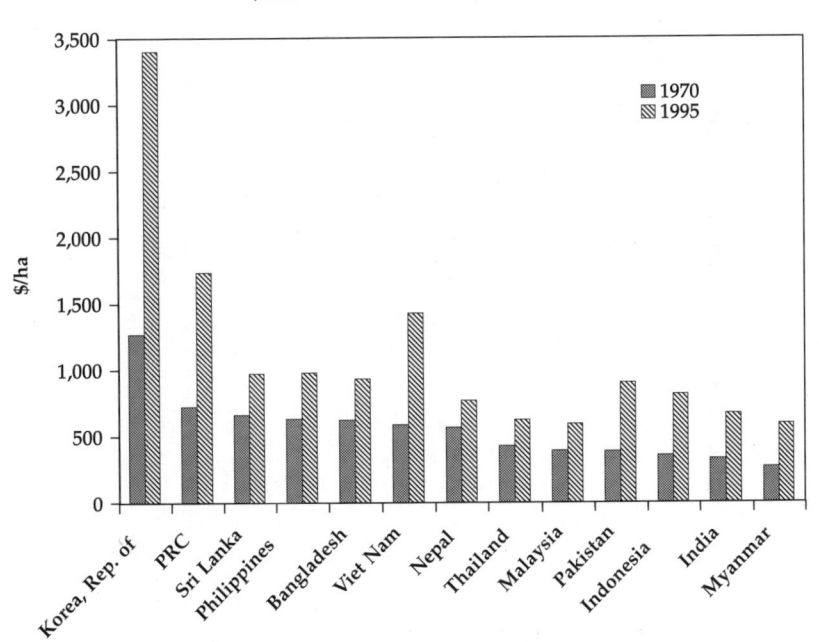

Figure V.2: Land Productivity in Asian Agriculture
(International 89-91 $/hectare)

Note: Net Agricultural Production is gross production minus feed and seed.
Source: FAO FAOSTAT. 1998 (Agricultural Production Indices, Land Use Domain).

Table V.7: Land productivity in Asian Agriculture (International 89-91 $/Hectare)

Year/Period	1970	1995	1967-82	1982-95	1982-89	1989-95	1967-95
	($/ha)		(percent per year)				
Bangladesh	617.3	931.3	1.39	2.12	1.61	2.86	1.73
PRC	720.4	1733.6	3.40	3.44	3.60	5.62	3.42
India	327.7	664.5	2.78	3.34	2.06	3.03	3.04
Indonesia	353.0	812.3	3.95	2.93	3.23	3.96	3.48
Korea, Rep. of	1270.1	3399.9	4.57	3.37	0.60	3.53	4.01
Malaysia	389.6	586.8	3.55	0.66	0.01	0.73	2.20
Myanmar	264.7	588.7	4.48	1.95	4.24	4.27	3.30
Nepal	564.4	764.8	0.80	1.60	4.33	-1.39	1.17
Pakistan	382.9	898.4	2.92	4.15	1.00	3.94	3.49
Philippines	631.8	977.2	2.04	1.65	-0.25	2.42	1.86
Sri Lanka	661.0	968.3	2.18	0.76	1.14	1.95	1.52
Thailand	425.1	617.9	1.56	1.57	4.64	2.07	1.56
Viet Nam	582.4	1426.3	2.72	4.40		4.12	3.49
AVERAGE	469.7	1058.6	3.03	3.55	2.81	4.43	3.27

Note: Net Agricultural Production is gross production minus feed and seed. Growth rates are 3-year centered moving averages.

Source: FAO FAOSTAT. 1998 (Agricultural Production Indices, Land Use Domain).

Figure V.3 shows the value for labor productivity, measured as aggregate agricultural output per worker, in 1970 and 1995; Table V.8 shows the rate of growth in labor productivity. During 1967–1995, labor productivity increased by 2.2 percent annually. The slower growth of labor productivity compared to growth in land productivity indicates that the Asian developing economies have generally adopted land-saving agricultural technologies. East Asian countries clearly outperformed South Asian countries in labor-productivity growth during the last 30 years. Annual growth averaged 2.7 percent in the PRC, 2.3 percent in Indonesia, 5.8 percent in the Republic of Korea, and 4.6 percent in Malaysia and a lower 2.0 percent in Viet Nam, 1.4 percent in the Philippines, and 1.5 percent in Thailand.

Figure V.3: Labor Productivity in Asian Agriculture
(International 89-91 $/worker)

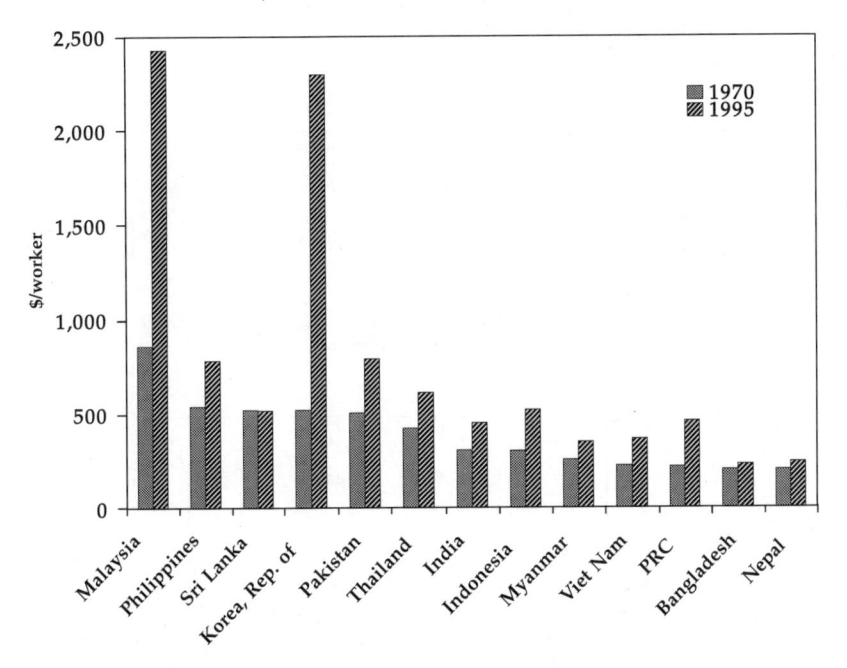

Source: FAO FAOSTAT. 1998 *(Agricultural Production Indices, Population Domain).*

Table V.8: Labor Productivity in Asian Agriculture
(International 89-91$ /Agricultural Worker)

Year/Period	1970	1995	1967–82	1982–95	1982–89	1989-95	1967–95
	($/ha)		(percent per year)				
Bangladesh	204.3	231.9	0.37	0.79	1.29	0.21	0.57
PRC	220.1	461.9	1.48	4.04	2.90	5.39	2.66
India	306.9	452.1	1.37	2.16	2.62	1.62	1.74
Indonesia	303.3	520.3	2.51	2.13	2.32	1.92	2.33
Korea, Rep. of	520.7	2,297.1	4.26	7.64	8.17	7.01	5.81
Malaysia	859.5	2,427.7	4.03	5.17	6.00	4.20	4.55
Myanmar	258.0	351.6	2.29	0.22	-1.84	2.68	1.32
Nepal	203.1	243.9	0.53	0.99	2.21	-0.41	0.74
Pakistan	505.1	790.1	0.84	2.84	3.44	2.14	1.76
Philippines	541.0	781.1	1.86	0.96	0.16	1.89	1.44
Sri Lanka	521.0	516.3	0.43	-0.58	-1.53	0.55	-0.04
Thailand	424.2	611.5	1.91	1.03	0.69	1.44	1.50
Viet Nam	225.4	368.5	1.66	2.37	1.59	3.29	1.99
AVERAGE	268.0	472.3	1.53	2.98	2.48	3.57	2.20

Note: Growth rates are 3-year centered moving averages.

Source: FAOSTAT. (Agricultural Production Indices, Population Domain) 1998.

In South Asia, on the other hand, annual growth trailed at 0.6 percent in Bangladesh, 1.7 percent in India, 0.7 percent in Nepal, 1.8 percent in Pakistan, and negative 0.04 percent in Sri Lanka. The relatively better performance in India and Pakistan was a response to the adoption of new (second- and third-generation) green-revolution technologies. The productivity performance of both countries slipped in the 1990s, however.

Labor productivity increased more rapidly after the green revolution and accelerated again at the beginning of the 1990s, particularly in the PRC, Sri Lanka, Thailand, and Viet Nam. Labor-productivity growth in the PRC and Viet Nam can be attributed mainly to the adoption of economic and institutional reforms during the late 1970s for the PRC and late 1980s for Viet Nam. Especially in the PRC, movement of labor between rural and urban areas was prohibited prior to reforms. As population continued to increase and agricultural labor productivity improved, a huge pool of surplus and underemployed labor developed throughout rural areas of the PRC. Following the reforms, rural workers were finally permitted to leave primary production and seek employment in various rural enterprises and even in urban-based industries. As a result, labor use in agriculture declined dramatically while agricultural output continued to rise. These dual developments led to a rapid growth in labor productivity in agriculture.

PATTERNS OF PUBLIC SPENDING ON AGRICULTURE

Government expenditures in agriculture have been estimated by Fan and Pardey (1998) and Pardey, Roseboom, and Fan (1998) using purchasing power parities (PPPs) as the basis for converting local currencies into U.S. dollars to make international comparisons. PPPs compare currency values by measuring the relative cost in local currencies of a detailed basket of traded and nontraded goods and services. These have the major advantage that they suffer less from distortions than market exchange rates, which are more susceptible to changes from policy-induced variations in currency values or sudden swings in financial transactions. PPP measures also capture the price of nontradables such as the labor, land, and facilities components that constitute a large share of government expenditures on agriculture; exchange rates are based only on a basket of traded goods and services (Craig, Pardey, and Roseboom 1991).

To convert expenditures denominated in current local currencies into international dollar aggregates expressed in base-year (1985) prices, current local currency expenditures were deflated to a set of base-year prices using each country's implicit GDP deflator. Then the 1985 PPPs reported by the World Bank (1997) were used to convert local currency expenditures measured in terms of 1985 prices into a value aggregate expressed in terms of 1985 international dollars. The results are shown in Tables V.9-V.12 and in Figure V.4, which summarize data presented in Fan and Pardey (1998).

Size and Trends in Public Expenditures

In 1993, India, at 35.9 billion public international dollars, ranked first in public expenditures on agriculture based on 1985 PPP values (Table V.9). Its expenditures were 16 percent higher than those of the PRC at $31 billion, the second biggest spender. Indonesia, Thailand, and the Republic of Korea spent between $4 and $6 billion on agriculture. The levels of public expenditures on agriculture in Bangladesh, Malaysia, Pakistan, and the Philippines were similar despite the dramatically different sizes of their agricultural economies. Public expenditures on agriculture were lowest in Myanmar, Nepal, and Sri Lanka.

Public expenditures on agriculture increased at 4.6 percent per year, on average. Growth in expenditures varied widely by period and country. It was highest during the 1970s, at 9.5 percent per year, slowed down during the 1980s, and was lowest at the beginning of the 1990s, at 0.6 percent annually, a period when almost half the countries experienced a real reduction in public spending on agriculture.

Table V.9: Government Expenditures on Agriculture in Asian Countries, 1985 US dollars (Purchasing Power Parity)

Year	1972	1975	1980	1985	1990	1993	1972-79	1980-89	1990-93	1972-93
	(million dollars)						(percent per year)			
Bangladesh	2,358	528	1,187	1,749	1,269	1,773	-3.56	-1.30	11.81	1.29
PRC	11,595	17,843	24,542	21,113	28,229	31,061	14.20	1.16	3.24	4.80
India	15,491	13,680	22,877	30,549	39,109	35,918	6.18	5.46	-2.80	4.09
Indonesia	1,436	3,020	5,026	4,351	6,157	5,958	15.68	1.36	-1.09	7.01
Korea, Rep. of	537	993	1,129	2,244	4,332	4,160	21.48	15.03	-1.34	10.24
Malaysia	348	458	1,264	1,851	1,830	1,693	11.43	3.00	-2.57	7.83
Myanmar	272	219	655	874	296	181	8.53	-8.13	-15.22	-1.93
Nepal	107	136	257	541	254	359	15.15	1.79	12.24	5.96
Pakistan	740	1,031	1,168	971	1,312	1,669	2.96	1.97	8.36	3.95
Philippines	416	1,145	729	604	1,409	1,694	12.68	7.19	6.32	6.92
Sri Lanka	627	449	589	2,124	614	596	-2.83	4.02	-0.97	-0.24
Thailand	902	767	1,850	3,181	3,190	4,513	8.26	3.54	12.26	7.97
AVERAGE	34,828	40,269	61,273	70,151	88,001	89,574	9.46	3.46	0.59	4.60

Note: Government expenditures in PPP US dollars was calculated in two steps: Government expenditures in constant (1985) local currency for each year was calculated, and then 1985 PPP exchange rates were used to convert local currency to PPP US dollars.

Source: Fan and Pardey 1998

Most East, Southeast and South Asian countries followed this pattern of slowing growth in public expenditures on agriculture over the last 20 years. There was a recovery in expenditures during the beginning of the 1990s, however, in the PRC and the South Asian countries of Bangladesh, Nepal, and Pakistan. Growth was largest in the East and Southeast Asian countries of Republic of Korea (10.2 percent), Thailand (8.0 percent), Malaysia (7.8 percent), and Indonesia (7.0 percent). The Republic of Korea's growth in public agricultural expenditures during the 1980s was mainly due to the rapid increase in government investments in rural development projects aimed at improving the quality of life of rural residents (APO 1991).

The size of national economies and their agriculture sectors varies widely throughout Asia. The PRC and India alone account for nearly three fourths of total agricultural GDP and government spending on agriculture in the latter part of the period examined. Thus, comparing expenditures in relative terms provides an alternative and, for many purposes, more meaningful perspective on public commitment to agriculture.

Government expenditure on agriculture expressed as a percentage of agricultural GDP is a useful comparative indicator of the size of the public commitment to agriculture (Table V.10). After increasing slightly in 1980, this ratio declined to 8.1 percent, on average, in 1993. In 1993, Thailand and the Republic of Korea had the highest public investment intensities, spending the equivalent of almost 20 percent of agricultural GDP on the agriculture sector. The governments of India, Indonesia, Malaysia, Sri Lanka, and the Philippines spent between 7 and 12 percent of agricultural GDP on the sector.

Malaysia had one of the highest agriculture intensity ratios during the 1980s but since the mid-1980s its spending intensity on agriculture has declined to below 10 percent. The Indian government has spent more intensively on agriculture than the PRC government during the past several decades. Especially in recent years, India's public spending on agriculture per dollar of agricultural output has been almost twice the amount spent in the PRC. Sri Lanka spent almost one third of its agricultural

Table V.10: Government Expenditures on Agriculture as a Percentage of Agricultural GDP

Year	1972	1975	1980	1985	1990	1993
	(percent)					
Bangladesh	6.8	1.3	2.7	3.6	2.3	3.2
PRC	5.7	7.9	9.1	6.0	6.5	6.3
India	10.4	8.0	12.3	13	13.9	11.7
Indonesia	4.1	7.9	10.4	7.3	8.2	7.2
Korea, Rep. of	3.7	5.6	6.5	10.6	19.9	18.7
Malaysia	5.7	6.1	13.5	14.7	10.7	8.1
Myanmar	4.6	3.5	8.6	8.3	3.3	1.7
Nepal	2.1	2.4	4.3	7.8	2.7	3.7
Pakistan	3.5	4.6	4.3	2.9	3.1	3.6
Philippines	3.1	6.8	3.6	3.0	5.9	7.0
Sri Lanka	10.7	7.1	7.7	26.9	8.1	8.1
Thailand	8.0	5.5	10.9	15.4	15.2	19.7
AVERAGE	6.9	7.0	9.3	8.5	8.9	8.1

Source: Fan and Pardey 1998.

GDP on the sector in 1985, but expenditure levels have declined since. Myanmar has drastically reduced its government spending on agriculture relative to the size of the sector, while the level of public resources Bangladesh has committed to agriculture has persistently been small when compared with the size of the country's agricultural economy.

The percentage of public expenditures on agriculture in total public expenditures provides some indication of the public priority afforded by the agricultural sector. As the Asian economies have become more industrialized, the relative importance of agriculture in the economy has declined. One might expect this to cause the share of overall government expenditures directed toward agriculture to shrink. However, this rule cannot be generally applied to Asian developing countries. Indeed, the rapidly growing economies like Indonesia, Malaysia, and the Republic of Korea have earmarked a reasonably stable share of their total government spending for agriculture over the past decade or so. In the Philippines and Thailand, the share of public agricultural expenditures in total government expenditures even increased during the 1980s.

On the other hand, the slower-growing economies of Bangladesh, Myanmar, Nepal, Pakistan, and Sri Lanka have dramatically reduced the shares of public expenditures going to agriculture. Between 1975 and 1980, the PRC and India committed similar proportions of public expenditures to agriculture, but the PRC's share has declined more than India's share in recent years (Table V.11).

Table V.11: Percentage of Government Agricultural Expenditures in Total Government Expenditures

Year	1972	1975	1980	1985	1990	1993
	(percent)					
Bangladesh	23.2	11	12.3	15.7	5.4	6.9
PRC	8.5	12.1	12.4	8.3	8.9	8.3
India	22.1	9.7	14.6	12.6	11.5	9.6
Indonesia	7.6	9.8	9.6	6.8	7.6	6.6
Korea, Rep. of	3.8	6.3	4.1	5.8	6.9	5.2
Malaysia	4	4.2	7.2	7.9	5.7	4.8
Myanmar	12.5	13.3	23.6	24.5	9.3	7.5
Nepal	13.7	15.5	16.4	22	8.5	10.5
Pakistan	5.7	6.7	5.4	2.9	2.6	2.6
Philippines	4.5	9	5.3	5.7	6	7.3
Sri Lanka	12.3	9	5.7	20	5.8	5.1
Thailand	7.8	5.9	8.1	11.7	10.4	10.4
AVERAGE	15.4	10.5	12.4	10.9	9.6	8.4

Note: Expenditures include those at both central and local government levels.
Source: Fan and Pardey 1998.

Public Expenditures on Subsidies and Investments

Public expenditures on agriculture can be allocated to current expenditures such as input subsidies or to investments such as research and development, irrigation, and rural infrastructure. The limited available data show that input subsidies often account for 20 to 60 percent of total Asian government spending on agriculture. Asian governments have intervened in the setting of fertilizer prices in support of a number of broad-based policy goals, for example. Many countries have protected domestic fertilizer production by restricting imports or maintaining import tariffs. In the Philippines, for example, domestic fertilizer prices were maintained well above world prices through the mid-1980s by a combination of import controls and subsidies to domestic fertilizer plants. More commonly, governments have subsidized farm-level fertilizer prices in support of several objectives, including income support for farmers and provision of incentives to increase the rate of adoption and level of fertilizer use, to increase crop production, and to balance other taxes against agriculture.

Fertilizer subsidies can become extremely costly to government treasuries. In Indonesia, subsidies have been maintained both for farmers and for the domestic fertilizer industry. The total fiscal costs for the two types of fertilizer subsidy were about Rupiah 670 billion (US$407 million) in 1986/87, representing nearly one sixth of total government development expenditures for agriculture and irrigation. Since 1986, Indonesia has been slowly phasing out fertilizer subsidies, but they remain large. Fertilizer subsidies in Bangladesh in 1983/84 accounted for about 14 percent of its budget allocation for agriculture. The economic costs can be even greater, as subsidies soak up funds that could be used for alternative investments and can induce the overuse of fertilizers above socially optimal levels. To the extent that subsidies are not fully funded to provide enough fertilizer to meet demand at the subsidized price, excess demand will be created, which can contribute to nonprice rationing, nonavailability of fertilizer, black markets, poor logistics, and untimely delivery of fertilizer.

Given the negative effects of subsidies, are there appropriate uses for them? Fertilizer subsidies to farmers may be cost-effective in stimulating farmers to adopt and utilize fertilizer appropriately together with new production technology. Temporary subsidies during the early stage of fertilizer adoption may be effective in overcoming the fixed costs related to adoption of new technology and in inducing farmer experimentation and learning during periods of rapidly changing technological potential. Such temporary subsidies should be phased out as adoption and appropriate use of fertilizer become widespread, however, as they actually have been in many Asian countries. At high levels of fertilizer use, the budgetary cost of the fertilizer subsidy becomes prohibitively expensive and the subsidy induces inefficient use of fertilizer beyond appropriate levels. Reallocation of expenditures on subsidies to more productive investments could have large payoffs (see Box V.1).

Box V.1: Subsidies versus Investments: Indonesian Food Crops

Fertilizer subsidies have been a cornerstone of Indonesian food production policy. In the late 1980s and early 1990s, the annual fiscal cost of fertilizer subsidies was over Rp 750 billion, ($455.5 million) almost the same as expenditures on irrigation development and seven times as much as expenditures on agricultural research and extension. An analysis of the dynamic food-crop supply response in Indonesia during 1969–1990, however, shows that the contribution of investment in technology to long-run output growth was far higher than the contribution of fertilizer subsidies. An examination of the historical sources of growth based on the model clearly shows the primary importance of technology in explaining growth in production of rice, corn, cassava, and soybeans. The share of output growth accounted for by public investment in research, extension, and irrigation is more than 70 percent for all four crops. Changes in relative prices due to fertilizer subsidies and output price supports also contributed to output growth, but the impact was minor compared to the contribution of technology. These results show that the fertilizer subsidies in Indonesia represent a serious misallocation of public resources. Given the high output response to public investment in technology, combined with the very low output response to fertilizer prices, the elimination of fertilizer subsidies and transfer of the resulting fiscal savings into investments in research, extension, and irrigation would have large benefits.

Source: Rosegrant, Kasryno, and Perez 1998.

As will be seen below and can be seen in the Indonesian example in Box V.1, public investment—in particular agricultural research expenditure—has been the driving force behind productivity growth in Asian agriculture. In the past several decades, a number of Asian governments have increased government spending on agricultural research by more than 5 percent per annum (Table V.12). In India, Malaysia, and Thailand government expenditures on agricultural research increased at more than 7 percent annually during 1972–93. The Philippines was the only country in which public spending on agricultural research declined in real terms during this period.

Nonetheless, the share of public agricultural research in total public expenditures on agriculture declined in several East Asian countries, in particular Indonesia, the Republic of Korea, Malaysia, and the Philippines (Figure V.4). In the largest countries, the PRC and India, the share has been fairly constant or slightly increasing. The expenditure share also declined significantly in Nepal. In the PRC and India, the share has been fairly constant or slightly increasing. Whereas Pakistan and Thailand invested about 11 percent of agricultural expenditures on research in agriculture in 1993, the level was 6 to 8 percent in Bangladesh, the PRC, and Sri Lanka, and only 3 to 5 percent in India, Indonesia, Republic of Korea, Nepal, and the Philippines.

SOURCES OF GROWTH IN ASIAN AGRICULTURE

Compared to the analysis of overall economic growth (see Chapter I), studies assessing the breakdown of agricultural output growth into factor accumulation and total factor productivity growth for Asian economies using comparable data sets across countries have been few. Fan and Pardey (1998) estimate the contribution of TFP to agricultural output growth, using comparable data available from international sources (Figure V.5).

The South Asian countries of Bangladesh, Nepal, and Sri Lanka, as well as Thailand in Southeast Asia, experienced

Table V.12: Public Investments in Agricultural Research in Asia, 1985 International Dollars

Year/Period	1972	1975	1980	1985	1990	1993	1972-79	1980-89	1990-93	1972-93
	(million dollars)						(percent per year)			
Bangladesh	50.2	58.3	104.7	122.7	143.9	136.7	9.53	2.68	-1.69	4.89
PRC	650.5	736.7	1,092.5	1,480.7	1,528.3	2,027.7	7.52	5.58	9.88	5.56
India	374.7	463.8	775.7	921.8	1,561.9	1,649.7	9.79	7.04	1.84	7.31
Indonesia	57.8	69.3	89.6	148.7	208.2	181.6	9.66	10.71	-4.45	5.61
Korea, Rep. of	43.2	44.2	59.8	79.3	108.7	135.3	4.49	5.74	7.59	5.58
Malaysia	35.2	60.9	99.5	147.3	145.9	173.3	14.87	4.26	5.90	7.88
Nepal	13.1	14.6	17.6	28.7	13.5	19.0	6.06	-1.04	12.24	1.79
Pakistan	66.7	90.3	121.8	194.1	205.4	187.6	8.48	5.95	-2.98	5.05
Philippines	53.2	45.7	37.6	28.7	57.9	51.6	-4.67	0.11	-3.75	-0.15
Sri Lanka	17.5	24.4	35.5	34.9	31.7	35.5	10.46	-2.33	3.89	3.42
Thailand	112.1	132.5	156.8	217.1	316.9	480.7	3.61	4.87	14.90	7.18

Source: Fan and Pardey 1998.

Figure V.4: Percentage of Agricultural Research Expenditures in Total Government Expenditures on Agriculture, 1972, 1980, and 1993

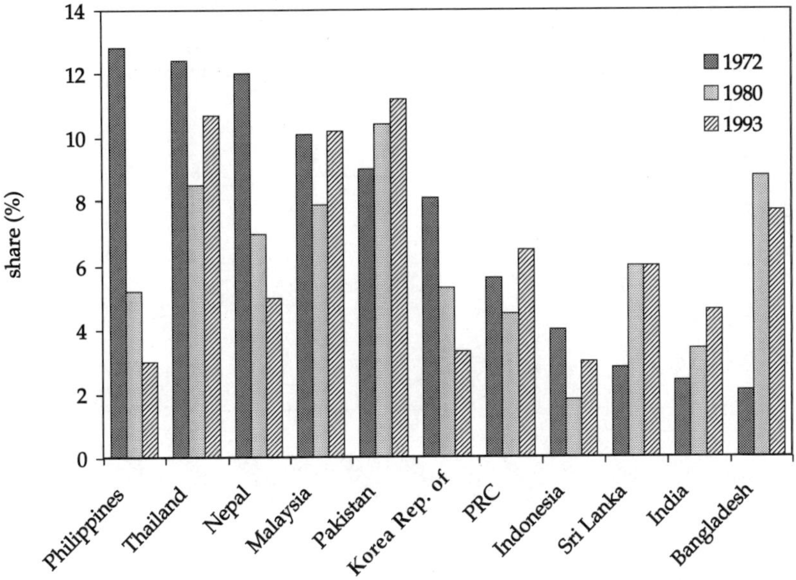

Sources: Pardey, Roseboom, and Fan 1998; Fan and Pardey 1998.

relatively low TFP growth during 1972–93, both absolutely and in proportion to agricultural output growth. It is noteworthy that the South Asian countries had low or negative rates of growth in agricultural research expenditures during the 1980s (Table V.12). In Bangladesh, agricultural TFP growth was slowed by relatively low investment in research, rural infrastructure, extension, and irrigation in the 1970s and dislocations arising from the civil war and separation of Bangladesh from Pakistan (Dey and Evenson 1991). In Sri Lanka, stagnation in TFP growth was also caused by intensifying civil war. Nepal is characterized by a difficult agroclimatic environment; moreover, the limited funding available for research was misallocated, with a heavy emphasis on crops that contributed relatively little to total area or value of production, like tobacco and sugarcane (Thapa and Rosegrant 1995). Thailand, on the other hand, experienced low TFP growth because it relied mainly on expansion of area to promote

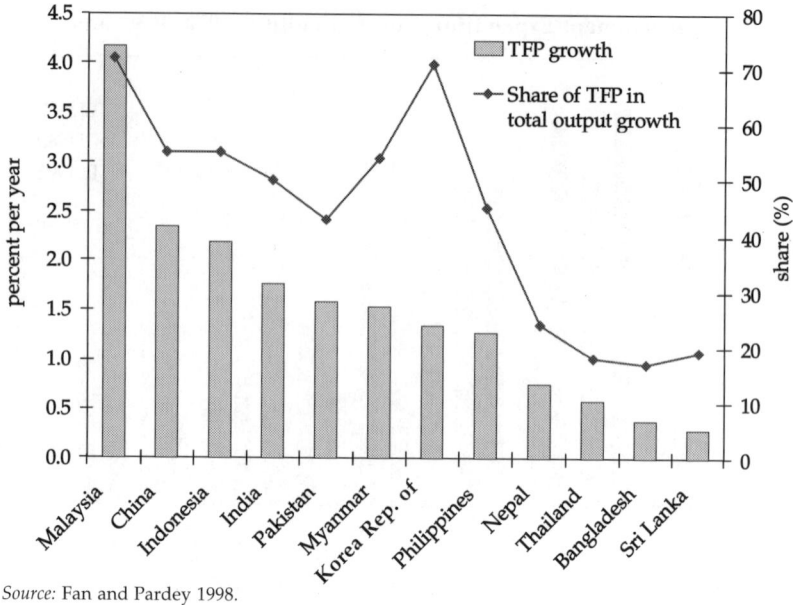

Figure V.5: Total Factor Productivity Growth in Asian Agriculture, 1972-93

Source: Fan and Pardey 1998.

agricultural growth during this period rather than on technological change driven by research. There was also little adoption of modern rice varieties because of their perceived relatively poor eating quality (Setboonsarng and Evenson 1991).

Explicit analysis of the sources of TFP growth in agriculture from country-specific studies bears out the importance of agricultural research and extension as a source of growth and illuminates additional important growth factors. In addition to being a primary source of TFP growth, the economic rates of return on agricultural research have been very high in Asian agriculture (see Box V.2).

The sources of growth in total factor productivity in agriculture can be understood through TFP decomposition analysis. TFP is estimated as a function of variables representing investments in research, extension, human capital, infrastructure, and other factors. Comparison of the sources of TFP growth in India and Pakistan is instructive.

Box V.2: Economic Rates of Return to Agricultural Research

In addition to being a primary source of TFP growth, investment in agricultural research has high economic rates of return. A ranking of country-level analyses on the estimated marginal internal rates of return to agricultural research for Asia shows that returns exceed 50 percent in 41 out of 65 studies and lie between 20 and 50 percent for 20 studies. Returns in Asian agriculture are considerably higher than in other developing and developed regions. For India, rates of return to agricultural research range from 40 percent (Evenson and Jha 1973) to 100 percent (Evenson 1987). Evenson, Pray, and Rosegrant (1999) find marginal internal rates of return diminishing over time: 59 percent during 1966–76 and 57 percent during 1977–87. For Pakistan, Pray (1978) estimates rates of return to agricultural research of 34–44 percent in 1906–56 and 23–37 percent in 1948–63. Azam, Bloom, and Evenson (1991) find a rate of 58 percent (1956–85) for Pakistan, and Nagy (1985) estimates a rate of 64 percent (1959–79) for Pakistan.

These high rates of return on agricultural research indicate that governments are underinvesting in agricultural research and that additional large investments in agricultural research would also have high payoffs. Although these past trends of economic returns to research cannot be simply extrapolated into the future, there is no evidence of diminishing rates of return over time. Some caution should be used in evaluating rates of return on research because of a host of potential errors in measurement and compilation that might inflate them. In particular, costs may be understated and benefits overstated; the use of truncated lags on the benefit stream in analyzing rates of return could bias rates of return upward (Alston, Craig, and Pardey 1998; Alston and Pardey 1996; Evenson, Pray, and Rosegrant 1999).

Even with a substantial downward revision in estimated economic rates of return to agricultural research, however, these rates would compare favorably with returns on other investments.

In India, public research and extension and private inventions have been the most important sources of TFP growth. For the period 1956–87, public research and extension account for 70 percent of TFP growth (Evenson, Pray, and Rosegrant 1999). In addition, expansion of roads has had a strong impact on TFP growth in India (Fan, Hazell, and Thorat 1998, see also Box II.1. in Chapter 2). The estimated effect of irrigation on TFP is also strongly positive, indicating that irrigation has a large impact on productivity, in addition to its value as an input. Moreover, the estimated effects of literacy are positive, showing the important impact of human capital development on productivity growth. For Pakistan, the factors with the highest positive impacts on productivity growth are research, share of modern varieties, rural literacy, and overall share of irrigation (Rosegrant and Evenson 1993).

A key result of the TFP decomposition analysis for both India and Pakistan is an understanding of the underpinnings of the substantial total productivity growth during 1955/56–65, before the rapid spread of modern varieties. For India, this was a period of rapid growth in investment in research and extension, strong growth in literacy, and very rapid growth in inventions in agricultural implements and inputs generated by private research and investment. Similarly, for Pakistan, heavy investment in research and rapid improvements in human capital, as shown by improvements in literacy and share of graduate personnel in research, induced strong growth in agricultural productivity even before the rapid adoption of modern crop varieties. These results show that strong productivity growth generated by research is feasible even when productivity is not embodied in the spread of modern varieties (Evenson, Pray, and Rosegrant 1999; Rosegrant and Evenson 1993).

In the early green-revolution phase (1965–75), TFP growth in India was sustained by rapid adoption of modern varieties and sharp increases in irrigation investment, despite somewhat lower rates of growth in investment in extension and public and private research. In Pakistan, TFP growth accelerated during 1965–75, due to sustained growth in irrigation development and explosive growth in the use of modern varieties, which more than offset the decline in growth of investment in research and human capital development (the latter reflected in a slowdown in the growth of literacy).

However, as the adoption of modern varieties slowed during the 1975–85 period, putting downward pressure on productivity growth, the patterns of productivity-enhancing public investment diverged in India and Pakistan. As a result, TFP growth during 1975–85 slowed only slightly in India, but turned negative in Pakistan. In India, investment in irrigation infrastructure dropped moderately, but was sufficient to sustain growth in the proportion of area irrigated, while expansion of the proportion of area irrigated in Pakistan virtually stopped. Growth in investment in public research, extension, and literacy in India actually increased relative to the previous decade. By contrast, the rate of growth in investment in research, technology, irrigation, and human capital development in Pakistan declined sharply relative to the previous decade.

Other factors contributing to the decline in TFP in Pakistan during 1975–85 were government policies taxing agriculture and increasingly serious waterlogging and soil salinity (Rosegrant and Evenson 1993; Azam, Bloom, and Evenson 1991). After 1985 there was a significant recovery in TFP growth in Pakistan as a whole and in the important Punjab region (Fan and Pardey 1998; Ali and Byerlee 1998). TFP growth in this latter period appears to be mainly due to labor-saving tractorization, at least in the Punjab, but sustainability of this growth appears questionable, because of the severe and growing negative impacts of resource degradation, particularly declining soil and water quality (Ali and Byerlee 1998).

More recently, TFP growth has slowed in India. Kalirajan and Shand (1997) show that both the level and contribution of TFP growth to overall output growth declined sharply between 1980–87 and 1988–90 in each of the 15 states that they analyzed. At the national level, TFP growth in Indian agriculture during 1975–80 was 1.71 percent per year and contributed 42 percent of agricultural output growth, but during 1986–90 slowed to 1.33 percent per year accounting for 30 percent of agricultural output growth (Desai 1994; Desai and Namboodiri 1997).

The decline in agricultural productivity growth was due in significant part to the relative neglect of investment in agricultural research during the 1980s, continuing into the 1990s (as shown in Table V.12), in favor of massive fertilizer and other input subsidies. Input subsidies grew from an already substantial 53 percent of agricultural plan expenditures in 1982/83 to 131 percent in 1992/93 (Ranade and Dev 1997). While the resulting growth in input use fueled fairly strong agricultural output growth during 1982–89, this input-led growth slowed after 1989 (see Tables V.1 and V.5). Given the importance of investment in research and extension, roads, and irrigation to agricultural output and growth in TFP, long-term prospects for productivity growth in India would be greatly enhanced by a shift of resources currently spent on input subsidies into these more productive investments. Input-led growth will not be sufficient without technological change (Kalirajan and Shand 1997; Kumar and Rosegrant 1994, 1997; Desai and Namboodiri 1997).

India also must deal with environmental problems that appear to have exerted a long-term downward pressure on TFP growth. Evenson, Pray, and Rosegrant (1999) estimate a negative coefficient for the time-trend variable in their analysis of sources of growth in TFP. They suggest that the negative sign could reflect negative effects from soil degradation (Antle and Pingali 1994) or cultivation intensity (Rosegrant and Pingali 1994), as well as other institutional or infrastructural factors. A lack of detailed environmental time-series data prevents a direct attribution of the negative trend to environmental degradation, but other evidence indicates that degradation is likely to be a significant contributor to the negative trend term. In areas of India (and elsewhere in Asia) where intensive rice monoculture has been practiced over the past two to three decades, there is considerable evidence of stagnant yields and/or declining trends in partial factor productivities, especially for fertilizers, and declining growth rates in total factor productivities (Cassman and Pingali 1995). In addition to these trends with respect to intensive monoculture rice, a similar slowing in partial factor productivity trends for the rice-wheat zone in India is reported by Kumar and Mruthyunjaya (1992) and Hobbs and Morris (1996) (see also Chapter VI).

The PRC's growth in both agricultural production and TFP has been extraordinary since the mid-1970s. Fan (1997) estimates TFP growth of 3.45 percent per year during 1972–1995, accounting for two thirds of agricultural output growth in the PRC. Productivity growth was particularly rapid immediately following economic reform, at 5.10 percent per year during 1979–84. The series of agricultural reforms resulted in sharply improved efficiency in agriculture as opportunities for off-farm activities became available, permitting a large transfer of inefficient labor out of agriculture; the growth in use of traditional inputs began to decline while total output continued to increase rapidly (Fan 1991). A range of analyses has documented the importance of economic reform in boosting agricultural productivity growth. Lin (1992) estimates that decollectivization accounted for about one half of total output growth in 1978–84. Fan (1991) finds that institutional changes accounted for two thirds of agricultural productivy growth during this period, and McMillan et al. (1989) estimate the contribution of reforms to productivity growth at 78 percent during 1978–84.

Huang, Rozelle, and Rosegrant (1995) estimate the simultaneous effects on growth in output of public investment in agricultural research, technology and irrigation, institutional innovations, pricing policies, and environmental factors, within a comprehensive dynamic analysis of sources of growth of agricultural production in the PRC during 1975–92. The results confirm that the economic reforms inititated during the establishment of the household responsibility system were an important source of agricultural growth, particularly during the reform period of 1978–84, but show that the one-time effects of those institutional innovations were largely exhausted after 1984.

Even after the efficiency gains from agricultural reform were mostly reaped, the PRC maintained remarkably high growth in agricultural output (5.54 percent per year) and TFP (3.91 percent per year) from 1985 to 1995 (Fan 1997). Huang, Rozelle, and Rosegrant (1995) show that most of this agricultural output growth since the mid-1980s has been the result of the expansion of agricultural research and investment in irrigation. Provincial-level analysis confirms that most of the growth in agricultural productivity in the PRC during 1984–93 was due to technological change, with relatively little continued contribution from efficiency changes induced by reform (Mao and Koo 1997). After a period of slower growth in the early to mid-1980s, investment in agricultural research and irrigation in the PRC has rebounded strongly, improving the prospects for continued productivity growth (Tables V.3 and V.12).

As with the case for India and Pakistan, however, Huang, Rozelle, and Rosegrant (1995) find that environmental degradation, including salinization, erosion, and natural disasters caused by the breakdown of the local environment, significantly reduced growth rates in production of rice, other grains, and cash crops in the PRC during 1975–90. Negative effects were largest for wheat, corn, and soybeans. These crops are much more likely to be grown in hilly and more ecologically fragile areas than paddy rice, which is grown more in plains areas and on terraces. These results suggest that increased attention should be given by policymakers to the adverse consequences of environmental stresses.

Private investment in agriculture consists mainly of investments by farmers aimed at increasing their long-term production capacity and private-sector investments in the crop research, input supply, processing, and marketing subsectors. The rapid growth in Asian agriculture has led to rapid income growth in rural areas and consequently to rapid growth in personal savings. This has created considerable capacity for increases in private agricultural investment within the rural sector. Private investments in agriculture have been relatively small in many Asian countries, however (Fan and Pardey 1998). At the same time, private investment can play an important role in agricultural productivity growth when government policies provide appropriate incentives (see Box V.3). The massive expansion of private-sector tubewell irrigation in India, Pakistan, and Bangladesh is the most successful example of private-sector irrigation development in the developing world

Box V.3: Private-Sector Research in Indian Agriculture

The private sector in India engages in considerable research and development (R&D) relevant to agriculture and has increased its investment quite rapidly over time. By the early 1990s, private-sector agricultural R&D was approximately one half the level of public-sector agricultural R&D. Private agricultural research accounted for more than 10 percent of TFP growth in India during 1956–87. The contribution of private research was highest during 1965–75, when it accounted for 22 percent of TFP growth. The contribution of private-sector industrial research declined, however, as India's trade and industrial policy turned inward and foreign technology was downplayed.

Many of the inventions produced in the private machinery, fertilizer, seed, and chemical industries have been embodied in inputs sold to farmers. The private firms undertaking R&D capture part, but not all, of the actual productive value of this technology in the form of higher prices. This partial capture of value means that private inventions will also provide social benefits in terms of productivity growth that are not directly appropriated by the private firms.

Policymakers need to be more aware of the expanding role of private-sector R&D in India, as well as of foreign suppliers of technology; Indian public policy toward private-sector R&D has to better recognize their favorable role in overall growth. Industrial policy and technology policy, including intellectual property-rights policy, will require careful evaluation and reform in order to encourage private investment in agriculture. Moreover, barriers to technology transfer should be removed in order to stimulate technology transfer and growth.

Source: Evenson, Pray, and Rosegrant 1999.

(tubewells are wells bored from the ground surface to the aquifer, from which water is pumped through a tubular casing using motor-driven or manually-operated pumps). A "groundwater revolution" in Bangladesh beginning in the 1980s was a key stimulant to rapid agricultural growth in the 1980s and early 1990s. Nearly 1.5 million hectares of land were newly irrigated after 1980, in large part from private installation of shallow tubewells spurred by deregulation of tubewell imports.

Public investments have been an important facilitator of private irrigation investment. Private tubewells in South Asia have grown most rapidly in areas with reasonably good roads, research and extension systems, and accessibility to credit and to electric or diesel energy; they have been concentrated in and around the command areas of large, publicly developed surface irrigation systems. Seckler (1990) notes three reasons for the complementarity between public and private irrigation investment in South Asia: deep percolation losses from the surface systems recharge the aquifers for tubewells; tubewells are often used together with surface irrigation water, which lowers pumping costs and concentrates these costs in periods of the highest marginal returns; and the tubewells ride piggyback on the infrastructure created for the surface systems.

It is likely that the share of private investment in Asian agriculture will increase in the future. As economies in the region continue to grow rapidly, the demand for agriculturally related technologies will increasingly move off-farm. Further increases in the use of inputs such as fertilizers, pesticides, and machinery will stimulate increased demand for new technologies and knowledge aimed at the input-supply sector. Rising incomes are leading to a rapid increase in the demand for processed agricultural products that in turn stimulates demand for post-harvest technologies related to the storage, processing, packaging, and marketing of agricultural produce (Fan and Pardey 1998). As shown in the case of expansion of private tubewells in South Asia, government policy plays a key role in stimulating private investment. Strengthening of property rights and removal of restrictions, regulations, and tax biases that penalize private investment will encourage the expansion of private investment in agriculture. In addition, properly targeted public investments can mobilize private investment.

While private investment in Asian agriculture is likely to grow, an important role for public investment in the sector will remain, because of incentive problems that discourage private investment in agricultural research and because of government desire to pursue equity or poverty-alleviation objectives. Agricultural research is in many cases long-term, large-scale, and risky, which means that most firms cannot carry out effective research and institutions may have to be set up on a collective (industry-wide or government) basis to achieve an economically efficient size and scale. The returns on new technologies are often high, but the firm responsible for developing the technology may not be able to appropriate the benefits accruing to the innovation—as in the case of improved open-pollinated rice and wheat varieties. The benefits of agricultural research often accrue to consumers (through reduction in commodity prices resulting from increased supply), rather than to the adopters of the new technology, so social returns may be greater than private returns to research. Appropriate government investments and policy interventions are therefore warranted, especially in areas with relatively low private incentives and relatively high social payoffs (Alston and Pardey 1996; Fan and Pardey 1998).

A number of developments could increase the role of the private sector in agricultural research in Asia. The appropriability of research results has increased, with hybrids being increasingly used and policy barriers to private-sector appropriation being reduced. Biotechnology innovations are likely to further the scope for private-sector involvement. In several countries the private seed sector has emerged as the dominant supplier of finished varieties for a number of crops. Policies to further increase private-sector involvement require continued attention; while many research activities require the long-term continuity of a public or semipublic institution, the potential for contracting for research should be explored more vigorously. Private-sector research, however, has generally shown little interest in solving the critical issues involved in increasing basic yield potential in wheat or rice varieties adapted to Asian agroclimatic zones or in developing hybridization procedures for additional crops. Moreover, there are some "orphan" commodities, usually tropical crops, fruits and vegetables, where the private sector makes no investments. Contracting of entire long-term research agendas to the private sector is therefore probably impossible and a significant and sustained public role in funding agricultural research will remain necessary (Binswanger 1994; Lipton 1994).

CONCLUSIONS

Strong agricultural growth in most Asian countries has been based both on rapid growth in input use and on productivity growth. The main sources of productivity growth have been public agricultural research and extension, expansion of irrigated area and rural infrastructure, and improvement in human capital. The rates of return to public research are high, showing the continued profitability of public investment in agricultural research, and strongly indicating that governments are underinvesting in research. The importance of productivity-driven growth is likely to increase relative to input-driven growth, because the growth rates in input use are declining as many regions in Asia reach high levels of input use. The public benefits from private research can be substantial, indicating that private firms capture only part of the real value of improved inputs through higher prices. Private investment in agriculture is likely to increase in importance, if policy reforms continue to create and improve the incentives for private investments by eliminating price distortions and strengthening property rights. However, continuing difficulties for private firms in recouping investments—particularly in research on many important crops—together with social objectives for public investment in agriculture, will continue to call for an important role for public investment in agriculture.

VI THE EVOLUTION OF CEREAL AND LIVESTOCK SUPPLY AND DEMAND: POLICIES TO MEET NEW CHALLENGES

INTRODUCTION

Recent signs indicate that the phenomenal green-revolution growth in cereal-crop productivity over the past 30 years, particularly for wheat and rice in Asia, is slowing, especially in the intensively cultivated lowlands. Slackening of infrastructure and research investments and reduced policy support partly explain the sluggish growth. Degradation of the lowland resource base due to long-term, intensive use also has contributed to declining productivity growth rates.

Fundamental changes in the diets of the populations in Asian developing economies are also a major factor in the evolution of cereal supply and demand. People with more money to spend on food shift from coarse grains to rice and from cereals to meat and other foods. Cereals that used to feed people now increasingly go to feed animals. Livestock consumption, which until recent years has been small relative to cereal consumption in Asia, has experienced extraordinary growth in demand in much of Asia that is beginning to exert great pressure for expansion in production. Although correction of some policies that have favored crop over livestock production could help, traditional forms of animal husbandry, with their low levels of productivity, cannot meet this new demand. More intensive alternatives hold the promise of greater production, but also a threat of increased environmental damage. In this chapter, the changing trends in cereal and livestock supply and demand are examined; the underlying factors that drive these trends are explored.

EVOLUTION OF CEREAL AND LIVESTOCK DEMAND

Rapid income growth, combined with increasing urbanization and changes in tastes and preferences, has caused a shift in diets from coarse grains to rice and secondary shifts from rice to wheat at the margin, as well as increased consumption of livestock and dairy products, vegetables, and fruits. Huang and Bouis (1996) show that urbanization induces structural shifts in diets: as populations move from rural to urban areas there is a wider choice of foods available in urban markets, as well as an exposure to a wider variety of dietary patterns of foreign cultures. Urban lifestyles place a premium on foods that require less time to prepare (inducing, for example, a shift from rice to wheat bread), as employment opportunities for women improve and the opportunity cost of their time increases. Moreover, urban occupations tend to be more sedentary than rural ones. People engaged in more sedentary occupations require fewer calories to maintain a given body weight. In addition, urban residents typically do not grow their own food. Thus, their consumption choices are not constrained by the potentially high-cost alternative of selling one food item at farm-gate prices (say, rice) to buy another food item at retail prices (say, bread), a choice faced by semisubsistence producers.

In addition, while changes in food demand patterns that cannot be attributed to increases in household incomes and changes in food prices may first be noticed in urban areas, as structural transformation proceeds to a more advanced level these same changes in food demand patterns may eventually occur in rural areas as well (Huang and Bouis 1996). The combination of these structural dietary changes with substantial income growth has brought about rapid changes in consumption patterns in Asian developing economies.

Trends in Cereal Demand

Table VI.1 shows historical growth in per capita food and (animal) feed demand and total demand for cereals in Asia, by region for the periods 1967–82 and 1982–95 (based on centered moving averages). The two subperiods roughly divide the period of 1967 to 1995 into a peak green-revolution period and a late green-revolution period. Growth in per capita cereal food demand declined substantially in Asian developing countries, from 1.2 percent annually during 1967–82 to 0.1 percent during 1982–95. The drop was most dramatic in East Asia, from 1.8 percent per year to –0.4 percent per year. In South Asia, on the other hand, growth in per capita cereal food demand remained virtually constant, at 0.5 percent per year. The fall in the growth rate of per capita food demand has been most rapid for maize, from 0.9 percent per year in 1967–82 to –2.9 percent annually in 1982–95, driven by the dramatic drop in East Asian food demand for maize. The overall decline was also substantial for rice and wheat, from 1.1 percent per year to 0.1 percent annually, and from 3.8 percent per year to 1.6 percent per year, respectively, during the same periods. Negative growth accelerated for other coarse grains, to 4.0 percent per year after 1982.

With both per capita cereal-demand growth and population growth slowing, growth in overall demand for all cereals has also declined over time, from 3.7 percent per year during 1967–82 to 2.8 percent from 1982 to 1995. Rapid shifts in dietary patterns—most apparent in East Asia—show up in a decline in the annual growth of total cereal demand from 4.2 percent during 1967–82 to 2.4 percent after 1982. This drop was mostly accounted for by a halving of the rate of growth of total wheat and rice demand and a reduction of the growth in total maize demand by nearly one third. Moreover, demand for other grains continued to decline sharply, from negative 1.2 to negative 2.3 percent per year over the two subperiods.

In South Asia, a dietary shift has been taking place from other coarse grains to increased consumption of rice. The annual rate of growth of total utilization for rice there remained virtually constant (and the rate of growth of per capita consumption of

Table VI.1: Annual growth in total utilization and per capita food and feed demand for cereal crops, by region, 1967–1995

	1967–82			1982–95		
	Total utilization	Per capita food demand	Per capita feed demand	Total utilization	Per capita food demand	Per capita feed demand
	(percent per year)					
Wheat						
East Asia	6.48	4.95	6.03	2.53	1.23	2.73
Southeast Asia	6.21	3.81	a	5.59	3.40	a
South Asia	4.91	2.27	a	3.59	1.49	a
Asia	5.84	3.75	6.21	3.59	1.60	4.36
Maize						
East Asia	6.01	1.78	6.36	4.42	-5.52	5.70
Southeast Asia	4.43	0.22	5.09	6.02	2.32	6.41
South Asia	1.32	-0.97	a	2.33	0.24	a
Asia	5.21	0.88	6.04	4.47	-2.91	5.27
Milled Rice						
East Asia	3.40	1.44	9.53	1.27	-0.16	5.45
Southeast Asia	3.47	1.10	2.96	2.42	0.49	2.86
South Asia	2.76	0.50	a	2.77	0.69	a
Asia	3.24	1.07	a	2.06	0.13	3.98
Other grains						
East Asia	-1.18	-4.40	2.15	-2.32	-7.54	0.89
Southeast Asia	8.15	a	a	1.76	-0.10	a
South Asia	0.59	-1.73	a	-0.58	-2.57	a
Asia	-0.25	-2.90	1.57	0.20	-3.97	2.80
All cereals						
East Asia	4.22	1.80	5.81	2.39	-0.44	5.11
Southeast Asia	3.81	1.16	4.32	3.32	0.88	5.47
South Asia	2.88	0.50	2.87	2.60	0.53	0.89
Asia	3.73	1.24	5.36	2.82	0.10	4.88

Source of basic data: FAO FAOSTAT 1998.

[a] Per capita demand is less than 2 kg. Growth rates are not computed.

rice as food increased slightly), while utilization of coarse grains declined each year after 1982. In Southeast Asia, growth in total cereal utilization of 3.3 percent per year during 1982–95 is driven by maize, which grew at 6.0 percent per year after 1982, closely followed by wheat, which grew at 5.6 percent annually in the same period. While per capita food demand and changing diets lie behind the growth in wheat consumption, it is strong growth in demand for animal feed that lies behind the figure for maize.

This crucial trend, the rapid growth in demand for cereals—in particular maize—as animal feed in Asian developing countries, emerges clearly from Table VI.1. Per capita demand for maize as feed grew at a very rapid 6.0 percent per year during 1967–82, and continued to increase at a strong 5.3 percent annually during 1982–95. Whereas annual growth in per capita feed demand for maize slowed slightly in East Asia (from 6.4 percent to 5.7 percent), it accelerated in the group of Southeast Asian countries (from 5.1 to 6.4 percent). Growth in Asian developing countries was also strong for per capita feed demand of wheat and rice, at 4.4 percent per year and 4.0 percent annually after 1982, respectively. This high growth is due to the rapid expansion of the livestock industry, especially in the more rapidly growing economies where consumption of livestock products has been expanding dramatically.

Growth in Demand for Livestock Products

Table VI.2 summarizes the historical growth in total and per capita demand for livestock products (referred to here as beef, pork, sheep and goat meat, and poultry; dairy products are treated separately in Chapter XII and the Appendix). Growth in meat demand has accelerated in developing Asian countries, with annual per capita food demand growth for all meats of 5.9 percent and total demand growth of a phenomenal 7.9 percent from 1982 to 1995, up from 2.4 percent and 4.6 percent during 1967–82, respectively. For the group of developing Asian economies as a whole, growth was fastest for poultry, with a per capita demand growth of 8.9 percent per year and a total

Table VI.2: Annual growth in total utilization and per capita food demand for livestock products, by region, 1967–1995

	1967–82		1982–95	
	Total utilization	Per capita food demand	Total utilization	Per capita food demand
	(percent per year)			
Beef				
East Asia	5.12	3.10	15.49	13.98
Southeast Asia	1.94	-0.46	3.92	1.96
South Asia	2.52	0.20	3.33	1.21
Asia	2.75	0.58	7.88	5.83
Pork				
East Asia	5.33	3.31	7.22	5.81
Southeast Asia	2.86	0.50	5.68	3.69
South Asia	4.15	a	2.91	a
Asia	5.00	2.78	7.04	5.08
Poultry				
East Asia	5.21	3.19	12.83	11.33
Southeast Asia	6.33	3.89	6.81	4.80
South Asia	4.08	a	10.14	a
Asia	5.50	3.27	10.92	8.86
Sheep and goat				
East Asia	5.95	a	8.96	a
Southeast Asia	5.49	a	4.89	a
South Asia	2.69	a	4.23	a
Asia	3.99	a	7.36	a
All meat				
East Asia	5.33	3.31	8.59	7.16
Southeast Asia	3.67	1.28	5.75	3.74
South Asia	2.77	0.44	4.19	2.06
Asia	4.59	2.38	7.86	5.86

Note: Total utilization for livestock includes food and feed demand, processing, waste and other uses.
Source of basic data: FAO FAOSTAT 1998.
a   Per capita demand is less than 2 kg. Growth rates are not computed.

demand growth of an extraordinary 10.9 percent per year in 1982–95. These figures basically double the respective annual rates of growth during 1967–82. Moreover, growth in per capita and total utilization of beef overtook growth in per capita and total pork utilization after 1982. Still, growth in total demand for pork and sheep/goat meat was also very rapid, at 7.0 and 7.4 percent per year, respectively, during 1982–95.

Subregional growth among livestock products was highest for beef in East Asia, a per capita food demand growth of a staggering 14.0 percent per year and total demand growth of an even higher 15.5 percent annually—a quadrupling and tripling, respectively, of the rates prevailing in 1967–82. Still, some caution is advised as regards figures for East Asia for the latter period: PRC livestock consumption (as well as production) figures, based on national food balance sheets, may have been overestimated in the 1990s (but not in the 1980s). Independent estimates based on surveys and feed use suggest that actual meat consumption in the early 1990s in the PRC was probably closer to 30 million metric tons in aggregate rather than the 34 to 45 million metric tons (1992–1995 values from FAO 1998) typically shown (Ke 1997; Delgado et al. 1998).

In Southeast Asia, per capita food demand for poultry grew rapidly at 4.8 percent per year during 1982–95, continuing the high growth trend during 1967–82 of 3.9 percent per year. Growth in total utilization grew even more rapidly, at 6.8 percent annually during 1982–95, up from 6.3 percent in the first sub-period. Per capita growth in pork and beef consumption also accelerated during 1982–95, to 3.7 percent per year and 2.0 percent per year, respectively. In South Asia, growth in total utilization of poultry more than doubled between the two sub-periods, from 4.1 percent per year to 10.1 percent per year. Growth continued at a relatively high level for beef (3.3 percent per year, from 2.5 percent annually in the first sub-period) and decelerated slightly for pork, from 4.2 percent per year to 2.9 percent annually after 1982.

EVOLUTION OF CEREAL PRODUCTION: TRENDS, CHALLENGES, AND POLICIES

Growth in cereal production in developing Asian countries slowed for all major cereals, as shown in Table VI.3, from 3.6 percent annually during the first sub-period to 2.4 percent per year thereafter. Maize took over from wheat, exhibiting a growth rate during 1982–95 of 3.9 percent per year, but wheat still showed higher rates of growth in production (2.9 percent per year) than rice (2.0 percent per year), while production growth was negative for other grains in both sub-periods.

The contribution of area expansion to growth in cereal production in Asia declined dramatically during the latter sub-period. For the whole region, cereal area growth was virtually stagnant after 1982, dropping from an already slow rate of 0.4 percent per year during 1967–82. Only maize area showed significant expansion after 1982, at nearly one percent per year. Wheat and rice area continued to expand after 1982, but at much

Table VI.3: Annual Growth in Cereal Crop Area, Production, and Yield, by Region, 1967–1995

	1967–82			1982–95		
	Area	Yield	Production	Area	Yield	Production
	(percent per year)					
Wheat						
East Asia	0.93	5.46	6.45	0.17	2.90	3.07
Southeast Asia	-1.45	5.30	3.72	0.36	-2.60	-2.20
South Asia	3.22	4.12	7.40	0.93	2.52	3.57
Asia	1.30	4.07	5.43	0.15	2.70	2.85
Maize						
East Asia	1.00	4.01	5.05	1.39	3.02	4.47
Southeast Asia	1.45	2.57	4.06	0.34	3.17	3.53
South Asia	0.63	0.88	1.51	0.62	1.68	2.31
Asia	1.09	3.48	4.62	0.95	2.94	3.93
Paddy Rice						
East Asia	0.57	2.78	3.36	-0.65	1.70	1.05
Southeast Asia	0.80	3.07	3.89	1.05	1.65	2.71
South Asia	0.71	1.88	2.60	-0.05	2.92	2.81
Asia	0.70	2.54	3.25	0.13	1.91	2.03
Other grains						
East Asia	-4.44	3.14	-1.43	-4.10	2.23	-1.96
Southeast Asia	5.40	2.85	8.49	-1.02	-0.26	-1.41
South Asia	-0.84	1.92	1.07	-2.35	1.85	-0.55
Asia	-1.76	1.63	-0.15	-1.38	1.05	-0.35
All cereals						
East Asia	0.00	3.84	3.84	-0.20	2.56	2.36
Southeast Asia	0.94	2.97	3.94	0.90	1.90	2.82
South Asia	0.71	2.72	3.44	-0.33	2.94	2.58
Asia	0.42	3.13	3.57	0.02	2.42	2.43

Source of basic data: FAO FAOSTAT 1998.

slower rates, and area planted to other grains (including barley, millet, oats, rye, and sorghum) declined sharply. Wheat area grew at only 0.2 percent per year during 1982–95, only about one tenth the rate of growth from 1967 to 1982, while rice area expansion was down from 0.7 percent per year to 0.1 percent. Area growth varied considerably across regions within Asia. East Asia (dominated by the PRC) and South Asia showed a decline in area planted to cereals after 1982, with declines in rice and coarse grain area offsetting the relatively large expansions in maize (East Asia) and wheat (South Asia). Cereal area continued to expand slowly in Southeast Asia, mainly due to growth in rice area.

The pattern of growth of cereal yields in Asia also shows a significant slowdown after 1982 (see also Figure VI.1). Wheat yield growth in Asia declined from 4.1 percent per year in the first sub-period to 2.7 percent in the second, on average, with wheat yield growth in East Asia slowing from 5.5 percent annually to 2.9 percent. Maize yield growth dropped from 3.5 percent annually in 1967–82 to 2.9 percent per year thereafter, on average. Annual yield growth recovered in South Asia, however, from 0.9 percent annually during 1967–82 to 1.7 percent during 1982–95. Rice yield growth exhibits a similar trend, with a drop in annual yield growth from 2.5 percent to 1.9 percent for the region as a whole. Declines in rice yield growth were substantial in most regions, but in South Asia, which was a relatively slow adopter of green-revolution rice technology, yield growth increased from 1.9 percent per year in 1967–82 to 2.9 percent annually in 1982–95. Rice yield growth in South Asia peaked in the late 1980s, at 3.3 percent annually, and slowed to 2.5 percent per year during 1989–95.

Challenges in Sustaining Cereal Production

The slowdown in cereal production growth in developing Asian countries since the early 1980s has been caused by declining world cereal prices and by factors related to the increasing intensification of cereal production. Green-revolution growth in

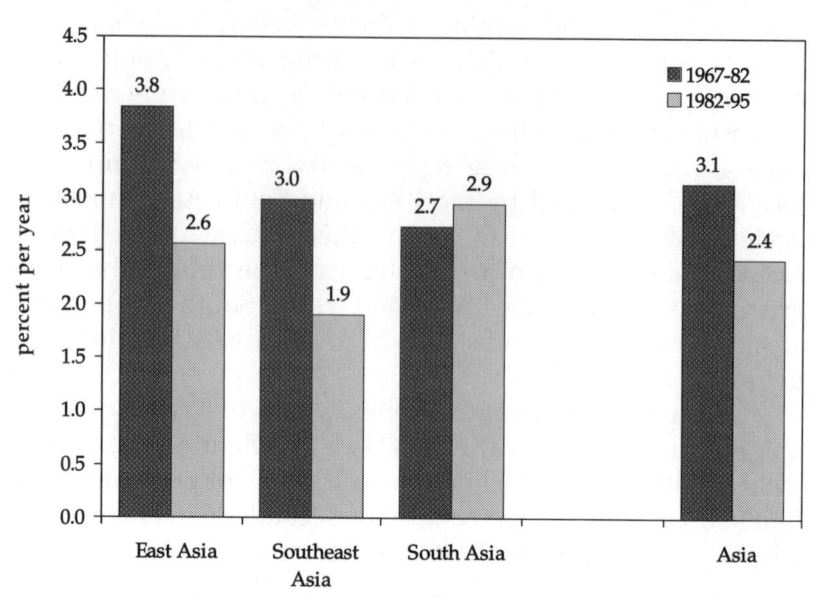

Figure VI.1: Annual growth in cereals yields, 1967–82 vs. 1982–95

Region	1967-82	1982-95
East Asia	3.8	2.6
Southeast Asia	3.0	1.9
South Asia	2.7	2.9
Asia	3.1	2.4

cereal-crop productivity resulted from an increase in land productivity; it occurred in areas of growing land scarcity and/or areas with high land values as a result of good market infrastructure and was associated with strong policy support. High levels of investment in research and infrastructure development, especially irrigation infrastructures, resulted in the rapid intensification of the lowlands. Consequently, the irrigated and the high-rainfall lowland environments became the primary source of food supply for Asia's escalating population.

Between 1982 and 1995, however, real world wheat prices declined by 28 percent, rice prices by 42 percent, and corn prices by 43 percent. Declining cereal prices caused a direct shift of land out of cereals and into more profitable cropping alternatives and have slowed the growth in input use and therefore of yields.

Probably more important in the long run, declining world prices have also caused a slowdown in investment in crop research and irrigation infrastructure, with consequent effects on yield growth (Rosegrant and Pingali 1994; Rosegrant and Svendsen 1993). Environmental and resource constraints have also contributed significantly to the slowdown in yield growth. The use of high levels of inputs and the achievement of relatively high cereal yields, particularly for rice and wheat, have made it more difficult in parts of Asia to sustain the same rate of yield gains, as farmer yields in these regions approach the economic optimum yield levels. At the same time, increased intensity of land use has led to the necessity for increasing input requirements in order to sustain current yield gains.

Moreover, Pingali, Hossain, and Gerpacio (1997) argue that the practice of intensive rice monoculture itself contributes to the degradation of the paddy resource base and hence to declining productivities. Declining yield growth trends can be directly associated with the ecological consequences of intensive rice monoculture systems such as buildup of salinity and waterlogging, declining soil nutrient status, increased soil toxicities, and increased pest buildup, especially of soil pests. Many of these degradation problems are also observed in the irrigated lowlands where wheat is grown after rice in rotation (Hobbs and Morris 1996). The nature and magnitude of environmental degradation is described in more detail in Chapter X.

Intensification *per se* is not the root cause of lowland resource base degradation, however, but rather the policy environment that encourages monoculture systems and excessive or unbalanced input use. Trade policies, output price policies, and input subsidies have all contributed to the unsustainable use of the lowlands (for more detail on trade and macroeconomic policies, see Chapter VII). The dual goals of food self-sufficiency and sustainable resource management are often mutually incompatible. Policies designed for achieving food self-sufficiency tend to undervalue goods not traded internationally, especially land and labor resources. As a result, food self-sufficiency in countries with an exhausted land frontier came at a high ecological and environmental cost. Appropriate policy reform, both at the macro as well as at the sector level, will go a long way toward arresting and possibly reversing the current degradation trends, but the degree of degradation in many regions will pose severe challenges to policy (Pingali and Rosegrant 1998).

Policies for the Crop Sector

Ironically, the very policies that encouraged increased food supply through intensive monoculture systems also contributed to the declining sustainability of these systems. Rice policies operated under two presumptions: (i) lowlands are resilient to intensification pressures and sustain productivity growth indefinitely; and (ii) modern technology provides a "silver bullet" solution to food-supply problems of Asia. Traditional farming systems were sustainable because of lower intensities of cultivation and because they benefited from a stock of farmer technical knowledge about the crop and paddy resource base, built up over millenia. Neither science nor farmer knowledge was able to predict the changes imposed by intensification and modern technology use on the biophysical resource base. Learning from the experience of intensification-induced environmental degradation, however, can help to bring this degradation under control, halt it, or possibly even reverse it through appropriate policy and management reforms (Pingali and Rosegrant 1998).

With the progression toward global integration, the competitiveness of domestic cereal agriculture can only be maintained through dramatic reductions in the cost per unit of production. New technologies designed to reduce significantly the cost per unit of output produced, either through a shift in the yield frontier or through an increase in input efficiencies, would substantially enhance farm-level profitability of cereal-crop production systems. Increasing input use efficiency would also contribute significantly to the long-term sustainability of intensive food-crop production and help arrest many of the problems.

Water and Irrigation Policy

Water allocation and water pricing have significantly contributed to the environmental costs of irrigated agriculture. For all practical purposes water has been, and still is, free and is therefore used beyond the social optimum. Two major degradation problems in intensified areas—salinity build-up and waterlogging and excessive reliance on monoculture rice—are directly related to the virtually free provision of water to farmers. When water is provided at no cost, there is no incentive to economize, making rice, a water-intensive crop, artificially more profitable than crops that use less water. Increasing water-use efficiency through opportunity cost pricing or market valuation of water would produce substantial environmental benefits and would not adversely affect yields. Yet this leverage for improving the sustainability of the resource base has been used only in isolated cases in Asia.

Comprehensive reform of water demand management for both groundwater and surface water is essential for increasing the productivity of water in agriculture, by saving water in existing uses and by improving the quality of water and soils. The most significant reforms will involve changing the institutional and legal environment in which water is supplied and used to one that empowers water users to make their own decisions regarding use of the resource, while providing correct signals regarding the real scarcity value of water, including environmental externalities. The combination of reforms that is appropriate will vary, depending on the location, level of institutional and economic development, and degree of water scarcity. Key elements of these reforms should include the establishment of secure water rights of users; decentralization and privatization of water-management functions; and the use of incentives, including markets in tradable property rights, pricing reform and reduction in subsidies, and effluent or pollution charges (Rosegrant 1997).

Principles for groundwater management reform are similar: successful approaches in the western United States, for example, have employed a variety of instruments to influence water demand, including pumping quotas (usually based on some notion of historical use), pumping charges, and transferable rights to groundwater. The governance structure in the groundwater basin establishes water rights, monitoring processes, means for sanctioning violations, representative associations of water users, financing mechanisms for administration and management, and procedures for adapting to changing conditions. Key elements for the success of this governance structure in the U.S. also make it highly appropriate for developing Asia: it is agreed upon and managed by the water users; it is responsive to local conditions; it operates with available information and databases, rather than requiring theoretically better but unavailable information; and it adapts to the evolving environment (Blomquist 1992, 1995; Rosegrant 1997).

Fertilizer and Pesticide Policy

Government intervention in the cereal market, especially through output price support and input subsidies, provided farmers with incentives for increasing cereal productivity. In addition to highly subsidized irrigation water, Asian farmers benefited from "cheap" fertilizers, pesticides, and credit (see Monke and Pearson [1991] for an example of the Indonesian price policy). The net result was that rice monoculture systems and rice-wheat systems were generally profitable through the decades of the 1970s and the 1980s, despite a long-term decline in real world rice and wheat prices through this period.

Input subsidies that keep input prices low directly affect crop-management practices at the farm level; they reduce farmers' incentives to improve input use efficiency, which often require them to invest in learning about the technology and how best to use it. In addition to inducing increased use—in many regions of Asia considerably in excess of the socially optimal level—subsidized fertilizer prices have tended to favor the use of nitrogen fertilizers over other nutrients, creating soil-fertility imbalances. The reduction and eventual removal of fertilizer-price subsidies would significantly improve the incentives for efficiency of fertilizer use (see also Chapters V and VII for a discussion of costs of input subsidies). Nonprice policies are also important in improving fertilizer-use efficiencies, including location-specific research on soil-fertility constraints and agronomic practices, improvement in extension services, development of improved fertilizer supply and distribution systems, and development of physical and institutional infrastructure (Desai 1986; Desai 1988).

The policy scenario for pesticides is similar to that of fertilizers. Pesticide subsidies provided during the early stage of production-technology adoption have led to indiscriminate pesticide use, resulting in ecological problems including disruption of the pest-predator balance, increased pest resistance, increased pest losses, and human health problems (Pingali and Rosegrant 1998). The design of policies is often based on farmer and policymaker perceptions of pest-related yield losses anchored around exceptionally high losses during major infestations, even when the probability of such infestation is low (Rola and Pingali 1993). For various environmental and human health reasons (see Pingali and Roger [1995] for detailed evidence in Philippine rice ecosystems), Integrated Pest Management (IPM) programs have been vigorously pursued. IPM research results have been translated into usable thresholds, resistant varieties, pest surveillance programs, and identification and definition of the role and importance of natural enemies.

Unless they are spatially and temporally flexible, however, IPM recommendations may overstate the case for applying pesticides and cause increases in applications in cases where a natural control ("do-nothing") strategy would be more effective. IPM will be successful only if farmers participate fully in adapting and using this technology. To make it more attractive, pesticides should never be subsidized since, as in the case of fertilizers, farmers would have no incentive to invest time in acquiring IPM skills. Removing all explicit and implicit subsidies on pesticides is essential to reduce pesticide use on farms. Accelerated IPM program implementation requires local political support and commitment, as well as funds for research and technology dissemination (Pingali and Rosegrant 1998).

Price and Trade Policy and Sustainability

Trade and macroeconomic policies are discussed in detail in Chapter VII, but it must be noted here that the macroeconomic setting has also contributed to unsustainable agricultural management practices. Agricultural price and trade policies in many Asian countries have often been both internally inconsistent and costly for long-term diversified growth. On the one hand, general trade and exchange-rate policies have penalized agriculture across the board, while on the other hand, crop-specific interventions such as output price protection and input subsidies have attempted to favor individual crops, particularly rice and wheat. Continued reform of macroeconomic and price policies to create a level playing field across agricultural commodities would assist in balancing demands on the natural resource base.

Agricultural Research and Extension

Chapter V stressed the critical importance of continued public investment in agricultural research and extension in maintaining productivity growth. Because public resources are scarce, it is also essential that available resources be used more efficiently. Greater emphasis should be placed on raising the competence of scientists and the quality of research management and on providing adequate operating funds and technical support. Oram (1990) suggests that the division of responsibilities and working relationships between the international and national research centers needs to be reexamined with a view to increasing efficiency. He suggests that decentralization of research regionally, to the farmer-participants and based on agro-ecological characterization, may be the most effective approach, because it provides better farmer input and feedback to upstream researchers and policymakers. Improvement in linkages between public agricultural research and small research-based firms and informal farmer research could have high payoffs. Farm-based research often specializes in choosing varieties specific to micro-environments and can be highly complementary with the formal research systems (Lipton 1994).

The importance of effective extension may be even greater in the future, because of the increasing need for efficiency in input use as opposed to input and crop-variety promotion. Technologies to implement IPM and to improve the nutrient balance and the timing and placement of fertilizer applications are highly complex, knowledge-intensive, and location-specific. Because new technologies are more demanding for both the farmer and the extension agent, they require more information and skills for successful adoption compared to the initial adoption of modern varieties and fertilizers. In addition, these improved technologies do not give as clear gains in yield and income as the initial adoption of new technologies. The increase in income from these technologies is in fact highly sensitive to the farmer's skills and efficiency in utilizing them (Byerlee 1987). As a result of the greater complexity of new crop technologies, increased investment in extension, education, and human capital is likely to have high payoffs.

To provide the necessary information dissemination and training for these new technologies, extension services will have to be upgraded. The poor performance of many extension services can be attributed to inadequate training, inappropriate organization, and lack of incentives. Possible options for reform are privatization of extension services, the training and visit extension system, and decentralization of existing public systems. Privatization of extension through contracting to private companies can introduce incentives for higher efficiency. But for much of Asian agriculture, private extension may be difficult to manage and monitor and may not generate adequate profits for the private company, because of the large number of widely dispersed small farmers cultivating many different cropping systems. Privatization of extension is more likely to be successful when extension is linked to delivery of a specific technology, such as hybrid corn, to larger, more homogeneous groups of farmers.

Similar to the case for research systems, decentralization of existing extension services could also help farmers cope with the additional complexity of efficiency-enhancing technology. Thimm (1990) recommends integrated national planning of research and extension to ensure the appropriate budgetary mix for proper operation, regular interdisciplinary evaluation of on-farm benefits of recommended technology, and establishment of a goal-oriented organizational structure that encourages a bottom-up flow from farmers to extension and research. The latter, when combined with adaptive, location-specific research, can facilitate the transfer of complex technologies.

EVOLUTION OF LIVESTOCK PRODUCTION: TRENDS, CHALLENGES, AND POLICIES

The most striking development in the evolution of livestock production is the rapid expansion during 1982–95 in production of poultry in Asia overall (up 10.8 percent per year) and in all regional subgroups (with growth rates ranging from 7.2 percent per year in Southeast Asia to 12.7 percent annually in East Asia). This production growth was only surpassed by the phenomenal growth in beef production in East Asia, mainly the PRC, at 16.0 percent per year during 1982–95, albeit from a relatively low base. These trends in number of livestock slaughtered, carcass weight per animal slaughtered, and meat production are summarized in Table VI.4. Moreover, as has already been noted, production figures for the early 1990s in the PRC might be inflated. Asian production growth was also very rapid for pork (7.1 percent annually) and sheep and goat meat, both again driven by East Asia.

Indeed, production growth accelerated substantially for each of the livestock products between 1967–82 and 1982–95. Growth almost tripled for beef production, from 2.8 percent per year in 1967–82 to 7.7 percent annually in 1982–95; nearly doubled for poultry production, from 5.5 percent per year to 10.8 percent per year; and accelerated substantially for both pork and sheep/goat meat production (from 4.9 to 7.1 percent per year for pork, and 4.0 to 7.2 percent per year for sheep and goat meat).

Growth in production across all these livestock products was driven by growth in numbers, in contrast to developed-country livestock production growth, where growth in carcass weight per animal slaughtered plays a larger role. But the extent to which numbers growth dominated did vary somewhat by product for the region. Growth in numbers accounted for 93 percent of poultry production growth in Asia, on average, during 1967–82, and for 90 percent during 1982–95. For beef production, growth in numbers accounted for 81 percent of growth in the latter sub-period, with the effect particularly strong in East Asia (91 percent). Moreover, numbers growth accounted for 84 percent of sheep and goat meat production and 77 percent of pork production growth (again, from 1982–95).

Table VI.4: Annual growth in livestock number, production, and yield, by region, 1967–1995

	1967–82			1982–95		
	Number	Yield	Production	Number	Yield	Production
	(percent per year)					
Beef						
East Asia	3.82	0.79	4.65	14.60	1.18	15.98
Southeast Asia	1.80	0.15	1.95	2.86	0.38	3.25
South Asia	2.08	0.56	2.65	2.33	1.12	3.48
Asia	2.26	0.49	2.76	6.25	1.37	7.71
Pork						
East Asia	3.03	2.16	5.26	5.64	1.52	7.24
Southeast Asia	2.55	0.31	2.88	4.47	1.17	5.69
South Asia	4.14	0.02	4.16	2.89	0.03	2.92
Asia	3.01	1.88	4.94	5.46	1.52	7.06
Poultry						
East Asia	4.91	0.26	5.18	10.95	1.55	12.68
Southeast Asia	5.96	0.31	6.49	7.53	0.32	7.18
South Asia	4.03	0.02	4.08	9.37	0.71	10.14
Asia	5.18	1.88	5.54	9.71	1.01	10.83
Sheep and Goat						
East Asia	5.70	0.23	5.94	8.34	0.35	8.71
Southeast Asia	5.55	0.11	5.65	2.71	1.91	4.68
South Asia	2.75	0.02	2.77	2.87	1.28	4.19
Asia	3.87	0.16	4.04	6.04	1.12	7.22

Note: Number is the number of livestock slaughtered, and yield presents the carcass weight per animal slaughtered.

Source of basic data: FAO 1998

Challenges for Increasing Livestock Production

Rapid growth in demand for livestock will put great pressure on production systems in Asia. For the livestock sector in Asia to continue its growth, some major changes in the way livestock is produced—by whom, on what scale, with what technology, and with what quality guarantees for the broader market—will have to take place. The low productivity of traditional smallholder animal husbandry will no longer suffice. Greater industrialization will be required. Some examples of a path for doing so, such as the growth of a commercial sector for poultry production in Thailand and the PRC, do exist. What they indicate about the need to invest in such a commercial sector, to promote a steady transfer of technology for the sector from developed to developing countries, and to institute serious quality control so that production can reach broader markets, has implications for how production of other meats must change in the years ahead. Still, this path does hold greater risk for the environment. Policies to help overcome these challenges must take that threat into account.

Livestock production is influenced by interventions in both the livestock and the feed-crop sectors. The emphasis here will be on the former, since the challenges facing maintenance of cereal yields in the face of environmental degradation and the policy remedies to overcome those challenges, described above, apply to cereal production for the feed sector as well as the basic staple-food sectors.

Traditional Livestock Systems: Limits to Productivity

Of the three major production systems that characterize Asian livestock systems—land-based grazing systems, mixed farming, and industrial farming systems—the first two have historically dominated Asian livestock production. Livestock was traditionally raised using resources that have few alternative uses, such as household food waste or land that was not fertile enough for cropping.

Delgado et al. (1999), based on Carney (1998), summarize the reasons why land-based grazing systems and mixed farming are important to the livelihood of the rural poor in Asian developing countries. First, livestock products are an important source of cash income, particularly for the poorest people. Second, these animals are an important asset, for women in particular. Third, livestock manure and draft power is important for the preservation of soil fertility and the sustainable intensification of farming systems. In addition, livestock allows the rural poor to exploit common property resources, such as open grazing areas. Moreover, livestock products represent a source of income diversification, especially in semi-arid systems characterized by only one cropping season per year. Finally, traditional livestock production is often the only source of income for the most marginal of the rural poor, such as pastoralists, sharecroppers, or widows. Moreover, considerable evidence shows that the rural poor and landless earn a larger share of their income from livestock production than better-off rural people. Thus, livestock production is one of the few rapidly growing markets that the poorest rural people can join even if they lack substantial amounts of capital, land, and training.

At the same time, permanent or seasonal nutritional stress has put at risk production from these traditional systems involving ruminants in pastoral and low-input mixed farming systems. In most Asian countries, it has been possible until now to alleviate such seasonal feed shortages through techniques that rely on feed grains only to a minor degree: conserving and storing forage, treating crop residues, and adding mineral nutrients to feed. But these coping strategies are now falling short as increased specialization in production reduces access to traditional feeds. Scarcity of traditional feeds is just part of a list of woes that Asian smallholders must overcome to boost productivity of their livestock systems, such as high incidence of disease and lack of access to better breeds.

Partial improvement of the system may not have the desired effect. For example, replacing local cattle with improved varieties will not solve the productivity problem without greater feed resources on the farm (see the complimentary volume in this Rural Asia series by Mingsarn and Benjavan [1999]). Even with access to better stock, moreover, higher production often relies on new management practices and farmers typically have limited access to education and training. At the levels of productivity characteristic of the traditional system, much more land would have to come under animal husbandry to meet growth in demand, land that is increasingly not available for this purpose (for more on this, see Chapter X). In short, the pressure that rising demand is now placing on traditional, resource-constrained production systems is rendering these systems increasingly unsustainable and less competitive for the large numbers of Asian poor who rely on them for significant portions of their income (Mingsarn and Benjavan 1999). Thus, there is little doubt that most of the growth in future livestock production will come from industrial farming systems.

Rapid growth in the demand for livestock products does provide a substantial opportunity for small-scale, rural producers, however. On the one hand, the true economies of scale in livestock production may not always favor industrial production and large-scale producers, once explicit and implicit government subsidies to large-scale producers are accounted for. The phasing out of subsidies and enforcement of pollution and health requirements for large-scale producers or the redirection of subsidies towards small-scale producers could keep a market share open for the poor, with substantial poverty-alleviation benefits. But economies of scale in production facilitate vertical integration from production to the delivery and marketing of these highly perishable products. To overcome this barrier, policymakers need to promote producer cooperatives or other organizational structures to coordinate small producers with processors and distributors. Failure to provide this and other policies targeted carefully towards small-scale producers would not only be a major missed opportunity for the rural poor to share in the gains to be made in the "livestock revolution" in Asia. It could also drive the rural poor out of livestock production (Delgado et al. 1999).

Increasing Industrialization of Livestock Production

Industrial livestock production has seen very high rates of growth in the last decade. Driven by rapid industrial and urban growth, the periurban poultry and pig industries, in particular, expanded. This growth was in response to increased demand as well as to technical changes that resulted in a more favorable feed-conversion ratio with higher returns on investment for these species. This structural shift to more intensive production also reflects the pressure on land from growing populations—part of the trend that has rendered expansion of land-intensive animal husbandry more difficult. These industrial systems are also much more productive and can be competitive in the international marketplace as well, as the case of poultry in Thailand indicates.

There, a commercial sector in poultry evolved from a system previously comprising smallholder producers, starting in the 1970s. Production rose from around 36 million birds in the mid-1970s to more than nine times that, or approximately 338 million birds, by the end of the 1980s. Furthermore, technology was close to United States standards, as measured by market age, average market weight, feed conversion ratio, and mortality, while production costs were actually lower in Thailand (primarily due to differentials in labor costs) (Schwartz and Brooks 1990). The rise of the commercial sector, with ten large firms (vertically integrated) accounting for about 80 percent of production, has meant the decline in importance of independent growers. The low production costs helped attract outside investment to help develop the industry (joint ventures fueled by Japanese investment, which also helped open Japanese markets to Thai chicken exports) (Henry and Rothwell 1996).

Such industrialization is now beginning in the PRC, although most production still comes from traditional animal husbandry. Indeed, the PRC emerged in the 1990s as a competitor to Thailand in chicken, with industrialization spurred by low production costs following market liberalization. Here, too, foreign investment (including technology) played a role. The result has been staggering growth rates in production, over 10 percent per year from 1990 to 1995. Increasingly, then, the commercial sector will need to become the source of production growth for livestock products in developing Asia.

Need for Technology Transfer and Quality Control

These two examples allude to, without underlining the importance of, continued technology transfer if developing Asia is not to lag far behind productivity trends elsewhere. Genetic potential for poultry meat production has vastly improved and continues to do so, but only a handful of companies worldwide supply the genetic material. In fact, the Thai poultry industry relies on imported stock (parent or grandparent) from the developed world (principally the US and UK); the infusion of foreign investment in the PRC also involved improved stock (Henry and Rothwell 1996). At the same time, adequate quality control will be needed to make sure higher production meets quality standards required for export.

Challenges for the Environment

As with intensive crop agriculture, the intensification of livestock production poses potentially severe environmental challenges. Production of livestock generates waste by-products that under some conditions can be recycled but, when animal concentrations are high, can become a serious pollution problem. Livestock and feed production use large quantities of water, not only as a direct input but also for waste disposal. The high concentration of industrial livestock production has the potential to produce organic discharges that can exceed the carrying capacity of the surrounding environment. The intensive production of pigs in the PRC has caused animal waste pollution problems that now need close attention. Malaysia is also experiencing environmental problems arising from the pig sector. Under proper management, these impacts can be alleviated and the environmental costs internalized and charged to producers or consumers so that environmental impacts are mitigated without threatening production. For example, the

Ponggol Pigwaste Plant in Singapore recycles wastewater at the cost of about 8–9 percent of the production cost of pork (Steinfeld et al. 1997). In Malaysia, aerobic treatment increases the cost of production of pork by 6 percent (Mingsarn and Benjavan 1999).

Livestock is a contributor to greenhouse gas emissions and might threaten biodiversity by encroaching on land used by wildlife. Greenhouse gases from livestock are generated from respiration, manure management, and livestock-related activities that utilize fossil fuels. High-intensity animal production has become the biggest consumer of fossil energy in modern agriculture (IPPC 1995). Livestock production is also a source of risk to animal and human health. Risks arise not only from uncontrolled endemic diseases, but also from those that appear when animal concentrations are high or when unconventional feeds are used. Human health can be affected by zoonotic diseases emanating from livestock production facilities and by food contamination resulting from underdeveloped or overwhelmed food monitoring systems (Delgado et al. 1999). Moreover, there are potential negative human and animal health effects when discharges that may contain heavy metals, such as copper, zinc, and cadmium, are not properly discarded, especially under a high-concentration and nutrient-surplus system.

Policies for the Livestock Sector

The ongoing transformation in livestock production creates tremendous policy challenges. A dynamic commercial sector must provide the source of growth and an appropriate policy framework to accompany this transition—one that recognizes the need for continued livestock technology transfer and the likely role of joint ventures in guaranteeing it, plus the regulatory issues, especially pertaining to quality control. At the same time, investment in breeding and research, plus reinforcement of the infrastructure to facilitate the production increases that will be forthcoming, are needed if livestock is to fulfill its potential as a major contributor to overall rural growth and poverty alleviation strategies in Asia (Delgado et al. 1999; Mingsarn and Benjavan 1999).

In broad terms, policies for the livestock sector are similar to those presented for the crop sector and agricultural production, especially the abandonment of distortional policies in favor of a market-oriented framework. While intensification of production means less land needed for production of livestock, environmental sustainability and human health will only be guaranteed by policies that monitor and prevent damaging by-products of the industrialization of livestock production.

Facilitation of Technology Transfer

As the Thai and PRC examples in the poultry industry described above make evident, the industrialization of the livestock sector in Asia implies the need to import the advanced technology already developed and in development. As they also show, the unimpeded flow of technology is just one aspect of the globalization that will affect the sector: joint ventures and competition for export markets are others. In demand-driven production systems in developing countries, such as in East Asia and periurban India, adoption of intensification technologies patterned on those in developed countries will probably be rapid. Where the livestock sector is demand-driven, adoption can and will occur through market forces. In particular, technologies for increased production of pork, poultry and dairy products will be largely transferable from developed countries. Technology transfer will drive a trend (similar to that for the crop sector described above) toward knowledge-intensive systems; smart technologies, supported by astute policies, can help to meet future demands while maintaining the integrity of the natural resource base. Better information on which to base decision-making is, therefore, urgently required (Mingsarn and Benjavan 1999; Delgado et al. 1999).

Even with the direct import of technology, however, it will be necessary to develop policies that facilitate a market-oriented environment; key research issues will remain in the livestock policy area for production, processing and marketing. Moreover, issues relating to livestock and the environment cannot be solved with technical innovations alone, but will require an enabling environment in which effective technologies can be introduced. In addition, the quality of extension services and of the livestock producers in general is of particular importance. In the PRC, for example, productivity increased by 41 percent in animal production, but low educational levels of the producers and weak extension systems hinder increased technology adoption (Liu 1995).

Livestock Research

In addition to direct transfer of technology, the Asian livestock industry can be expected to benefit substantially from progress in biotechnology. Embryo technology could lead to the replacement of conventional breeding. This could lower the cost of importing individual animals and remove the challenges of their environmental adjustment. Biotechnology could also be used to identify genetically superior individuals for particular traits, like resistance to specific diseases. Specifically tailored vaccines could increase the effectiveness and reduce the risk of industrial livestock production. There is also considerable potential for increasing control over endemic diseases (Mingsarn and Benjavan 1999). Further livestock research in Asian developing countries should also focus on the development and adaptation of nonconventional feeds; improved storage, preservation, and processing technologies; improved breeding programs; and better disease controls (Devendra, Smalley, and Li Pun 1998).

Integration of Crop and Livestock Policies

Although technology will be adopted in demand-driven areas, in regions where demand is growing less quickly, such as in most of South Asia, technology uptake will be slower; important pockets of stagnant technology are likely to remain. In order to facilitate this transformation, livestock and feed (that is, crop-sector) policies that are closely related in these economies should be established in an integrated way with regard to achieving livestock objectives (Rae 1992). Historically, though, livestock-sector and crop-sector policies have often pursued diverging objectives in Asian developing countries, with negative effects on livestock production. In the Philippines, for example, while livestock has historically received high levels of protection through import tariffs, the price of corn and soybeans has simultaneously been greatly increased through tariffs and import quotas, the net result being a tax on the livestock sector (Dimaranan, Rosegrant, and Unnevehr 1996).

Livestock policies include subsidies for credit, breeding stock, and slaughter services. Moreover, in some Association of Southeast Asian Nations (ASEAN) countries, production units must be licensed and environmental regulation might constrain production activities. Tariffs have been levied on imported medicines, baby chicks, and feeds. And both tariffs and quantitative controls have been imposed on the importation of livestock products. Research and development policies focus on breeding improvements, extension and training programs, animal inspection, and vaccination and artificial insemination programs. Moreover, livestock production is affected by policies related to the investment or regulation of slaughterhouses (Rae 1992).

Rae (1992), citing a study by Kasryno et al. (1989), states that livestock policies in Indonesia have forced domestic prices below border values by 20 percent to 40 percent, at the same time that other policies raised the price of feeds by 20 percent to 75 percent above border values. Dairy products, however, receive government support that more than offsets the policy-induced increase in feed prices. In Malaysia, product-price supports in the dairy industry offset the slight distortions in the feed prices, and the effective protection rates in poultry and pig protection are close to zero, since the effects of product and feed policies basically offset each other. In Thailand, distortions in feed prices contribute little to effective protection rates. The Government controls the pork market via ownership of slaughterhouses and influences overall domestic processing and trading. Both beef and dairy production are protected through tariffs on imports of beef and quantitative import controls on dairy products, reducing the incentives for efficient production (Rae 1992).

Policies Protecting the Environment

Policies to maintain the nutrient balance in nutrient-deficit mixed farming in developing countries should focus on providing incentives and services for technology uptake to preserve and enhance crop-livestock integration. For nutrient-surplus industrial and mixed farming systems, a mixture of regulations to control animal densities and waste discharge and incentives for waste reduction is required, including the removal of subsidies on concentrate feed and on inputs used in the production of feed.

Policies to address environmental degradation in grazing areas of developing countries need to include elements of institutional development through local empowerment and property-rights instruments. Incentive policies may help to reduce grazing pressure in the semi-arid zones through, for example, the introduction of full cost recovery for water and animal health services or for grazing fees. Similarly, taxation of pasture and cropland in rainforest areas can discourage forest conversion.

CONCLUSIONS

Intensification of agricultural production, combined with wrong incentive structures from past or present policies, has caused substantial environmental degradation in the crop sector. With rapidly increasing demand for feed, pressure for livestock production could cause similar or even more severe environmental degradation. Policies that mitigate or even reverse negative environmental effects in the crop sector and help preempt larger problems in the livestock sector include the removal of trade, macroeconomic, and price distortions on input and output markets (see also Chapter VII), and the establishment of price incentives and regulations to reduce the production of environmental externalities in both sectors. Modernizing the aging livestock production systems in many Asian countries will require substantial investments to improve the feeding potential, ensure a suitable animal environment, and provide other modern production and processing technology. Moreover, the increasing importance of new, knowledge-intensive technology in both sectors requires a market-friendly environment for the adoption and adaptation of new technologies, the removal of restrictions on technology imports, and, in particular, a better and more decentralized research and extension system.

VII Impacts of Trade, Macroeconomic, and Price Policies on Agriculture

INTRODUCTION

Trade and macroeconomic policies are important determinants of the rate of overall economic growth. Broad-based economic growth, based on favorable trade, macroeconomic and price policies, can in turn have substantial benefits for agriculture, including the creation of domestic markets for agricultural commodities and the generation of capital for investment in agriculture. Indeed, in much of Asia, particularly East and Southeast Asia, the broad performance of macroeconomic policies, including fiscal discipline, adequate incentives for savings and investment, and an outward-looking trade policy, was overwhelmingly conducive to economic growth, at least until the early to mid-1990s (see also Chapter I).

In addition to setting the economic environment for overall growth, these policies also have a profound impact on the performance of individual sectors of the economy such as agriculture. Trade, macroeconomic, and price policies can affect agriculture by either taxing (disprotecting) or subsidizing (protecting) the sector. Trade restrictions have a direct effect on the domestic prices of tradable (often agricultural) goods, and have an impact on the real exchange rate which, in turn, affects the domestic prices of tradable goods in relation to locally produced goods. For example, import duties and quotas raise the domestic price of import-competing production in relation to exportables (including many agricultural commodities) and therefore encourage a shift away from export production. The same policy instruments reduce the demand for imports, which lowers the price of foreign exchange so that the domestic prices of tradable goods fall in relation to nontradable goods and hence indirectly bias production incentives against both import-competing and export goods. Protection for industrial import substitutes thus penalizes the domestic production of agricultural goods in the following ways:

- the rise in the domestic price of the protected industrial output reduces the relative price of agricultural products;
- the cost of industrial inputs (fertilizer, pesticides, and farm equipment) for agricultural production increases; and
- the induced appreciation in the real exchange rate renders agricultural exports and import-competing products less profitable than nontradable goods (Bautista 1993).

Alternatively, trade and macroeconomic policies can protect agriculture or specific commodities within the agricultural sector. Agricultural protection, by raising domestic food and agricultural prices above world prices, penalizes consumers and introduces inefficiency by attracting excess resources to production of the protected commodity or sector and by rendering unprotected sectors less competitive. Protectionist policies require large government fiscal outlays to farmers to pay price supports and subsidies and may also encourage excessive use of agricultural chemicals, thus damaging the environment. Input subsidies distort input allocation decisions and compromise scarce government resources that could be used for directly productive investments. In many cases, agricultural protection also represents an inefficient transfer of income from consumers and taxpayers to farmers, because the fiscal costs of protection are higher than the benefits received by farmers and because income is redistributed to wealthier, large farmers, who receive the greatest share of benefits. In addition, price stabilization policies significantly have influenced agricultural production in many Asian developing economies (see Box VII.1).

Despite the high costs of either taxing or protecting agriculture through trade and macroeconomic policies, a common "developmental" path, consisting of a period of taxation of agriculture followed by increasing agricultural protection, has been adopted by most countries as they develop and industrialize. Will this historical pattern of declining taxation and increasing protection of the agriculture sector be repeated in the developing countries of Asia? What impact might present-day trade and macroeconomic policies on agricultural taxation and protection have on the playing out of this historical pattern in the Asian developing countries? Might developments in international trade negotiations and institutions or developments in the structure of agricultural trade provide countervailing forces to increasing agricultural protection?

HISTORICAL PATTERNS OF AGRICULTURAL PROTECTION

As economies grow, they tend to shift from taxing agriculture to protecting agriculture relative to other sectors. This shift occurs at an earlier stage the weaker an economy's comparative advantage in agriculture and more rapidly the higher an economy's growth rate and the faster the decline in its comparative advantage in agriculture (Anderson and Hayami 1986). This pattern has been observed in Western Europe and North America and, more recently, in Japan; Taipei,China; and the Republic of Korea. In Japan, agricultural protection was zero or negative in the 19th century, became positive in the early years of the 20th century, and rose sharply after the 1950s. The Republic of Korea and Taipei,China have followed the Japanese pattern, but at an accelerated pace. Over a 20-year period, beginning in the early 1960s, the Republic of Korea and

> **Box VII.1: Price Stabilization Policy**
>
> In addition to direct interventions to influence agricultural price levels, many Asian countries have attempted to reduce price fluctuations from season to season and from year to year. Price stabilization—implemented through domestic procurement and imports—concentrates primarily on the main imported staple cereals. Export crops, such as rice in Pakistan and Thailand, are regulated by a public monopoly or through private traders in a system of variable taxation of exports. Rice and wheat trade was monopolized by the governments of Bangladesh, Indonesia, Pakistan, and the Philippines, as well as by the centrally planned economies of the PRC and Viet Nam. For import-competing cereals, domestic procurement and distribution were used to alleviate seasonal instability and to provide a floor price for farmers, whereas imports were used to deal with year-to-year fluctuations.
>
> Although quantitative evidence is limited, it is likely that successful rice price stabilization in the early stages of economic take-off in Southeast Asia had significant benefits. During this period—characterized by a dominance of rice in diets, a large number of smallholder rice farmers, and a highly unstable world rice market—severe swings in domestic rice prices could have triggered general inflation and political instability that would have reduced public and private investment. Successful price stabilization may have had a positive macroeconomic impact under these circumstances.
>
> In the longer run, however, as the share of agriculture in the economy has declined, price stabilization in Asia has provided only small benefits—and at a high cost. Public management of imports has been characterized by bureaucratic rigidity that has caused delays and inefficiencies in adjusting
>
> (continued next page)

Taipei,China shifted from slightly taxing agriculture to heavily assisting it.

The historical rise in levels of agricultural protection in industrializing countries is linked to the changing role of

Box VII.1 (continued)

imports to variations in domestic production and prices and by forecasting errors in estimating import requirements. Both these results have often increased price variability instead of reducing it. The budgetary costs of managing public stocks and open-market operations have been very high and the lack of operational transparency has encouraged corruption and rent seeking. Public intervention in domestic markets has squeezed the normal seasonal price spreads, driving private traders from the market and reducing market efficiency.

By the late 1980s, many Asian countries were realizing that price stabilization was an ineffective and unsustainably expensive strategy for achieving stable prices. In Pakistan, the government abolished its monopoly in rice in 1988/89; in Bangladesh, the rice monopoly in external trade ended in 1992/93. The Philippines has proceeded slowly, opening up import trade in wheat, flour, and animal feeds to the private sector, but retaining the National Food Authority monopoly on international trade in rice and maize. BULOG in Indonesia, probably the most effective public agency in Asia in stabilizing grain prices (albeit at very high budgetary and market-efficiency costs), is in the process of a significant reduction in its stabilization role. In Viet Nam, a complete liberalization of prices was introduced in the agricultural sector in 1989 and restrictions on imports and exports of important products, such as rice, were relaxed. In 1997, additional implicit and explicit restrictions to internal trade were removed and export quotas were decentralized.

(Balisacan 1998; Timmer 1997; Goletti, Minot, and Berry 1997; Anderson and Roumasset 1996; Islam and Thomas 1996; Dollar 1994).

agriculture during economic growth. This changing role was described in some detail in Chapters I and III. But the economic policy making of agricultural protection also has political aspects. First, the declining importance of food in household budgets as incomes grow reduces the political pressure for low food prices. Consumer interest in lower food prices becomes highly diffuse; individual consumers do not gain sufficiently from low food prices to motivate strong lobbying efforts. Second, the decline in the relative size of the agricultural sector in employment and production makes it easier for farmers to organize politically to seek special support and less costly politically for the government to provide this support. Farmers increase their effectiveness in lobbying and gain widespread sympathy, especially when the decline of the sector is rapid. In addition, there is a tendency for growing economies (particularly those of densely populated countries) to lose their comparative advantage in agriculture and become net food importers. This development provides greater scope for protecting farmers through "covert" policy instruments, such as import controls, that do not require budget outlays and a political rationale for agricultural protection in the name of food security (Lindert 1991; Anderson and Hayami 1986).

TRADE, MACROECONOMIC, AND PRICE POLICIES IN ASIA

In most Asian developing countries, the indirect effects of trade and macroeconomic policies have caused overvaluation of the real exchange rate, which in turn has lowered effective agricultural prices, taxing the agricultural sector (Bautista 1990). In the Philippines, for example, overvaluation of the exchange rate arising from the protection of domestic industry lowered agricultural prices by 30 percent during the 1980s (David 1989). Policies in Pakistan, Sri Lanka, and Thailand similarly induced overvaluation of the real exchange rate by 15–25 percent during the 1980s (Bautista 1990). Nevertheless, there was a downward trend in the degree of overvaluation in the real exchange rate in most Asian countries throughout the 1980s, followed by a gradual real appreciation in the 1990s that again penalized agriculture and other tradable sectors (see Chapter IX for more detail on real currency appreciation in the 1990s).

To compensate at least partially for the discrimination of trade and exchange-rate policies against agricultural production, most East and Southeast Asian countries have provided subsidies for agricultural inputs (fertilizer, credit, and irrigation) and have targeted direct price supports to favored crops, particularly rice and wheat. These selective subsidies and price supports have fallen far short of fully offsetting the pervasive taxation caused by the trade restrictions. In addition, they distort economic incentives and penalize consumers, while imposing costs on farming activities that receive lower subsidies and on the environment, the latter by encouraging overuse of inputs. However, as David (1990) shows, there was a gradual but significant reduction in direct price support for most food crops in East and Southeast Asia after the mid-1980s: many countries reduced levels of commodity-price protection and moved toward a policy of following the long-term world price in setting domestic agricultural prices.

In Indonesia, for example, taxation of agriculture through currency overvaluation has been minimal. The average overvaluation during 1980–90 was about 7 percent, and when overvaluation reached higher levels (16 percent in 1984 and 14 percent in 1987), the rupiah was devalued to restore balance (Müller 1995). The slight overvaluation was balanced by heavy fertilizer subsidies that kept fertilizer prices at about one half of world prices; by moderate direct price protection of rice and other food crops; and by heavy price protection for favored commercial crops such as sugar, soybeans, and milk and dairy products in order to promote crop-diversification strategies. Beginning in the mid-1980s, policies shifted toward a gradual reduction of fertilizer subsidies and agricultural price protection (Gonzales et al. 1993). By 1994, nominal protection rates were low for virtually all agricultural commodities, estimated at –5 percent for rice, 5 percent for corn and soybeans, –2 percent for cassava, 20 percent for milk, 14 percent for poultry products, 1 percent for beef, and 5 percent for other livestock products (San and Rosegrant, 1998). The reduction in protection of rice led to the elimination of producer subsidies for rice farmers. The annual value of subsidies to rice production peaked at $1.1 billion in 1985 and was –$71 million in 1990 (Müller 1995).

Compared to other Southeast Asian countries, Malaysia's trade and macroeconomic policies have been relatively neutral toward agriculture, with the exception of rice, which has been highly protected. Exchange-rate distortions due to trade and macroeconomic policy have been minimal, with overvaluation averaging only 6 percent from 1980 to 1990 (before the beginning of the real appreciation in the 1990s due to massive short-term capital inflows (see Chapter IX). In contrast, direct rice-price interventions to support farm income and encourage production have resulted in nominal protection rates in excess of 100 percent. Important export crops such as rubber and palm oil, on the other hand, have been subject to export taxes to generate government revenues and to finance research, marketing, and replanting programs. The level of taxation, however, has declined from 25–30 percent in the early 1980s to 5–10 percent in the 1990s.

In the Philippines (as in many Asian countries), providing income support to producers and maintaining affordable prices to consumers were primary (but conflicting) objectives of direct price policy. The policy was intended to prevent or smooth out sharp changes in food prices in response to changes either in world prices or domestic supplies. Low and stable rice prices were considered important for keeping the rate of inflation low, for reducing upward pressure on wages, and for improving the profitability of the industrial sector. In practice, price interventions have had varied impacts for different commodities. Maize has received high protection (averaging 62 percent in 1990–92) through import tariffs and restrictions due to a strong emphasis on import substitution. Rice has fluctuated between being protected during periods of higher imports and being taxed during the years that rice was exported. The export crops, sugar and copra, have been highly taxed, but the level of taxation has declined somewhat since the mid-1980s. Although direct price support favored some crops, an overvalued exchange rate implemented in support of import-substituting industrialization heavily taxed the agricultural sector. The Philippine exchange rate was overvalued by about 25 percent in the early 1980s, but trade policy reforms implemented after 1986 significantly reduced the overvaluation to around 10 percent in late 1990.

In Thailand, while maintenance of low and stable food prices for civil servants and the urban poor was considered important, the primary objective of price policy was to maximize foreign-exchange earnings from agricultural exports by maintaining a high and stable export price. Before the early 1980s, therefore, price interventions followed the classic pattern of taxation of major export commodities, including rice, rubber, and cassava, and protection of crops targeted for import substitution, including cotton, soybeans, and palm oil. Trade liberalization beginning in the early 1980s, however, has essentially eliminated the taxation of rice, rubber, and maize and has greatly reduced the protection for other crops such as sugar, for which the nominal protection rate declined from over 90 percent in 1985 to –8 percent in 1990. Exchange-rate policies in Thailand resulted in an average overvaluation and taxation of agriculture of about 20 percent during 1980–84, but improvements in current-account balances reduced the overvaluation to about 10 percent in the late 1980s.

Before the PRC began its economic transition process in 1978, the chief role of agriculture was to provide cheap food, capital, and labor for industrial development. Agricultural products were priced below international market prices, thus transferring rents from agriculture to the industrial sector. The implicit taxation of agriculture was enforced by mandatory procurement quotas, a state grain monopoly, strict foodgrain acreage controls, administrative prescription of agricultural inputs, restrictions on nonagricultural activities, and prevention of rural-urban migration. This strategy of squeezing the agriculture sector failed, however, due to the cumulative misallocation of resources and low productivity in agriculture (Stone 1988; Wiemer 1994).

Since 1978, the PRC has been in transition between taxing and subsidizing agricultural production. During the first reform period of 1978–84, state procurement prices for agricultural products were raised and rural markets were opened; the household responsibility system was also introduced (see also Chapter VIII). Interregional trade was authorized in 1984. In the second phase of reforms, during 1985–93, the agricultural pricing and marketing systems were liberalized and rural off-farm activities were encouraged. By 1993, more than 90 percent of all agricultural produce was sold at market-determined prices. The average nominal rate of protection during 1978–93 ranged from –27 to –68 percent for wheat, from –11 to –56 percent for rice, and from –21 to –57 percent for maize, based on a comparison of quota prices with international prices. Since 1994, however, this high taxation has dropped substantially. Recently, in the face of rising grain prices, declines in production, and increasing inflationary pressure, there have been some signs that the government, at least at the provincial level, is attempting to regain control. In 1995, the governor's responsibility system was introduced, whereby provincial governors are responsible for balancing grain supply and demand and stabilizing grain prices (Fan and Tuan 1998). So far there is no clear indication of a shift to increasingly protectionist policies.

Compared to the more export-oriented growth model of East and Southeast Asia, most of South Asia has pursued a strategy emphasizing import substitution and self-sufficiency in capital-goods production and neglect of agriculture and exports. Direct agricultural-price policies have often compounded the negative impacts of indirect trade and macroeconomic policies in this region, with the result that agriculture has been much more heavily taxed than in Southeast Asia. In Pakistan, for example, some protection was provided by direct price supports for wheat, ordinary rice, and cotton in the 1960s, but the overvaluation of the rupee outweighed the direct protection of these crops, while increasing the price-induced taxation of basmati rice (Dorosh and Valdés 1990). In the 1970s and early 1980s, direct effects of trade policies turned negative and became the dominant form of taxation for wheat, basmati rice, and ordinary rice, because the exchange-rate effect (the distortion caused by the appreciation of the real exchange rate) was smaller than in the 1960s. For cotton, trade policies had only a small direct effect on domestic prices, but domestic prices remained significantly lower than equilibrium free-trade prices because of the indirect effect of exchange-rate appreciation and overall trade policy.

Beginning in the mid-1980s, large declines in the world prices of cotton and rice resulted in significant reductions in implicit taxation, because domestic prices of these commodities were not allowed to fall precipitously. Thus, the five major agricultural products in Pakistan (wheat, basmati rice, ordinary rice, cotton, and sugarcane) were consistently taxed from the 1960s to the mid-1980s. Dorosh and Valdés (1990) estimate that the combined effect of trade and exchange-rate policies and agricultural-price policies reduced wheat production by 24 percent and basmati rice production by 52 percent during 1983–87 compared to what they would have been without government intervention. In the absence of direct and indirect price interventions, farm incomes from these five major crops would have been 40 percent higher during that period.

Import substitution policies became the cornerstone of Bangladesh's development policy and remained so until the onset of policy reforms in the 1980s. High export taxes and export restrictions, as well as fixed and multiple exchange rates, characterized the external trade regime. The import-control regime and the exchange-rate policy combined to cause an overvalued exchange rate until the mid-1980s, imposing an implicit tax on exportables. The reform process that started in the 1980s included the gradual removal of trade distortions, the introduction of a flexible exchange rate, and the elimination of agricultural input subsidies. Appreciation of the exchange rate during the second half of the 1980s offset the protection provided by direct trade policies for wheat, cotton, and potatoes and increased the implicit taxation of rice, lentils, jute, and tea. Nevertheless, the overall degree of taxation of the agricultural sector had declined by early 1990s compared to the late 1970s. The implicit protection rate of the agriculture sector as a whole averaged –18.1 percent between 1986/87 and 1990/91. Average nominal protection of rice ranged from –26.3 percent during 1976/77–1980/81 to –0.5 percent during 1981/82–1985/86 to −10.3 percent during 1986/87–1990/91. The total nominal protection rate for wheat declined sharply from −21 percent during 1981/82–1985/86 to −0.3 percent in 1986/87 (Rahman 1994).

In India, agriculture has been heavily taxed through both direct and indirect price policies, with only partial compensation through input subsidies that have introduced additional distortions in production incentives. Protection of industry and overvaluation of exchange rates have taxed agriculture and export restrictions on agricultural commodities kept domestic agricultural prices below world prices. Overall taxation of the agricultural sector was estimated at 30 percent during 1970–84 (Ranade and Dev 1997). Even in the state of Punjab, which received substantial input subsidies, Bhalla and Singh (1996) estimate aggregate measures of support of −32.64 percent for the period of 1981/2 to 1992/3.

India instituted economic reforms during 1991 that were necessitated by an unmanageable fiscal deficit and a balance-of-payments crisis. These reforms included liberalization of industrial and trade policy, relaxation of foreign-exchange controls, and reduced trade- and investment-licensing requirements. The reforms were concentrated in the nonagricultural sector, but the liberalization had a positive effect on agriculture by reducing the anti-agricultural bias of protectionist industrial policies. Overall net taxation of agriculture has been cut to just above 9 percent during the post-reform period, although important crops, such as rice, still face taxation rates in excess of 30 percent. In addition, state trading in agricultural commodities is still pervasive.

The Indian government has attempted to compensate farmers for high taxation through indirect and direct price policy by heavy subsidies to fertilizer, power, credit, and irrigation. Total input subsidies grew dramatically starting in the early 1980s, although fertilizer subsidies were reduced during the economic reform process initiated in 1991. Input subsidies were equivalent to 53 percent of plan expenditure in 1982/83, and 131 percent in 1992/93 (Ranade and Dev 1997). Very conservative estimates of losses due to subsidies in 1989/90 range from 0.5 percent of GDP for irrigation; 1 percent of GDP for electricity; and nearly 1 percent of GDP for fertilizers (Srinivasan 1994). By 1995/96, following the introduction of economic reforms, fertilizer subsidies dropped to 0.6 percent of GDP (Joshi 1998).

The record of trade and macroeconomic policies and their impact on the agricultural sector in Asia has thus been highly complex. Policymakers have pursued inherently conflicting objectives, such as maintaining high prices for farmers and low prices for consumers, and have simultaneously taxed and protected agriculture. On balance, trade and macroeconomic policies have until recently produced a substantial bias against agriculture, with the degree of taxation higher in South Asia than in East and Southeast Asia. Economic reforms since the mid-1980s have significantly reduced this bias, however. The evolution in policies can best be described as being toward a reduction in distortions and toward increased liberalization in both direct and indirect policies, rather than the shift from taxation to protection that is implied by the historical pattern of agricultural policies. Developments in international trade institutions and structural changes in international trade also appear to lead to continued liberalization of policies rather than increased protectionism, as seen below.

GATT AND WTO

The General Agreement on Tariffs and Trade (GATT) served as the multilateral framework for international trade relations for nearly fifty years, following the decision of the United States not to ratify the charter of the International Trade Organization that was proposed at the Havana conference of 1947–1948. Although technically the GATT was not an organization but a multilateral agreement, it nonetheless succeeded in substantially reducing barriers to world trade through eight successive rounds of multilateral negotiations held under its auspices. As a result of the Uruguay Round (UR) agreement in December 1993, the World Trade Organization (WTO), a formal international organization, subsumed the GATT. As of September 1997, the WTO had 132 member countries and 34 observer countries, all but four of which have applied to join the organization. The Kyrgyz Republic became the 133rd member of the WTO in December 1998. The PRC; Taipei,China; and the Russian Federation are the largest trading countries that are still negotiating accession (see also Box VII.2, on the PRC and the WTO). In addition to creating the WTO

Box VII.2: The PRC and the World Trade Organization

China was a founding member of GATT, but withdrew in 1950 after the formation of the People's Republic of China (PRC). It has been seeking to rejoin the international body, now the WTO, since 1986. The benefits of accession by the PRC to WTO are likely to be large, both for the PRC and for the rest of the world.

Bach, Martin, and Stevens (1996), in a general equilibrium analysis, find that the gain to the PRC from accession to the WTO would be $22 billion, $17 billion allowing for tariff exemptions. Anderson et al. (1997) estimate that accession of the PRC (and Taipei,China) to the WTO would add 30 percent to the estimated global real income gains from the UR and that aggregate world trade could increase by 13 percent instead of 10 percent by 2005. These results do not include additional (dynamic) gains from increased foreign investments, increased respectability of the PRC abroad, gains to all WTO members from generally greater liberalization and transparency in the PRC, and a "peace dividend" in the form of reduced tensions between the PRC and some of its major trading partners.

The negotiations over PRC accession to the WTO cover all aspects of multilateral trade negotiations, including agriculture, the customs system, import licensing, industrial subsidies, predictability and transparency of the legal system

(continued next page)

and correcting many of the weaknesses of the GATT, the UR agreement brought agricultural trade back into GATT disciplines. Agreement was also reached on measures to strengthen the protection of trade-related intellectual property (TRIPs) and to liberalize trade-related investment measures (TRIMs). Moreover, the dispute-settlement mechanism has been strengthened (Srinivasan 1998).

The inclusion of agriculture in the UR agreement is an important breakthrough. The provisions seek to provide a

---

Box VII.2 (continued)

and legislative reform, the Agreement on the Application of Sanitary and Phytosanitary Measures, the Agreement on Technical Barriers to Trade, non-tariff barriers, State trading, TRIMs and TRIPS. Other stumbling blocks include

- the need for strict application of WTO rules to this former centrally planned economy (in the light of other transition economies lining up for accession);
- the status (developing or developed economy) of the PRC and the related fear of the PRC's size;
- implications for the textile and clothing sectors;
- the extent of market access offered by the PRC; and
- enforcement of intellectual property rights.

To overcome these stumbling blocks, the PRC needs to include a strong commitment to WTO obligations, provide greater trade-policy transparency, and agree to a compromise on the issue of its developing-economy status in WTO. WTO members also should replace their "moving goalposts" of demands for the PRC with a clear set of requirements within a specified time frame. The return of Hong Kong, China, one of the most liberal WTO members, to the PRC in July 1997 might accelerate the latter's accession into the organization.

(Anderson 1996a; Anderson et al. 1997; WTO 1998)

framework for the long-term reform of agricultural trade and domestic policies, to achieve increased market orientation in agricultural trade, and to strengthen the rules governing agricultural trade in order to improve predictability and stability for both importing and exporting countries (GATT 1994).

The most important component of the agricultural agreement is the requirement that all nontariff border measures be replaced by tariffs that initially provide the same level of protection. Tariffs resulting from this "tariffication" process and other agricultural tariffs are to be reduced for developed and developing countries over a period of 6 and 10 years, respectively, by an average of 36 percent and 24 percent, with minimum reductions for each tariff line of at least 15 percent and 10 percent. Many loopholes and exemptions limit the scope of agricultural trade liberalization under the UR agreements, however. For example, loopholes in the process of tariffication have allowed countries to bind tariffs at levels well above the existing average tariff (so-called "dirty tariffication"). Countries are also allowed to retain import restrictions until the end of the implementation period under certain circumstances.

Developing countries are afforded special treatment—through the "rice clause" negotiated by Japan and the Republic of Korea to continue managed rice imports—that in effect exempts those countries from the liberalization of agricultural commodities that are predominant staples in their traditional diet (Ammar 1997; Srinivasan 1998). The UR on agriculture also restrains domestic subsidies to agriculture, but measures that have a "minimal" impact on trade (so-called "green box" policies), such as income support to agricultural producers that is decoupled from production, are permitted to continue. Agricultural research, disease control, infrastructure, and food-security policies are also among the green box policies. The main farm programs of the United States and the European Union were exempted and Japan was allowed to increase its direct payments to farmers. Moreover, developing countries were exempted from commitments to reduce input subsidies provided to "low-income or resource-poor producers." Again, these significant exemptions will limit the immediate effectiveness of subsidy-reducing policies.

Given the high levels of bound or base-level tariffs resulting from dirty tariffication, the fairly long implementation periods, the permitted exemptions, and the exclusion of some trade-distorting policies from elimination, it is likely that the extent of liberalization of agriculture under existing rules will be small. Indeed, available quantitative estimates of the growth in trade and in incomes with full implementation of the liberalization agreed to in the UR show notable but modest benefits (Srinivasan 1998; Anderson 1996a). Hertel et al. (1995) estimate global welfare gains from UR agreements on agriculture and manufacturing of $260 billion per year, or 0.4 percent of global GDP, by 2005. The largest gains are for developing countries, particularly those that are more open or that liberalize to a greater extent. The East and Southeast Asian developing countries are therefore projected to be substantial gainers, with an increase in GDP of 4.7 percent by 2005. Francois, McDonald, and Nordstrom (1995) also find that Asian developing countries will benefit significantly more from the UR agreements than other regions, with the PRC receiving a boost in GDP annually of 4.0 percent, other developing East and Southeast Asian countries receiving a 3.2 percent increase, and South Asia receiving a 3.1 percent increase, based on 1992 real prices.

Despite the modest quantitative benefits estimated by modeling, integration of agricultural trade into the WTO is a major achievement and the most significant benefits cannot be captured in quantitative models. The dynamic gains to developing countries in Asia are likely to be substantial, including the improvement of access to international technology and capital; strengthening of investor confidence, employment and productivity growth; and encouragement to unilateral trade liberalization programs. The political benefits in Asia (and elsewhere) may be especially large, because the UR agreements will help to break up the existing constellation of political forces that, as described above, has tended to make increasing agricultural protection an inevitable part of economic growth.

Anderson (1996b) argues that multilateral trade agreements can alter the political equilibrium in favor of more open trading policies. A country that reduces trade barriers and allows more imports may harm its import-competing producers, but if its trading partners simultaneously lower their barriers to the first country's exports, the producers of those exports will benefit. Exporters may benefit more than the import-competing industries lose, so that the liberalizing politicians also become net gainers in terms of electoral support—in both countries (Grossman and Helpman 1995).

Political gains from trade negotiations involving exchange of market access will tend to be greater, the larger the number of countries involved and the broader the products and issues coverage of the negotiations. Expansion of membership in the WTO compared to the GATT has dramatically increased the scope for exchange of market access among countries. In addition, it is easier for politicians to placate those harmed by reform when similar producers in other countries are being simultaneously asked to make sacrifices (Anderson 1996b). This has been especially important in the case of farm-policy reforms during the UR (Tyers and Anderson 1992).

Thus, the inclusion of agriculture under the GATT/WTO umbrella is a major step forward, but the limited degree of real agricultural liberalization leaves much to be done in the years ahead. Governing rules should be tightened by reducing or eliminating the special and differential treatment given to both developed countries and to developing countries in Asia and elsewhere. Special exemptions weaken the political benefits of the WTO, because they reduce the scope for governments to use the agreements to justify domestically the adoption of politically unpopular but economically desirable reforms. Elimination of farm-export subsidies should be considered in the next comprehensive round, to bring agriculture into line with nonfarm products, for which export subsidies are banned under the GATT. Dirty tariffication, which set bound tariffs well above actual tariffs, could permit protection levels to be raised considerably, if countries should choose to use tariffs to insulate their domestic markets from fluctuations in international markets. The next multilateral negotiations should aim to bring those bound tariffs down at least to applied rates (Anderson 1996a).

REGIONAL TRADE AGREEMENTS

The long and often frustrating UR negotiation, culminating at one point in the collapse of the "final" negotiating session in Brussels in December 1990, was a significant factor in the revival of interest in regional trade agreements (RTAs) that seek to liberalize trade among countries within a region. Prominent RTAs include the North American Free Trade Agreement (NAFTA) and the Southern Common Market (Mercado Comun del Sul, or MERCOSUR) in the southern cone of South America. Negotiations have begun to develop a free (or at least liberalized) Asia-Pacific trade area based on the Asia Pacific Economic Cooperation (APEC) forum. In Asia, ASEAN formed the AFTA in 1992 (see Box VII.3) and the member states of SAARC signed a framework agreement for SAPTA in 1993 (see Box VII.4).

The proliferation of RTAs raises serious questions as to whether these agreements strengthen or weaken the multilateral trading system and the prospects for continued trade liberalization. Modeling exercises show that RTAs generally boost interbloc trade and generate slight positive income gains (see Boxes VII.3 and VII.4). As would be expected due to the smaller number of countries that liberalize trade within a regional bloc, the estimated quantitative benefits of RTAs are significantly lower than for global liberalization under the GATT/WTO.

There is substantial disagreement, however, regarding the broader political and economic implications of RTAs. Josling, Tangerman, and Warley (1996) argue that RTAs are a positive force for liberalization of agricultural trade and international trade generally and that regional liberalization efforts complement multilateral trade negotiations. The complementarity between regional free trade and multilateral liberalization is achieved through the impact of the partial

Box VII.3: ASEAN (Association of South East Asian Nations)

The ASEAN is the longest-established regional group for economic cooperation within Asia. The association was created in 1967 with the participation of Indonesia, Malaysia, Philippines, Singapore, and Thailand. Brunei joined in 1984, Viet Nam in 1995, the Lao PDR and Myanmar in 1997, and Cambodia in 1999.

The initial objective of ASEAN was to foster peaceful national development of its member states through cooperation. In 1977, a limited program of preferential trade arrangements was adopted by ASEAN member states, followed by the establishment of an ASEAN Free Trade Area (AFTA) in January 1992, whose full implementation is scheduled for 2002.

ASEAN has been an important forum for political and diplomatic cooperation among the states, but has been relatively ineffective on trade issues, due to the very limited trade concessions given to partners and the lack of common external tariffs or a unified stand on trading issues.

Two recent studies by DeRosa (1995) and Lewis and Robinson (1996) find that the AFTA is trade-creating on a net basis. AFTA contributes comparatively little to higher economic welfare in ASEAN countries, however, except possibly the two highest-income and particularly open ASEAN countries, Malaysia and Singapore, as each would contribute a greater share of the increased intraregional demand for manufactured goods previously supplied by developed countries outside the region. Moreover, as Singapore has no tariffs to remove and Malaysia has very few, preferential liberalization in ASEAN is asymmetric. Most-favored-nation liberalization on the part of ASEAN members would raise trade to a much higher degree. For a regional trading block to take off in East Asia, it is likely that Japan would have to play a catalyzing role.

Recent arrangements among the ASEAN members to accelerate and deepen tariff reductions could in the future lead to faster growth of intragroup trade. A more integrated and unified ASEAN would also command more attention within the Asia-Pacific Economic Cooperation (APEC) forum and possibly at the global level.

(DeRosa 1995 and 1998; Panagariya 1997; Frankel and Wei 1997).

Box VII.4: SAARC (South Asian Association for Regional Cooperation)

SAARC was created in 1985 as a follow-up to the South Asian Regional Cooperation (SARC) forum, including Bangladesh, Bhutan, India, Maldives, Nepal, Pakistan, and Sri Lanka, to complement existing bilateral and multilateral cooperation.

During the Seventh Summit in 1993, a Framework Agreement was signed on a South Asian Preferential Trading Arrangement (SAPTA). Under SAPTA, trade cooperation is to be phased in through reduction of tariff rates on an item-by-item basis, reduction of tariffs on a sector-by-sector basis, and establishment of a South Asian Free Trade Area (SAFTA).

Overall trade cooperation efforts have been rather small, however, due to frequent political disputes. Although the SAARC countries have, in recent years, unilaterally liberalized their trading regimes, numerous trade barriers remain. The commodities for which tariffs would be reduced under SAPTA constitute only a small proportion of intra-SAARC trade for each member country. Due to limited trade complementarities, trade is unlikely to pick up significantly after removal of trade barriers. Sri Lanka has the most liberal trade regime of the SAARC countries and would benefit substantially from a removal of restrictive trade policies in the other SAARC countries.

Srinivasan and Canonero (1995) find that SAARC would substantially promote trade for the smaller members, as their initial levels of trade are relatively small, and India and Pakistan would be large enough partners to make a difference in trade for them. Sengupta and Banik (1997) find that intra-SAARC trade could be increased by 30 percent (by 60 percent if informal trade is taken into consideration) if the association were to establish a free trade area, as has been proposed.

Trade expansion within SAPTA and the realization of the envisioned regional integration in the form of a SAFTA would require a serious political commitment to intraregional trade, but would clearly reinforce the pace of economic change in the region.

(Rajapakse and Arunatilake 1997; Rahman 1997; Khan 1997)

opening of borders for national agricultural programs. According to Josling, Tangerman, and Warley (1996), this complementarity operates at four different levels. First, there will be pressure to implement policy instruments that do not distort competition within the RTA. Second, there will be pressure to harmonize national regulations and standards in order to reduce internal transaction costs and diminish the potential for trade conflicts. Third, trade policies, even where nominally decided at the national level, will have a tendency to conform to the rules specified by the RTA; the tendency for external policies to converge in free-trade areas is reinforced by the probable development of coordinated policies toward third-country trading partners in negotiations. Fourth, there is an inevitable weakening of national policy instruments as trade opportunities expand within the regional bloc.

On the other hand, Srinivasan (1998) and Bhagwati (1997) are highly critical of the movement toward regional trading agreements. They argue that RTAs will push the world toward trading blocs rather than toward a multilateral system. By excluding nonbloc members, RTAs are inherently trade-diverting and thereby weaken supporters of an open, multilateral system, while simultaneously creating new interest groups opposing multilateral liberalization. To the extent that trade diversion does take place under RTAs, new interests will oppose further liberalization. Furthermore, given limited resources, the attention of policymakers will be distracted from the global multilateral system when RTAs are under discussion or negotiation. Krueger (1995) notes that the last point is of particular concern. When attention should center on the formation and strengthening of the WTO, it is instead diverted to proposals such as the creation of a more powerful APEC and the addition of new members in NAFTA. Even if, in the long run, RTAs evolve toward multilateral liberalization, the distraction of attention during WTO's formative phase is a significant cost.

According to Bhagwati (1997), the only compelling arguments in favor of active pursuit of RTAs would be the failure of global multilateral trade negotiations, such as the GATT and the WTO, or the transformation of RTAs into true common markets that would eliminate investment and migration barriers as well as trade barriers and would generate much greater benefits. He argues that the Asian members of APEC should push for active engagement with WTO rather than for transformation of APEC into a regional trade agreement, and should use APEC as a forum to coordinate positions and policies at the WTO and upcoming multilateral negotiations.

The debate over the benefits and costs of RTAs is unlikely to be resolved in the near future. Regional trading agreements in Asia are likely to be moderately trade-creating and to produce small income benefits. Regional trade arrangements may also enforce discipline on national trade policies, allowing nations to speed up liberalization and ultimately producing a self-reinforcing process toward open markets. But RTAs also pose the danger of creating trade blocs that would strengthen antitrade liberalization groups and distract attention from the ultimate goal of global multilateral trade liberalization.

STRUCTURAL CHANGES IN INTERNATIONAL AGRICULTURAL TRADE

Policy Reforms in Developed-Country Agriculture

In addition to progress on multilateral and regional trade negotiations, fundamental changes in the structure of agricultural trade will influence trade and macroeconomic policies in the Asian economies and the impact of these policies on agriculture. The United States and Western Europe have historically provided large subsidies and high levels of protection to agriculture. However, unilateral agricultural policy reform that has been completed in the United States and is under negotiation in the European Union will fundamentally change developed-country farm policies, moving them away from the paradigm of market intervention and price manipulation in favor of direct payments to farmers tied less to output.

In the United States, the Federal Agriculture Improvement and Reform (FAIR) Act of 1996 dramatically altered the support system that US producers had relied upon since the Great Depression. The Act replaces the long-standing, crop-linked, deficiency-payment/supply-management program that covered wheat, rice, feedgrains, and upland cotton with a program of fully decoupled, temporary contract payments based on land acreage. Payments are capped at about $36 billion over 1996–2002 and are scheduled to decline over the 7-year period. The FAIR Act also eliminates the Acreage Reduction Program, gradually eliminates dairy-price supports, and modifies US peanut and sugar programs.

During 1993–95, the Common Agricultural Policy (CAP) of the European Union went through significant reforms that initiated a shift from direct price supports to decoupled income supports by lowering intervention prices. In July 1997, the European Commission initiated Agenda 2000, which seeks to broaden the shift from price supports to income supports under the CAP, to continue the reduction in intervention prices, and to reduce export subsidies on agricultural products. These policy reforms in previously highly protected agricultural markets should reduce tensions in international markets and encourage the development of a more liberalized agricultural trade system.

Changes in the Nature of Agricultural Trade

Josling, Tangerman, and Warley (1996) note that the nature of trade in agricultural goods is also changing over time, in ways that tend to expand its importance in the multilateral trade system and to increase the relative political strength of interests that favor liberalized trade policy. Agricultural trade has historically been dominated by bulk products, such as cereals, milk, meat, and sugar. Trade in bulk commodities is still based largely on comparative advantage and policy-induced distortions and these bulk products have not generally attracted international investment.

But the share in world trade of differentiated products with higher value added such as processed foods, including fruits and vegetables, is now becoming much larger in both developed and developing countries. Producers of these goods can develop comparative advantage by investment from abroad and can build markets through quality and name recognition. Government price-support policy has traditionally been less important in this trade and is in fact often detrimental to processors who have to deal with high-priced raw materials. Government regulations regarding quality and food-safety standards are more significant. Growing mobility of technology, management, and marketing skills is making it possible to develop the agricultural and food sector through international investment. Foreign processing firms are searching for reliable sources of raw materials at low cost and are increasingly looking abroad. Changing consumer tastes, developed through urbanization and contact with other cultures or introduced by immigrants, have expanded markets for commodities previously considered nontradable. All of these developments point to the globalization and further liberalization of agricultural trade (Josling, Tangerman, and Warley 1996).

International Price Variability

What will be the impact of the agricultural trade liberalization and the changing structure of agricultural trade on international price variability? Some observers have argued that the increasing openness of trade regimes could usher in a period of increasing price variability that could destabilize farm incomes and harm poor consumers who still spend a large share of income on food (Timmer 1997). But with the changing structure of international agricultural trade, a number of countervailing forces will come into play that on balance could actually reduce international price variability.

Policy reforms in North America and Western Europe may indeed have increased the likelihood of increased variability in international prices. Until recently, the developed countries have borne most of the costs of maintaining food stocks, largely as a byproduct of domestic farm-support programs. However, as North American and European governments scale back price-support programs in favor of direct payments to farmers, they no longer need to buy and hold large reserves. In 1996, the United States and European Union together held less than one half the stocks they held in 1993. This policy-induced reduction in stocks could mean larger price fluctuations in the future, because fewer supplies may be available to the market to dampen price changes when production varies and private-sector stocks may not make up for the difference in reduced public stocks.

Given that the lack of transparency and consistency in government stockholding and trade policies were often sources of instability in the past, however, a reduced involvement by governments in stock management and a more transparent trade policy could contribute to increased stability in the future. Moreover, market liberalization initiatives—including those completed under the UR agreements and those resulting from future multilateral trade negotiations, regional trading arrangements, and other unilateral initiatives—should contribute to stability in international markets by inducing greater adjustments to demand/supply shocks in domestic markets. Protectionist policies in the past shifted the impact of fluctuations in domestic production to the world market, increasing international price variability and insulating the domestic market from external fluctuations. The elimination or reduction of protectionist policies is therefore expected to help reduce price instability in the world market. Improved market integration should also smoothe the effect of international trade on available supplies: the elasticity of demand facing any group of producers would increase, as would the supply elasticity facing any group of consumers. Increased stability should follow in an economy that responded faster and more strongly to shocks as consumers and producers made rational decisions aided by improved information and price transparency (FAO 1996; Islam and Thomas 1996).

Although it is too soon to assess the full impact of trade liberalization and structural change on international price variability, an analysis by Sarris (1998) of inter-year price variability of cereals from 1972 to 1996 concluded that no increase in interyear variability in world cereal markets had taken place. Recent price changes do not appear to herald anything unusual and are not outside the range of normal annual variations. Furthermore, analysis of the intrayear cereal price variability showed that there does not seem to be a tendency for the coefficients of variation of monthly seasonal prices to increase over time; if any, the tendency is towards a decline in variability. It is unlikely that international price instability will cause significant problems for the continuation of the trade liberalization process.

CONCLUSIONS

In much of Asia, gradual reform in the mid-1980s and early 1990s of trade and macroeconomic policy regimes that have historically penalized agriculture have recently improved its competitive position. This long-term policy evolution was temporarily interrupted in the early 1990s, when the competitive position of agriculture and other tradable sectors began to erode due to the dramatic appreciation in real exchange rates in East and Southeast Asia (and a significant but smaller appreciation in South Asia); this erosion resulted from macroeconomic policies and the massive influx of short-term foreign capital. The sharp depreciation of currencies in several countries in East and Southeast Asia during the financial and economic crisis that began in 1997 eliminated effective taxation of agriculture caused by real exchange-rate overvaluation. This development should provide a significant stimulus to agriculture in these countries (these issues are discussed in detail in Chapter IX).

Despite the setbacks caused by the highly variable incentive environment for agriculture during the 1990s, continued reform of trade and macroeconomic and price policies would create a level playing field across economic sectors and across agricultural commodities and provide a stimulating environment for agricultural exports. These conditions would certainly provide further incentives for efficient agricultural growth. The inclusion of agriculture in the UR agreements and the creation of the WTO should provide a supportive international environment for further national agricultural liberalization. It should also provide a framework for reform of agricultural trade and domestic policies, by strengthening the rules governing agricultural trade in order to improve predictability and stability for both importing and exporting countries. International agricultural product differentiation and international investment and technology transfer will further encourage agricultural trade liberalization and encourage agricultural growth.

VIII Economic Reform in Asian Transition Economies

INTRODUCTION

A number of economies in Asia, including the PRC, the Lao PDR, Mongolia, Viet Nam, and the Central Asian Republics of Kazakhstan, Kyrgyz Republic, and Uzbekistan, have in recent years moved, at varying speeds and with varying success, from centrally planned to market-oriented economies. The transition process involves some or all of the following policy reforms:

- macroeconomic stabilization to reduce budget deficits;
- external-trade liberalization to remove trade barriers and rationalize the exchange-rate regime;
- liberalization of prices;
- legalization of nonstate enterprises and removal of legal discrimination based on type of ownership; and
- transformation of the system of property rights, leading to privatization of state-owned enterprises and establishment or extension of private property (Parker, Tritt, and Woo 1997).

The PRC began the process of economic transition in 1978; the Lao PDR and Viet Nam followed in 1986. The Central Asian Transition Economies (CTEs) and Mongolia began their transitions in 1991 following the collapse of the Soviet Union. In the East and Southeast Asian Transitional Economies (ETEs)—the PRC, the Lao PDR, Viet Nam—the opening of the economy initiated periods of rapid growth. The PRC's GDP grew at a rapid rate of 9.9 percent per year between 1978 and 1996, compared to a pre-reform growth rate of 5.2 percent per year; the Laotian GDP grew at 5.5 percent per year between 1986 and 1996, compared with annual pre-reform growth of 5.1 percent, and the Vietnamese GDP grew at 7.1 percent annually between 1986 and 1996, compared to the 3.8 percent per year during the pre-reform period (Table VIII.1).

Agricultural-sector growth has been an important force for economic growth in some transition economies, increasing in both the PRC and Viet Nam during the reform process compared to pre-reform performance. Annual growth in real agricultural GDP averaged 5.1 percent in the PRC during 1978–96, compared to the 2.2 percent in the pre-reform period, and 4.9 percent in Viet Nam during 1986–96, compared to pre-reform growth of 3.8 percent. In contrast, the Central Asian transition economies, together with Mongolia (which is geographically in East Asia, but in terms of transition experience, closer to Central Asia) experienced huge economic declines at the onset of the transition. Real GDP contracted sharply in all the CTEs during the early transition period. There were considerable differences in the growth performances among these countries, however, with the Kyrgyz Republic experiencing the sharpest drop from positive 7.5 percent per year to negative 10.9 percent annually. By 1996, output in the Kyrgyz Republic was only 56 percent of the 1991 level; in Kazakhstan, 60 percent; and in Uzbekistan, 84 percent. In Mongolia, on the other hand, the decline in output was relatively less.

Still, agricultural value added in 1996 was only 65 percent of the 1991 value in Kazakhstan, 87 percent in Uzbekistan, and 95 percent in the Kyrgyz Republic. In Mongolia, on the other hand, agricultural output actually increased at 2.33 percent per year in the post-reform period. The economic decline was accompanied by hyperinflation, reaching over 1,000 percent annually in the CTEs. The economies of Central Asia and Mongolia have nevertheless apparently turned the corner, with positive economic growth resuming and hyperinflation coming under control. Real GDP growth in Kyrgyz Republic was 5.6 percent in 1996 and was estimated at 6.0 percent for 1997; in

Table VIII.1: Comparison of Transition Economies in Asia

	PRC	Lao PDR	Viet Nam	Mongolia	Kazakhstan	Kyrgyz Republic	Uzbekistan
Pre-Reform							
Growth in real GDP (percent per year)	5.24	5.05	3.81	2.20	-0.92	7.54	4.71
Growth in real ag. GDP (percent per year)	2.18	n.a.	3.80	3.18	n.a.	5.47	3.91
Post-Reform							
Growth in real GDP (percent per year)	9.91	5.52	7.14	-0.46	-9.86	-10.87	-3.48
Growth in real ag. GDP (percent per year)	5.11	5.37	4.93	2.33	-8.12	-0.95	-2.86
At beginning of reform							
Agriculture as % of GDP	28.1	60.6 ('89)	36.1	14.1	26.7 ('92)	37.8	37.1
Industry as % of GDP	48.2	13.4 ('89)	35.3	34.2	44.6 ('92)	36.0	35.7
GDP per capita, con. 87 US$, PPP	700.0 ('80)	715.7	833.9 ('89)	1568.5	3986.8	2703.0	2448.6
Employment in agriculture as % of employment	75.1	78.7	72.1	32.0 ('90)	22.2 ('90)	32.1 ('90)	34.9 ('90)

Note: Pre-reform growth rates are 1966–77 for the PRC; 1984–85 for Lao PDR and Viet Nam; and 1987–90 for the rest; post-reform growth rates are 1978–96 for the PRC; 1986–96 for Lao PDR and Viet Nam; and 1991–96 for the rest. Individual values are for reform begin (PRC: 1978; Lao PDR and Viet Nam: 1986; rest: 1991) unless indicated otherwise.

Source: WDI (World Bank), for Kazakhstan and Lao PDR post-reform ag. GDP and for Mongolia pre- and post-reform: 1998e.

Kazakhstan, growth was estimated at 0.5 percent in 1996 and at 2.0 percent in 1997; and in Uzbekistan, growth was estimated at 1.6 percent in 1996 and at between 2.4 percent and 5.2 percent for 1997 (World Bank 1998e). In Mongolia, growth was 2.6 percent in 1996 and projected to exceed 3 percent in 1997 (World Bank 1998c).

In this chapter are explored the underlying factors that have influenced the performance of the transition economies of Asia, with special reference to the agricultural sector. What determined the relative success of the PRC, Lao PDR, Viet Nam, Mongolia, Kazakhstan, Kyrgyz Republic, and Uzbekistan in the transition process, and what policy lessons can be drawn from the transition process? Which underlying factors, especially agricultural factors, influenced their performance? Some of the important issues that can influence the outcome of the transition process include

- the impact of strategy, policy, and sequencing of reforms compared to exogenous factors and country-specific initial conditions such as agrarian structure and pattern of industrialization;
- the speed with which the reform package was implemented;
- the pace and method of reform of state enterprises, the tools and degree of privatization, and the creation of property rights to establish positive incentives;
- the impact of trade liberalization and the integration into the world economy; and
- environmental and social policy (Parker, Tritt, and Woo 1997; Green and Vokes 1997; Rana 1995a).

INITIAL CONDITIONS AND EXOGENOUS SHOCKS

Perhaps the biggest initial difference between the ETEs (PRC, Lao PDR and Viet Nam) on the one hand and the CTEs and Mongolia on the other was in the motivating force behind the transition process. Reform in the ETEs was driven by internal developments—primarily dissatisfaction with economic performance—and was undertaken by essentially stable governments. The beginning of reform in the PRC followed the poor performance of the economy, and agriculture in particular, during the Cultural Revolution of 1966–76. Economic and agricultural growth was similarly discouraging in the Lao PDR, following the establishment of the People's Democratic Republic in 1975, and in Viet Nam, following reunification in 1975. Both countries began market-oriented reforms in the mid-1980s with the Jin Tanakan Mai, or New Economic Mechanism, in the Lao PDR and the Doi Moi, or renovation policy, in Viet Nam (Green and Vokes 1997).

Reform was forced on the CTEs and Mongolia by the twin shocks of the breakup in 1990 of the Soviet-run Council for Mutual Economic Assistance (CMEA) trade and financial arrangements, followed by the collapse of the Soviet Union itself in 1991. The breakdown of the Soviet-based economic system caused the termination of payment mechanisms for cross-border transactions, the loss of traditional markets, the end of massive budget transfers from the central government of the Soviet Union, a sharp deterioration in the terms of trade for traditional exports, and hyperinflation that discouraged savings and investment and distorted price signals (Green and Vokes 1997; Parker, Tritt, and Woo 1997). These massive initial shocks drove an economic downturn comparable in magnitude to the Great Depression in the United States in the 1930s.

Fundamental differences in initial economic structure also have had a profound influence on the performance of the transition economies. In the CTEs and Mongolia, the state sector accounted for 80 to 90 percent of GDP and of employment at the time of the Soviet Union's demise. The state sector was much smaller in the Lao PDR and Viet Nam at the onset of economic reform (similar in fact to those of other, noncentrally planned developing countries), accounting for 15 and 22 percent of GDP, respectively, and less than 10 percent of employment. The state sector in the PRC was intermediate in size, accounting for about 57 percent of GDP and 45 percent of employment. Moreover, the PRC economic planning system, while centralized and hierarchical in principle, was much less extensive than in the CTEs and Mongolia (Rana 1995a, 1995b). The dominance in the CTEs and Mongolia of an inefficient state sector that was heavily dependent on subsidies and economic linkages with the Soviet Union has been a severe constraint on resumption of economic growth.

Directly related to the dominance of the state sector is the relatively large importance of the industrial sector in the CTEs, as compared to some of the ETEs, like the Lao PDR and Viet Nam. At the beginning of the reform process, industry accounted for about 45 percent of GDP in Kazakhstan and 36 percent in the Kyrgyz Republic and Uzbekistan (Table VIII.1). The high level of industrialization posed a transition problem in the CTEs because the industrial sector was extremely inefficient, depending on subsidies and intra-Soviet Union economic relationships. Analyses of the causes of economic decline and recovery in Eastern Europe and the former Soviet Union have found that the degree of "overindustrialization" was an important explanatory factor for economic performance during the early years of transition and reform. Overindustrialization captures the degree of industrialization at the beginning of the transition relative to the degree of industrialization typical of a market economy in the same range of GDP per capita (De Melo et al. 1997). The lower degree of overindustrialization in Uzbekistan compared to the other countries of the former Soviet Union and Eastern Europe, for example, has been identified as an important factor explaining its relative success during the early transition years (Taube and Zettelmeyer 1998; Zettelmeyer 1998).

The structure of the agricultural sector itself has been a further defining difference between the CTEs and Mongolia on the one hand and the ETEs on the other. The critical differences are in the scale and management of production and the nature of the technology utilized. In the CTEs, agriculture was organized into sovkhozy (state farms) and kolkhozy (collective farms). These state enterprises were in general very large, with those in Kazakhstan averaging 35,000 to 40,000 hectares, for example. Heavy capital investment was made in rural roads, irrigation, and large-scale farm machinery, but much of this investment was highly inefficient (Green and Vokes 1997). Agricultural-production decisions were driven by command-and-control agricultural polices similar to those of the Soviet Union. In this system, the State owned and exploited the land and determined the amount and composition of agricultural production according to the overall needs of the economy for food and foreign exchange. It also set the production share accruing to the government (which was virtually 100 percent until 1990) under the state order system; determined the requirements of chemical inputs for collective and state farms; and established farm, input, and consumer-goods prices separately from production quotas, often after planting. Finally, it restricted the movement of labor within the rural sector and between the rural and urban sectors. At the same time, the government compensated farmers for the difference between production costs and planned income, either through budgetary transfers (the state farms) or through input subsidies (the collective farms). State production orders, highly subsidized fixed input prices, and fixed output prices determined artificial profits pre-set for each farm, with essentially no flexibility for decision-making at the individual farm level.

As a result of these policies, agriculture in the CTEs faced a serious crisis well before the beginning of economic transition. The command-driven cotton monoculture that dominated much of Central Asia had caused devastating environmental problems, including excessive water use, high water losses, and associated waterlogging, increased soil salinity, declining water quality, and the deterioration of the Aral Sea and its immediate vicinity (see Box VIII.1).

Crop yields are very low compared to yields in developing countries with similar agroclimatic environments. In 1987, 7 percent less cotton was produced on 15 percent more land compared with the 1976-1980 averages in the five Central Asian Republics (Frederick 1991). Although the inherent faults in agricultural policy and the first signs of the approaching crisis became evident in the 1970s, they were simply ignored under the prevailing command-administrative system of management.

> **Box VIII.1: The Aral Sea Disaster**
>
> The Aral Sea region is a notorious example of the wide-reaching negative consequences of the command-and-control policies exercised in the CTEs. The rapid increase in irrigated areas since the 1960s—particularly for cotton cultivation—in this arid Central Asian region had devastating economic, ecological, and health consequences.
>
> The Aral Sea level is determined by the inflow of two feeding rivers, the Amu Darya and the Syr Darya. In the1960s, this inland lake was the world's fourth largest, but due to extensive irrigation water withdrawals from the rivers, inflows that were once 72 km^3 and 37 km^3 per year, respectively, have dropped to a combined total of less than 10 km^3, on average, during the last decade.
>
> As a result, the surface area of the Aral Sea declined by almost 50 percent and 35 million people lost access to the use of the lake for its water, fish, reed beds, and transport. The area surrounding the Aral became completely desolate, the animal and plant life has been significantly decimated or destroyed, and desert storms now carry tons of toxic wastes over many kilometers.
>
> (continued next page)

The CTEs thus entered the transition period with a large-scale, low-productivity, poorly managed, and highly inefficient and inflexible agricultural sector.

In the ETEs, agricultural performance was also relatively weak under the pre-reform institutions, but the structure of agriculture was more conducive to rapid output response to economic reform. Agricultural production was organized into state farms and collectives, with the latter predominating, but was never as fully communalized or centrally controlled as in the CTEs. Moreover, operational farm sizes in the ETEs were very small and therefore more easily adaptable to changing relative prices and opportunities during the transition.

> Box VIII.1 (continued)
>
> Mismanagement and excessive use of irrigation water have contributed to the transport of salts with the drainage water, which, together with the rapid rise of the groundwater level, has led to the progressive salinization of soils. More than 30 percent of irrigated areas of the Central Asian Republics have a medium to high degree of salinization; land losses due to salinization are estimated at 1 million hectares over the past 30 years. Water quality declined dramatically, due to the limited waste-dilution capacity of the rivers, high salt concentrations flushing out from irrigated areas, and runoff of agricultural chemicals (including defoliants from cotton harvesting).
>
> In the lower reaches of the Syr Darya river, morbidity has increased 20-fold over the last 20 years. Infant mortality in a number of districts exceeds 110 per 1,000, that is, three times the average for the former Soviet Union. Out-migration increased and both living standards and life expectancy declined considerably in the region, especially after the collapse of the Soviet Union. The cotton bias led to a gradual decline of agricultural productivity and production and contributed to the general economic decline in the region.
>
> *Source:* Golubev 1993; Cai 1999.

Although in the PRC, for example, collective farms accounted for more than 95 percent of total cultivated land area, the farms were arranged in much smaller units of about 50,000 communes with, on average, 2,033 ha; and these, in turn, were disaggregated into 750,000 production brigades with about 136 ha and approximately 5 million production teams of about 20 ha, which was the actual unit of production. In addition, private plots, limited to 5–10 percent of the teams' land, were distributed to individual households. Moreover, private activities (including animal raising, handicraft and other activities) accounted for 30–50 percent of total household income. Only a small part of the land was cultivated through state farms (Wong 1996). In the Lao PDR, collectivization efforts, begun in 1976, were undertaken on a much smaller scale and even at their peak affected less than half of the cultivated area, due to strong farmer resistance and declining productivity. Many of the cooperatives consisted of little more than traditional labor exchange groups renamed to benefit from the lower taxes and other subsidies accorded by the Government (Vokes and Fabella 1996; Green and Vokes 1997).

In Viet Nam, the collective system was well established in the north, but cooperatives had considerable autonomy from government administrative control. Farm households were allowed to work 5 percent of their land as private plots, but in practice the area under private plots was often considerably larger. Prior to reunification in 1975, agriculture in the south was similar to smallholder agriculture in other Southeast Asian countries. Efforts to collectivize agriculture in the south beginning in 1976 were met with strong farmer resistance. As a result, only 5.9 percent of all farmers were organized in collective farms in the Mekong Delta, for example. Whereas in the north cooperative farm work was carried out by production teams using communally owned equipment, in the southern collective farms, production activities were mainly carried out on a family-farm basis, and collective effort was only made for obtaining inputs and for marketing outputs (Pingali and Xuan 1992). Although in principle the system was very close to that of the Soviet Union, planners actually had much weaker control over the economy in many ways. Agriculture was incompletely collectivized and extensive parallel markets, with wide differentials between free-market and state prices, were tolerated (Fforde and De Vylder 1996). After 1980, limited reforms were implemented throughout the economy through a contract farming system that permitted individual families to farm the land following land preparation.

As shown in Table VIII.1, the share of agriculture in total employment in the ETEs was also very high at the beginning of the economic transition process. When the transition process initiated rapid industrial growth, the large reservoir of relatively unproductive labor in agriculture provided low-cost labor to the growing industrial sector without damaging agricultural growth.

Thus, prereform agriculture in the ETEs was small-scale, relatively labor-intensive, and relatively decentralized compared to that in the CTEs and Mongolia. These structural differences constituted a significant advantage as the countries entered the transition process. With small-scale agriculture, production techniques and cropping choices can adapt rapidly in response to changing relative prices and trade liberalization, even without well-functioning financial markets. The decentralized nature of agriculture further facilitates output response to economic reform and encourages the development of small-scale trading and rural services and industry. Labor released from agriculture during the reform process was a crucial enabling factor for industrial growth. Reform of the capital-intensive, centralized collective farms in the CTEs has proven much more difficult. Recovery and positive supply response to reform is dependent on changes along the entire processing and marketing chain; weaknesses anywhere in the chain can slow the response (Rana 1995a; Green and Vokes 1997).

AGRICULTURAL REFORM

Starting from these very different preconditions, agricultural reform has been a key factor in the process of economic transition in the ETEs, but has been slow and ineffective in the CTEs. The PRC was the path breaker in agricultural reforms that served as a model for Viet Nam and the Lao PDR. Agricultural reform in the PRC began in 1978 with the reorganization of production and the improvement of incentives. The so-called household responsibility system was an important component in the reform process. Under this system, labor teams' lands were divided and contracted to individual households, which became responsible for production and allocation decisions, as well as for delivering taxes and quota and above-quota sales to the state. Although land ownership remained with the village, land use contracts were usually set for 15–25 years, which provided enough security to establish incentives for farmers to invest in farm improvements and resulted in productivity increases. After initial successes, this system spread to over 90 percent of all production teams. Whereas in 1978 the average team size was about 20 ha, after the decentralization, the average farm size was about 0.4 ha—a de facto return to private farming. Moreover, in 1993, a new policy was adopted to extend land leases for an additional 30 years when existing contracts expired. Thus, the fundamental reform that induced rapid productivity growth in PRC agriculture after 1978 has been the combination of the household responsibility system and improved security in land rights (see also Chapter V).

Nevertheless, the reform of property rights in the PRC has not fully secured these rights; a number of studies have pointed to continued weakness in property rights as one reason for the recent slowdown of growth in the farm sector (Ahuja et al. 1997). The household responsibility system individualized the claim to residual income, but continued to vest land ownership in the collective, thus discouraging farmers from making long-term investments in the land (Wen 1995; Choe 1996). Empirical analysis indicates that continued land-tenure insecurity has adversely affected medium- and long-term investments in farms in the PRC (Yao 1995; Feder et al. 1992). These results indicate that granting absolute individual land rights could boost agricultural productivity (Prosterman, Hanstad, and Ping 1996; Crook 1994).

Complementary reform measures further accelerated output growth in the PRC agriculture sector. Beginning in 1979, price incentives were improved by reductions in procurement quotas, increases in quota and above-quota prices, and the liberalization of rural markets and private trade. In early 1985, the mandatory quota system was changed to a contract procurement system, in an attempt to reduce the government's financial burden and to increase the role of markets in grain production and distribution. The contract system reduced farmer incentives, however, and included demands for contract management and enforcement of contracts that could not be met by existing institutions. Thus, by the end of 1985, the contract system was abolished and mandatory quotas were re-established. Grain procurement prices were sharply raised between 1986 and 1989, but these were accompanied by continued increases in farmer autonomy. As a result, farmers diversified into more profitable activities such as fruit production, fisheries, and rural nonfarm enterprises. In the 1990s, agricultural policy has continued to waver between liberalization of output markets and concerns for the impacts of such liberalization on consumers. In 1993, both procurement and ration (selling) prices of grain were liberalized, resulting in rapid increases in the market prices for grains between 1993 and 1995, which, in turn, induced the government to re-impose some administrative restrictions on grain markets (Lin 1997).

Despite these highly beneficial agricultural-sector reforms, broader economic policies have continued to penalize agriculture in the PRC, as in the nontransition economies of East and Southeast Asia. Trade and exchange-rate policies systematically depress the domestic prices of agricultural produce below international parity prices, while increasing the prices of manufactured goods.

Agricultural reforms in the Lao PDR and Viet Nam followed a path similar to that of the PRC, beginning at a later date but then proceeding rapidly. In the Lao PDR, rural reforms were initiated in 1986, when collectivization was de-emphasized and a return to family farming encouraged. Long-term leases and inheritance rights to leases were granted. Given that collectivization was not far advanced, these reforms mainly formalized existing usufructuary rights. Agricultural marketing was liberalized in 1987, allowing the private sector to compete with state enterprises. Price and exchange-rate reforms were introduced in 1988, removing price controls on all but eight strategic commodities. The official exchange rate was unified and depreciated, improving agricultural competitiveness (Hamid 1995).

In Viet Nam, agricultural reform began in 1986, and by 1988 the process of decollectivization had also shifted most decision-making power back to farm households. The system of agricultural-commodity procurement at below-open market prices was also eliminated in 1988. In 1989, virtually all prices were deregulated, so that all commodity prices were market-determined and direct subsidies nearly eliminated. State agencies still control much of large-scale wholesale and international agricultural trade, but local markets are largely in private hands. As in the Lao PDR, price reforms were accompanied by currency devaluation, which further increased the competitiveness of agriculture (Hamid 1995; Green and Vokes 1997).

The agricultural reform experience in the CTEs stands in stark contrast to that of the ETEs, with reform measures buffeted by the devastating economic crises following the breakup of the Soviet Union. Of the three countries reviewed here, the Kyrgyz Republic has undertaken the most comprehensive agricultural reforms, while reforms have been slowest in Uzbekistan. The reform experience in agriculture and other sectors in Mongolia is summarized in Box VIII.2. In the Kyrgyz Republic, the state mandatory sales system had been eliminated by 1995, most state and collective farms had been converted to joint-stock farms, input and output markets had been liberalized and the maximum period of land use rights had been extended from 49 to 99 years. Private farms now account for one third of cropped area. For crops where private smallholder farming activity had been most active, such as vegetables and potatoes, these reforms have improved incentives and boosted production. Production also recovered somewhat in the grain sector after 1994.

Of the utmost importance in agricultural reform in the Kyrgyz Republic, as well as the other CTEs, will be the creation of genuine incentives in the ownership structure of the at least nominally reformed state and collective farms. In principle, the "shareholders" in joint-stock farms should be motivated to improve productivity in order to maximize dividends on their shares, but this incentive structure has in fact rarely been implemented even in the nominally reformed joint-stock farms. In most cases, neither dividends nor shares have practical significance, so there is no true sense of ownership. In essence, the shareholders continue working under the direction of the previous collective-farm manager. More thoroughgoing reform will be required to introduce appropriate incentives into large-scale farms in the region. An alternative model, adopted in much of Eastern Europe, is the dismantling of the state and collective farms and the distribution of land and assets to individuals, creating several hundred private farms in place of one collective farm (Brooks and Lerman 1994). This process has already been undertaken to a limited extent in the CTEs.

In Kazakhstan, more than 90 percent of the state and collective farms have been at least privatized by conversion into joint-stock companies or cooperatives. On many or most of these farms, however, the legal status of land titles is ambiguous and many farms have not actually been restructured (ADB 1997b). Nevertheless, there has been an increase in activity on household plots, individual farms, and small agricultural enterprises. In 1990, private farming accounted for less than 10 percent of crop production and about 40 percent of livestock production, but by 1995, its share had increased to more than 25 percent and 65 percent, respectively (De Broeck and Kostial 1998).

In the Soviet era, Kazakhstan was the largest grain exporter to other parts of the Soviet Union, averaging around 10 million metric tons per year in the 1980s. During the early stages of transition, the Kazakh government imposed trade restrictions, such as export tariffs on wheat, in an attempt to maintain domestic food availability. State procurement in agriculture was initially maintained, but broke down during the transition as output was sold outside the state order system. The state procurement system and direct government subsidies to agricultural credit and inputs were virtually eliminated by 1995, with direct government subsidies to agriculture declining from 10–12 percent of GDP prior to independence to about 2 percent of GDP in 1995.

Simultaneously with Kazakhstan's cutback in direct subsidies, there was a sharp contraction in credit to agriculture: by the end of 1995, bank loans to the agricultural sector were less than 1 percent of GDP, down from 5 percent in 1993. A final blow to agriculture was the rapid deterioration in the terms of

> **Box VIII.2: The Reform Experience in Mongolia**
>
> In the pre-reform period, the Mongolian economic structure more or less mirrored the situation in the Central Asian Republics. Agriculture, mainly livestock production, was organized into state-owned farms and cooperatives comprising, on average, 400,000 ha and 23,000 ha, respectively. This sector contributed only 14 percent to GDP before the reform process, compared to the 34 percent of the industrial sector. A wide range of social services was provided by the Government, resulting in very high social indicators compared to countries with similar per capita income.
>
> At the outset of the reform process, Mongolia experienced possibly the largest contraction in external economic relations of all formerly Soviet-dependent economies. Trade volumes with former CMEA countries dropped sharply and Soviet credits, averaging about 37 percent of GDP annually, abruptly stopped.
>
> Mongolia's initial approach to reform was partial and often diffuse, ranging from some more radical measures, like the world's first voucher privatization program, to the maintenance of government control over trade, foreign exchange, and resource allocation. Significant reforms were
>
> (continued next page)

trade following price liberalization and hyperinflation. In 1992/93, agricultural prices increased by only half as much as prices in other sectors; in 1993, the prices of agricultural inputs increased 18.8 times while output prices increased 7.8 times. The agricultural terms of trade have continued to deteriorate in subsequent years (De Broeck and Kostial 1998).

A fundamental question remains as to whether the previous size and composition of the agricultural sector in Kazakhstan (and much of the rest of Central Asia) is viable in a market-oriented economy. The extensive grain and livestock agriculture practiced in Kazakhstan was as much the result of

> Box VIII.2 (continued)
>
> realized within three years in all economic sectors, however. All of the smaller enterprises and more than half of the large state enterprises have been sold via auctions or insider buyouts via vouchers. Agricultural cooperatives have been completely eliminated and the assets have been distributed to the members, but about half of the state farms have remained in public hands. Whereas the early reform period was characterized by high inflation and lack of monetary control, the situation stabilized in the mid-1990s. Price liberalization and the dismantling of the state order system were implemented gradually, responding to the needs of the growing private sector.
>
> Despite some lopsided policy measures, the Mongolian example is often cited as a successful one for the transition to a democratic market system. This was possible because, among other reasons, (i) Mongolia's socialist ideology was less entrenched; (ii) early parliamentary elections helped ensure macroeconomic reforms; and (iii) the support of private business led to a gradual reform in the other economic sectors. However, almost 30 percent of the population still lives below the poverty line, social indicators have worsened in the post-reform period, and foreign investment in this remote economy has been sparse.
>
> (Boone et al. 1997).

politically motivated policies, such as the Soviet Virgin Lands campaign of the 1950s, as it was the result of economic considerations. In addition to agroclimatic limitations to productive agriculture in much of the area cropped prior to independence, the geographic isolation of Kazakhstan (with consequent high transportation costs) is a serious constraint on production for external markets. Even an efficient, reformed agricultural sector in Kazakhstan could be considerably smaller than the pre-independence agricultural sector.

Uzbekistan is the largest agricultural producer among the CTEs and agriculture has played an important part in the relatively strong performance of the Uzbek economy during the transition period. Agriculture in Uzbekistan accounts for 35 to 40 percent of GDP and contributes about 60 percent of Uzbekistan's export earnings—largely due to cotton exports — and is responsible for 40 percent of total employment. Agricultural reform has been slow, however, and severe barriers to sustained growth remain. Uzbekistan has nominally reformed much of the sovkhoz and kolkhoz sector: by 1994 most sovkhozy had been converted to kolkhoz joint stock companies, without much real change in management or improvement in production incentives. Property rights to land and security of tenure remain uncertain and the transition to autonomous cooperative management or private farming has been slow. Most rural people live and work on about 2,000 former sovkhozy that have been transformed into kolkhozy. Farm workers on kolkhozy have been allocated small household plots, averaging 0.2 ha, to supplement household food supplies and incomes. The government has also recently given about 20,000 private farmers the right to use land allocated from collectives and former state farms. These private (or dekhan) farms range in size from 5 to 100 ha, with an average farm size of about 20 ha. Household plots and dekhan farms are an important source of agricultural growth; about one half the value of agricultural production, mainly fruits, vegetables, and livestock products, come from household plots that account for only 15 percent of irrigated land (ADB 1998a).

Most agricultural commodities have been freed from the compulsory state procurement system, but officially sanctioned trade associations exercise strong control over the disposition and price of output of most commodities. Moreover, the most important commodities, cotton and wheat, remain under rigid government quotas and state orders that levy a burdensome tax on the incomes of farms and their workers. The government subsidizes the prices of farm inputs in partial compensation for the low returns on output, but these subsidies have been declining. The bulk of the subsidies are the free distribution of water, subsidized credit, the free use of land, and cut-price electricity, agrochemical inputs, and fuels. These subsidies further distort production incentives and do not compensate for the taxes on output. It is estimated that in 1997 the net impact of government pricing, procurement, subsidy, and credit policies on cotton and wheat alone constituted a transfer of 4 percent of GDP at the official exchange rate and 8 percent of GDP at the unofficial exchange rate. As a result of these policies, the terms of trade have shifted dramatically against agriculture, with the agricultural sector losing 60 percent of its purchasing power between 1990 and 1997 (ADB 1998a).

In each of the CTEs, significant agricultural reforms are still necessary to correct serious remaining policy and institutional inefficiencies and establish a base for long-term agricultural growth. Probably the most crucial reforms are the development of the legal framework for property rights, the establishment of private farms, and genuine privatization of state and collective farms. Remaining production quotas and state orders for agricultural output should be eliminated and replaced by market prices. Trade and macroeconomic reform to open the international trade regime and institute a market-based exchange rate would also benefit agriculture.

The ETEs have progressed much further than the CTEs in agricultural reform, but still face important issues. Although the long-term leasehold systems that currently prevail in these countries provide considerable security, fully defined property rights in land would provide better security and production incentives for agriculture. As with other developing countries in Asia, there remains a persistent tendency to tax agriculture through exchange-rate and trade policies. Agricultural growth would be further stimulated by the removal of these biases against agriculture. These macroeconomic and trade policy issues were discussed in more detail in Chapter VII.

STATE ENTERPRISE REFORM

One of the most important, and also the most difficult, reforms in the move from centrally planned to market economies is the reform of state-owned enterprises (SOEs). Including the state and collective farms discussed above, these SOEs controlled virtually all agricultural production in the CTEs and maintain a heavy role in production today despite early transition reforms. In the ETEs, state enterprises have been relatively more dominant in the nonagricultural sector. In both the CTEs and ETEs, however, state enterprises controlled (and in many of the transition economies continue to control) a significant share of agricultural input and output marketing, agricultural processing, and international trade in agricultural commodities. In addition, the large implicit and explicit subsidies to industrial and service-sector SOEs exact a tax on the agricultural sector, in the same manner that biased trade and macroeconomic policies do (see Chapter VII). Moreover, the inefficiency of the SOEs in agriculture and other sectors has been a significant drag on the process of agricultural transformation.

At the onset of the transition process, the state sector was highly inefficient because of the legacy of central planning. The central plan allocated output targets, inputs, and investment to the SOEs, which were operated to meet physical production targets and protected from competitive pressures. Financial performance was irrelevant because profits and losses were redistributed among firms. As a result of these incentives, SOEs were inefficient and unproductive; massive subsidies to cover operating losses were a key source of inflationary pressure. The transformation of these enterprises is therefore essential to creating an efficient market economy. The strategies followed in the ETEs have again been quite different from those in the CTEs. While the strategy has not been identical in the PRC, Viet Nam, and Lao PDR, the broad plan of these countries has been to corporatize the state-enterprise sector, attempting to introduce financial discipline and market incentives to the SOEs while encouraging the development of a parallel private sector, and then downsizing the state-enterprise sector. The PRC's experience illustrates both the successes and the limitations of this approach.

Beginning in 1978, the PRC has pursued a dual-track transition that created parallel market and central-plan tracks in both the price system and enterprise ownership. A relatively free market was introduced, while state supply was maintained at the lower plan price. Plan prices, and the proportion of the economy subject to plan prices, were gradually adjusted until prices and incentives converged. As a result, a nonstate sector of private and semiprivate enterprises, foreign joint ventures, and community-based rural enterprises has developed alongside the SOE sector (Parker, Tritt, and Woo 1997). The state-enterprise system itself has undergone significant reforms aimed at improving incentives to respond to market signals. Direct local-government control over SOEs has been reduced, but the local government often retains heavy influence through control over financing, allocation of workers, and appointment of workers. A variety of mechanisms has been tried to introduce market-type incentives to SOEs, but the predominant form has been the enterprise contract, which specifies profit targets and remittances to the government. The establishment, particularly in the coastal areas, of Special Economic Zones (SEZs), with far greater institutional autonomy, has further encouraged innovation and investment. Strong responses to the partial reforms in the SOEs, and to an even greater degree to the more comprehensive reform in the SEZs, have been key forces behind the rapid economic growth in the PRC during the economic reform period.

Despite the success of these reforms, significant problems remain in the state-enterprise sector that could dampen future growth in the PRC. The enterprise contracting system has failed to eliminate the negotiability of state-enterprise financial relations with the government, to harden taxes through clear stipulation of tax rates, or to make taxation uniform and transparent. As a result, soft budget constraints protect state enterprises from the market environment. For example, during the economic slowdown in 1989–90, state enterprises were able to pass most of their losses to the government budget and banking system, causing severe inflationary pressures.

A number of factors make comprehensive state-enterprise reform difficult. Profitability is not directly linked to enterprise constraints because of continued input and output price distortions; the financial relationship between enterprises and the government remains distorted and subject to rent seeking on both sides. SOEs also perform a variety of social responsibilities including provision of housing, transport subsidies, and other social services. These responsibilities make SOEs uncompetitive with nonstate enterprises, creating the demand for further government subsidies. This complex set of relationships must be reformed before budget constraints for state enterprises can be hardened (Rana 1995b). The evolution of the PRC economy nevertheless continues to exert pressure for further reform of state enterprises. The corporatization and management of enterprises through price signals generated by stock markets is becoming increasingly accepted, and more and more state enterprises are being sold to nonstate companies (Cao, Fan, and Woo 1997).

In Viet Nam, the reform and downsizing of the state-enterprise sector has been more rapid, beginning with radical reforms in 1989 that cut all government budget subsidies to SOEs and eliminated 5,000 firms, while exposing SOEs in some sectors to private-sector competition. These dramatic reforms boosted the output of SOEs and in just three years, revenues from these enterprises increased from 6 percent to 11 percent of GDP (World Bank 1996a). A second round of restructuring began in 1991, with the number of SOEs falling from around 12,000 to 6,600 by 1995 due to closures, consolidation, and sale of smaller enterprises. This reform and restructuring reduced the proportion of loss-making SOEs from two thirds to less than 10 percent (Green and Vokes 1997). Nevertheless, reform of the larger state enterprises has been slow, and the remaining SOEs still benefit from a range of protective and distortionary measures, such as exchange-rate controls and land policy, that hinder free entry and competition and bias state enterprises toward inefficient and capital-intensive production.

The Lao PDR also moved quickly, beginning in the late 1980s, to reform the state sector: most small and medium-scale enterprises were sold, leased, converted into joint ventures, or liquidated by 1994. The large SOEs have not been privatized, however, and contracting systems similar to the PRC model have had only mixed success (Green and Vokes 1997; Rana 1995a).

Progress and problems in reforming the state-enterprise sector in the CTEs in many ways mirror those in the reform of the large state and collective farms described above. Many small- and medium-scale SOEs have been privatized and others have been closed, but large-scale SOEs have in many cases survived essentially intact from the Soviet period. The Kyrgyz Republic has again moved most rapidly among the CTEs, privatizing more than 6,000 enterprises, mainly through cash and voucher auctions, thus significantly reducing the size of the state sector. By 1996, SOEs accounted for only 47 percent of industrial output (ADB 1997a). Government plans are for phased privatization of the remaining large SOEs that are particularly concentrated in the energy, telecommunications, transport, and mining sectors. In Kazakhstan, although a few large enterprises have been privatized since 1996, the remaining large-scale SOEs have neither been corporatized nor privatized and are a significant source of inefficiency and fiscal instability. Direct subsidies to loss-making Kazakh SOEs in 1996 accounted for 2–3 percent of GDP (ADB 1997b). Uzbekistan has basically completed the privatization of small state enterprises. The process of corporatization of medium- and large-sized enterprises has been much slower, however. By 1996, only 19 percent of medium-sized and 17 percent of large-sized companies had been partially privatized or incorporated (ADB 1997c).

Fundamental challenges remain for reform of the state enterprise sector in the CTEs. With little reform of remaining SOEs, corporate governance structure is weak due to the lack of an appropriate legal and regulatory framework, financial discipline, or incentive structures. The lack of financial discipline and the poor incentive structure have perpetuated the inefficient managerial and operational practices of the centrally planned system. In order to strengthen market-based incentives, continued progress toward privatization should be accompanied by adoption of effective corporate structures, removal of government subsidies and enforcement of hard budget constraints, introduction of transparent enterprise accounts, and development of bankruptcy implementation procedures (ADB 1997d). Finally, it is essential also to open the economy to global markets in order to improve SOE access to technology, increase competitive pressures, and upgrade labor, management, and capital.

TRADE, MACROECONOMIC, AND STABILIZATION POLICIES

The combination of trade liberalization, real currency devaluation to market-based exchange rates, and macroeconomic restraint and stability have been essential to successful economic transition and economic and agricultural growth. Transition economies that have aggressively pursued trade liberalization have experienced dramatic growth—in exports and imports—that has been an important contributor to output growth. Trade expansion has depended primarily on the output response to the liberalization of trade regimes that gave domestic producers the incentive to compete in global markets and to gain access to new technology (De Menil 1997). Macroeconomic stability is particularly difficult to achieve in transition economies, as it involves sharp cuts in subsidies and other public spending on the one hand and the adoption of both creative and judicious ways to increase state revenues to finance budget deficits on the other. Successful fiscal and exchange-rate stabilization is important, however, not only to stimulate trade, but also to control inflation and to maintain incentives for investment. High inflation distorts relative price incentives and creates uncertainties that inhibit savings, investment, and growth.

The PRC and Viet Nam have pursued somewhat contrasting but relatively successful reforms of trade, exchange-rate, and stabilization policies. The PRC followed a gradual, but nevertheless far-reaching, approach to trade and exchange-rate policy reform. Beginning in 1978, quantitative trade controls by planning authorities were gradually relaxed, the central government's monopoly of foreign trade was eliminated, foreign-exchange controls were loosened, and local authorities granted broad exemptions from central authority and control. By 1988, only 45 percent of exports and 40 percent of imports were under central control. The number of foreign national trading companies expanded from about a dozen in 1979 to more than 5,000 in 1990. Beginning in 1979, export companies were allowed to retain a share of foreign-exchange earnings and retention rights were expanded from 15–40 percent in 1980 to 70 percent in 1988. Moreover, in 1985, foreign-exchange swap centers increased the flexibility in foreign-exchange trading.

Perhaps the most important boost to trade, however, was the granting of broad local exemptions from central regulation by local governments, primarily through the SEZs and Open Economic Zones. The extent and influence of these exemptions is shown by the difference between statutory import duties and effective tariff rates. In 1992, the average rate of import duty was 42.8 percent, while tariff revenue as a percentage of actual imports was only 5.6 percent (De Menil 1997).

The PRC has also been generally successful with a gradual approach to stabilization policies, but periodic inflationary periods indicate continued structural imbalances. The PRC has enforced hard budget constraints on the agriculture sector and on the rapidly growing private sector in industry and has maintained positive real interest rates for these sectors of the economy. However, as noted with respect to state enterprises, soft budget constraints and subsidized credit to the SOEs have encouraged excess investment; SOE deficits caused severe inflationary episodes in 1986–87 and 1994–95. Implicit subsidies to SOEs through negative real interest rates and noncollection of bad debts have been estimated at 3 to 4 percent of GDP; bad enterprise debts may account for 20 percent of bank portfolios. The government has contained these inflationary periods through direct administrative controls, including ceilings on bank lending, direct prohibitions on investment, and price regulations. But administrative controls will become less effective as economic reforms continue. Continued successful stabilization will require further management reform and hardening of budget constraints on the SOEs and the banking sector (World Bank 1996a).

The reforms in Viet Nam were described above, but it is worth reiterating here that a large share of the steps to liberalize and stabilize the economy were undertaken in a very short span of time, combining aspects of an Eastern European "big bang" reform with the gradualist PRC approach. Faced with continued poor economic performance and hyperinflation, the government undertook radical reforms to liberalize and stabilize the economy in the space of six months, at the beginning of 1989. Price controls, state procurement, and rationing were eliminated for most goods. The official exchange rate was sharply decreased and parallel exchange markets were unified. Devaluation was accompanied by sharp increases in real interest rates, which became positive, and fiscal deficits and the growth of domestic credit were cut dramatically (De Menil 1997). State-enterprise reform was also initiated at this time, but this reform has continued on a more gradual path. As in the PRC, accelerated reform in this sector would help reduce inflationary pressures.

The process of trade liberalization and fiscal stabilization in the CTEs was initiated during a period of massive economic decline, collapse of trade, and hyperinflation. Despite these severe conditions, the Kyrgyz Republic has followed a policy of sustained stabilization and liberalization since 1994. Inflation was reduced from over 1,000 percent in 1992 to about 20 percent in 1997. This was accomplished by dramatic cuts in budget deficits (from 14 percent of GDP in 1995 to 7 percent of GDP in 1996), reductions in money-supply growth, and liberalization of deposit and interest rates. The government has established a foreign-exchange market with a floating exchange rate and a fully liberalized exchange regime with no restrictions on capital or current accounts. Nontariff trade barriers have been removed and all export taxes have been eliminated (World Bank 1998a).

Kazakhstan has also brought inflation down from nearly 2,000 percent per year to below 40 percent in 1996 and 11 percent in 1997, by introducing tighter credit and monetary policies and by reducing the budget deficit to 2–3 percent in 1996 and 1997. Exchange-rate stabilization and structural reform of the trade regime have stimulated growth in exports (World Bank 1998b; ADB 1997b).

In Uzbekistan, trade liberalization (like agricultural and state enterprise reform) has been relatively slow and fitful. After the initial go-slow approach to market-oriented reforms following release from the Soviet system, fiscal and monetary stabilization and the acceleration of structural reform beginning in 1994 coincided with an improved economic situation. But in response to the twin economic shocks in 1996—poorer than expected cotton and wheat harvests and sharp declines in world prices of cotton and gold, Uzbekistan's primary exports—the government relaxed monetary and fiscal policy and tightened trade and foreign-exchange restrictions (World Bank 1998d; ADB 1997c). Although some fiscal tightening was resumed in 1997 in response to inflationary pressures caused by the relaxation of fiscal policy, the government remains cautious about the pace of trade liberalization and stabilization.

As can be seen from this brief summary, the pace of trade liberalization and economic stabilization has varied widely across the CTEs. A number of cross-country econometric analyses have been undertaken to seek to explain the performance of the CTEs, other former Soviet countries, and Eastern Europe since 1990. All of these countries experienced large output contraction after the break up of the CMEA and the collapse of the Soviet Union. The analyses indicate that the speed and comprehensiveness of liberalization and stabilization reforms in the CTEs are positively related to an earlier resumption of growth and to the rate of growth achieved when growth resumed (Fischer et al. 1995; Taube and Zettelmeyer 1998; Zettelmeyer 1998). These studies also address the "Uzbek growth puzzle"—the relatively strong performance of the Uzbekistan economy despite slow economic liberalization and stabilization reform. The results show that Uzbekistan's favorable performance did not occur because of the cautious approach to reform, but rather in spite of it. The variables which have driven Uzbekistan's relatively good performance— cotton exports, energy self-sufficiency, and low initial industrialization—have more than offset the detrimental effects of trade, macroeconomic and structural policies (Zettelmeyer 1998; Taube and Zettelmeyer 1998).

PROVISION OF SOCIAL PROTECTION DURING TRANSITION

An important challenge in economic transition is to protect the poor during the reform process. Even a successful transition from a centrally planned economy to a market-oriented economy will increase income inequality and leave some people behind in the growth process. Greater disparity of wages, income, and wealth is to a certain extent inevitable during the transition, because allowing wages to be determined by the market, which is necessary to create incentives for productivity, will also increase inequality. Increased inequality can increase poverty, but if economic growth is sufficient and social safety nets are adequately developed, poverty can be reduced despite increasing inequality. The rapid economic growth in the ETEs due to economic reform has in fact dramatically reduced poverty, but significant pockets of poverty nevertheless remain (see Chapter II). The problem in the CTEs has been far more serious because transition was begun in the midst of a severe contraction of economic output and the withdrawal of massive funding from the Soviet Union to support social services.

At the outset of the transition process, the ETEs were characterized by relatively low (or regionally varying) social indicators, widespread poverty, and lower levels of income and state social-service provision compared to the CTEs (Graham 1997). However, lack of labor mobility and the cooperative work structure together ensured some level of community solidarity. The reform process in the ETEs coincided with real economic growth coupled with labor absorption by the rapidly growing economy, resulting in an absolutely reduced incidence of poverty. Labor absorption was facilitated by the relatively smaller size of the state-owned industrial sector, which did not require large layoffs, at least at the onset of the reform process. Thus, increases in inequality have been given less attention. On the other hand, in Mongolia, an economy that formerly received large transfers from the Soviet Union, inequality and poverty have become more pronounced (Green and Vokes 1997; Graham 1997).

The problems of increasing inequality and large pockets of rural poverty have yet to be addressed in a comprehensive manner. In the case of Viet Nam, for example, Van de Walle (1998) argues that the development of a reliable, effective system of redistributive transfers and safety nets will be crucial if Viet Nam is to make a successful transition to a market economy. The author concludes that Viet Nam's poverty-reduction program and safety net could improve through

- strengthened institutional structures and policies, including national norms for a consistent identification of the poor across regions;
- survey and other instruments for measuring and monitoring local needs and program performance;
- integration and coordination among subprograms;
- welfare-maximizing redistribution of resources across space; and
- increased resources and attention to help households and communities deal with covariant risk.

Under the Soviet system, the CTEs were characterized by wide-ranging social-protection systems. Programs included subsidized housing, electricity, and water; relatively high nonwage benefits; and other social services, such as health and education provided at a minimal fee or for free. As a result, social indicators such as life expectancy and literacy were relatively high compared to the per capita incomes in these countries of between US$2,500 and US$4,000. Following the disintegration of the Soviet Union, however, these services were not fiscally sustainable: living standards eroded severely in some parts of the CTEs and both poverty and inequality increased substantially. Thus, unlike the ETEs, the CTEs experienced sharp drops in social indicators in the first years of the transition process (ADB 1997d).

Grootaert and Braithwaite (1998) find that social protection systems in the former Soviet Union and Eastern Europe have been inadequate to meet the challenges of transition, being both costly and poorly targeted. The authors identify the working poor, especially workers with little education or outdated skills, as the largest group affected by transitional poverty. The newly unemployed, who live from social transfers or other nonearned income, are most affected. Poverty rates in the transition period are higher in the former Soviet Union than in Eastern Europe, due to the existence of some form of social safety net in the latter. In the short to medium term, creating employment in the informal sector may generate a larger payoff than creating jobs in the formal (still to be privatized) sector, so programs to help (prospective) entrepreneurs should take center stage in poverty-alleviation programs.

Graham (1997) identified female-headed households and the urban unemployed in Mongolia, single and elderly pensioners and the very poor in Viet Nam, and the aging rural population in the PRC as groups that need to be targeted by safety nets. The creation of viable institutions to provide basic social services and insurance to a large part of the population will also be crucial in confronting poverty. In the CTEs, the universal social support system of the Soviet era is not viable, so targeted assistance programs have to receive much larger attention and will be crucial to the successful implementation of the reform process.

Whereas gradual reform might lead eventually to comprehensive social reforms, radical approaches can support a better redirection of resources to the poorest, because they would allow decreased opportunities for opposition from more vocal groups, for example. Targeted assistance programs as well as general social reforms must be implemented as both an integral and complementary part of the macroeconomic reform program, with comprehensive stakeholder participation in the process, so that successive governments have a stake in their continued implementation and beneficiaries perceive the transition process as socially equitable and thus support it.

ENVIRONMENTAL PROTECTION DURING TRANSITION

The centrally planned countries of East and Central Asia put little emphasis on the environmental impacts of economic growth. The Aral Sea disaster stands out as an example of the results of the pursuit of output growth without taking into account the medium- and long-term environmental consequences. Thus, at the onset of the reform process, all the transition countries were burdened with a legacy of environmentally damaging mismanagement reaching over all economic sectors; these problems were especially severe in the CTEs. The transition period offers a significant opportunity to improve environmental quality in the CTEs. At the beginning of the reform period, the collapse of the Soviet Union and the resulting industrial contraction effectively cut industrial pollution rates in the CTEs. There are at least early indications that the recovery in industrial output will not be accompanied by equivalent increases in pollution because of more effective environmental regulation and enforcement (World Bank 1996a).

In contrast, rapid economic growth in the ETEs contributed to increased pollution rates and other environmental degradation through both an inefficient and highly polluting state-owned sector and the newly developed, strongly growing private sector (Esty 1997). Continued growth in agricultural production in the ETEs raises environmental questions related both to the intensification of existing cropland and extensification that requires continued land clearing in increasingly sensitive areas. Forest degradation is considered one of the largest environmental problems as the result of arable land expansion, fuelwood consumption, commercial logging, shifting cultivation and fire damage. Other environmental damage includes water and air pollution and the elimination of wetland ecosystems (Marcoux 1996). These problems are not confined to the transition economies and will be addressed in more detail in Chapter X, which discusses future resource and environmental challenges for agricultural growth in Asia.

However, transition periods do provide an excellent opportunity for market reforms and for the establishment of sound rules of environmental protection for the transforming economies. First, changes in relative prices should promote more efficient use of energy and natural resources. Privatization and reduced state interference will encourage management to improve the operating performance of existing plants, while replacing old equipment with cleaner production technologies. Well-designed environmental standards, regulations, and enforcement procedures can contribute to this process. A clear separation of enterprise ownership and management from environmental regulation should encourage realistic environmental standards (World Bank 1996a). Foreign direct investment and international cooperation—such as the Aral Sea water and environmental management programs—should assist in improved environmental quality.

LESSONS AND FUTURE DIRECTIONS FOR REFORM

The debate over the relative merits of rapid ("big bang") reform and gradual reform is almost beside the point—both approaches have been successful when appropriate policy reforms have been successfully implemented. However, the forces triggering the transition—internal or exogenous—and the initial conditions at the onset of the crisis are important determinants of the early output response upon initiation of the transition process, regardless of the speed of reform. In the medium and longer term, policies that are conducive to investment and openness are the primary determinant of economic and agricultural growth.

The emerging evidence from economic reform in Central Asia shows that more rapid and comprehensive reform can be conducive to a quicker economic turnaround and faster growth (albeit perhaps at a higher social cost). The PRC, on the other hand, has had great success with a gradual approach; success was arguably best in the agricultural sector, however, where reform was fastest, and much less so in the state-enterprise sector, where reform has been slowest. Viet Nam undertook highly successful and relatively rapid economic liberalization and stabilization reforms in 1989, but has encountered difficulties in the slow-reforming state-enterprise sector. Gradual reform typically allows for a more comprehensive reform process, but a lack of urgency for thorough renovation can bring about inappropriate and incomplete reforms or lead to a stop-and-go process, such as the PRC experienced during its reform process. Nevertheless, political and social stability during the transition process may be as important to success as economic stability and may dictate a more gradual approach to reform than is optimal from the point of view of efficiency.

The agricultural sector has played a pivotal role in determining the early pace of reform and constitutes an important backbone for the acceleration of economic growth in transition economies. The rigid, transfer-dependent, centralized and collectivized agriculture structure in the CTEs fell apart with the collapse of the Soviet Union and paralyzed growth prospects in the initial reform years, despite the adoption of market reforms. The central government's grip on agriculture was much less pronounced in the ETEs; in particular the higher prevalence of smallholder agriculture contributed to a smoother transition in the agriculture sector and thus in the overall economy. The existence of surplus labor that could be released to the industrial sector was also favorable for economic growth during the transition process. Moreover, rapid growth in the agriculture sector helped cushion the adverse impacts of initial dislocations arising from reform in the nonagricultural sectors, providing for a degree of income stability at the onset of the reform process.

Finally, the agriculture sector provides positive stimuli to other economic sectors through

- its linkages with related industries, like food-processing;
- the provision of social (and sometimes environmental) stability during the transition process, as a large

segment of the population tends to be employed in the agriculture sector at the onset of the reform process;
- the provision of food security and savings in foreign exchange on food imports; and
- the creation of foreign exchange through the export of cash crops.

Because of the crucial role of agriculture in the transition process, increased attention must be paid to the acceleration of market-oriented reforms in the agriculture sector of the CTEs. Required reform measures include a comprehensive land reform and restructuring of both state and collective farms and removal of state intervention in and liberalization of input and output markets, as well as the rebuilding of banking and financial services for agriculture.

A "dual system" with state enterprises coexisting with the private sector (as in the PRC) can provide initial stability and a degree of social protection while market institutions develop. State enterprises that provide broad welfare benefits and social services to the vulnerable and needy groups of society or that provide employment to a large number of people might have to be protected at the outset of the transition process as a safety-net component. In addition, during the early stages of reform, dismantling of state enterprises is politically difficult and potentially harmful in the absence of market-oriented institutions and human capital. It may be more appropriate at such an early stage to focus on the corporatization of state enterprises, while simultaneously promoting the growth of the nonstate private sector and the integration of the economy into the global economy. The dual-track transition in the PRC is one example where private enterprises are gradually overtaking state-subsidized companies in scale and scope. However, state enterprises that exist under "decentralized socialism," that is, with marketed products but centralized property rights, can lead to increased opportunities for corruption, high inefficiencies, and an incentive structure that together maximize monetary transfers from the principal to the agents (Parker, Tritt, and Woo 1997). Therefore, in the medium to longer term, the continued existence of state enterprises with soft budget constraints, special foreign-exchange rights, and other preferences severely undermines the development of competitive markets. It is therefore essential in both ETEs and CTEs to quicken the pace of reform of state enterprises, moving beyond the improvement of corporate governance to true privatization of the state sector.

The existence or early creation of an economic, legal, and social enabling environment that is responsive to policy reforms and conducive to openness and investments is essential for all types of reform measures undertaken during the transition process. Macroeconomic stabilization and trade liberalization are important components of the policy reform process. Openness to global markets and international trade allows an economy to develop according to its comparative advantage, to catch up technologically, and to adapt its labor force and capital stock continually to changing factor endowments (Thomas and Wang 1997). Trade liberalization directly boosts export (and import) growth, instills competitive, market-oriented behavior in the transition economies, and increases the living standards of the people. Far-reaching trade liberalization has been a key component of the success of the East Asian transition economies; more rapid and comprehensive reform of trade policies would significantly benefit the Central Asian transition economies.

IX THE FINANCIAL AND ECONOMIC CRISIS IN EAST AND SOUTHEAST ASIA

INTRODUCTION

After two decades of rapid economic growth, many of the countries of East and Southeast Asia entered a period of serious economic and financial crisis in 1997. Between mid-1997 and the spring of 1998, the currencies of four Southeast Asian nations (Indonesia, Malaysia, the Philippines, and Thailand) and of the Republic of Korea fell 40–80 percent against the US dollar, precipitating a financial and economic crisis whose full impacts on the countries and the global economy are still unfolding. As a consequence of the crisis, the gross domestic product (GDP) for East and Southeast Asian countries fell sharply in 1998. Indonesia's GDP declined by 14 percent in 1998; Republic of Korea's by 6 percent, Thailand's by 9 percent, and Malaysia's by 7 percent, while the Philippines had barely positive growth (Severino 1999).

While the initial impact of the crisis was felt mainly in urban areas, where banks closed their doors, factories shut down, and hundreds of thousands swiftly became unemployed, the crisis has also had significant consequences for the agricultural sector and for rural residents in the affected countries. The events raise important questions about macroeconomic policies and the fundamental soundness of governance in Asian economies, the role of international capital and domestic financial markets in Asian development, and the incentives for and role of the agricultural sector in modernizing Asian economies. The policy and governance lessons that are drawn from the experience of the economic crisis will have profound implications for the future development of rural Asia.

CAUSES OF THE CRISIS: TWO VIEWS

The causes of the financial and economic crisis are complex and will be only briefly summarized here. The two main schools of explanation for the East Asian economic crisis are the "fundamentals hypothesis" and the "financial panic/contagion hypothesis". The fundamentals school argues that the crisis was primarily due to poor financial-market performance and weak financial and corporate regulatory oversight, combined with a deterioration of macroeconomic fundamentals in the region beginning in the early 1990s. The financial panic/contagion hypothesis, while noting that there was deterioration in economic fundamentals in the region, places more emphasis on the role of panic by international investors and inappropriate initial policy responses by countries and by international rescue programs that deepened the crisis beyond what the macroeconomic fundamentals warranted. This school of thought emphasizes the importance of finding the right regulatory framework to reestablish investment confidence. In fact, these alternative theories are not incompatible: the evidence indicates that both a significant deterioration in the macroeconomic fundamentals in the affected countries and investor panic contributed to the onset and depth of the crisis.

The lead-up to the crisis in the most severely affected countries—Indonesia, the Republic of Korea, Malaysia, Philippines, and Thailand—was characterized by an interlinked set of developments beginning in the 1990s. These developments included

- significant real appreciation of currencies and exchange-rate misalignment that diminished these countries' international competitiveness;

- large and growing current-account deficits financed with the accumulation of foreign debt in the form of short-term foreign-currency-denominated and unhedged liabilities;
- excessive short-term lending by international investors in the 1990s, followed by a sharp reversal of the short-term capital flows in 1997, which set off a round of competitive devaluations; and
- overborrowing and overlending in the financial sector, much of it to risky and low-profitability projects.

In addition to the large foreign debt, the East and Southeast Asian economies had a traditionally deep structure of domestic debt due to high internal savings rates, making these economies more vulnerable to foreign shocks (Wade 1998). The massive withdrawal of short-term international capital caused a widespread financial crisis and real or de facto bankruptcies for banks and firms that could not repay the large amounts of foreign-currency-denominated debt. The exchange-rate crisis made things worse, as the currency depreciation increased the real burden of the foreign-currency-denominated debt. Domestic interest rates increased sharply, further contracting the availability of credit to the economy. The withdrawal of international capital, together with the virtual collapse of domestic bank capital, was an enormous contractionary shock to the real economies of the crisis countries.

Development of the Crisis

How did the scenario play itself out? Neither the causes nor the outcomes of the financial crisis are uniform across the affected countries. Following are some of the common elements and some of the differences across the countries. More detailed treatments are available in Corsetti, Pesenti, and Roubini (1998a, 1998b) and Radelet and Sachs (1998). McLeod and Garnaut (1998) provide both a detailed comparative analysis and several country case studies.

Real Appreciation of Currencies

In retrospect, perhaps the leading indicator of looming troubles was the steep appreciation of the real exchange rate (the ratio between prices of tradable and non-tradable goods), which diminished the international competitiveness of the crisis economies. According to Radelet and Sachs (1998), appreciation of the real exchange rate was over 25 percent between 1990 and 1997 for the four Southeast Asian economies and 12 percent for the Republic of Korea (compared to 30 percent between 1987 and 1997). The primary reason behind the real appreciation was the massive inflow of foreign (mainly short-term) capital during this period. External forces also contributed to the real appreciation: the crisis economies linked their nominal exchange rates, in varying degrees, to the US dollar, which appreciated after 1995 relative to other major currencies. The real appreciation undermined the competitiveness of tradable-goods industries, including the rapidly growing export industries.

Other economic developments further weakened competitiveness. In Thailand, for example, there was extraordinarily rapid growth in real wages, due to the drying up of the pool of relatively low-productivity unskilled labor in agriculture; skilled labor for industry also became a bottleneck, because the educational system did not produce enough persons with needed skills (Wade 1998). Between 1990 and 1994, the real wage in Thailand increased by over 9 percent per year (Warr 1998).

The decline in international competitiveness in the crisis economies was shown by the dramatic fall in growth rates of exports across these economies in 1996. Export growth rates in nominal terms fell from an average of 24.8 percent in the five countries in 1995 to 7.2 percent in 1996 (Radelet and Sachs 1998). Indonesia's export position was least affected, experiencing a drop from 13.4 percent in 1995 to 10.4 percent in 1996. Export growth in the Republic of Korea fell from 30.3 percent to 5.3 percent; in Malaysia from 26.0 percent to 6.7 percent; and in the Philippines from 31.6 percent to 17.5 percent. Thailand experienced the most dramatic decline, with a drop in export growth from 25.1 percent in 1995 to −1.3 percent in 1996 (Garnaut 1998a). This collapse of export growth and the resulting loss of investor confidence in the stability of the Thai baht was the proximate trigger for the speculative outflow of short-term foreign capital from Thailand that precipitated the financial and economic crisis in the region.

Large Current-Account Deficits

Concurrently with the real exchange-rate appreciation, the current-account balances in the region deteriorated. During the 1985-89 period, the current accounts of the five crisis economies were relatively strong, averaging just 0.3 percent of GDP. The Republic of Korea and Malaysia had current-account surpluses of 4.3 percent and 2.4 percent of GDP, respectively, and the largest deficit was Indonesia's 2.5 percent (Radelet and Sachs 1998). Beginning in 1990, however, current-account balances weakened, with a further worsening during 1995 and 1996. In Thailand, the current-account deficit reached 8 percent of GDP in 1996; in Malaysia, 7.6 percent during 1995–96; in the Philippines, 3.7 percent; and in Indonesia, 3.5 percent. In the Republic of Korea, the current account deficit jumped from 1.8 percent in 1995 to 4.8 percent in 1996 (Garnaut 1998a).

Large Short-term Capital Inflows

Massive private foreign capital inflows into the East Asian economies, motivated by the strong economic growth prospects and resulting high investor returns in the region, also were the driving force behind the real appreciation of the currencies and growing current-account deficits. Large capital inflows were also driven by

- the liberalization of capital markets in the early to mid-1990s;

- the deregulation of domestic financial systems that facilitated foreign investment;
- the relatively poor financial-sector supervision that allowed domestic banks to take on substantial foreign-currency and -maturity risks;
- stable nominal exchange rates; and
- special incentives provided to foreign investment.

A strong element of "moral hazard" appears to have contributed to the huge expansion in foreign capital inflow. Moral hazard is the possibility that the expectation of future government support will induce an undesirable change in behavior, in this case inducing foreign investors to be more willing to make risky investments because the potential losses were seen as being effectively underwritten by the governments, or by international agencies such as the IMF.

In addition to the large private inflows, Stiglitz (1998) argues that the large amounts of international public lending to the crisis countries were undertaken without due consideration of the standards for risk assessment. Similar to the case for private investment, underlying this lending strategy may have been the assumption that it would be guaranteed by either a direct government bailout of the borrowing institutions or an indirect bailout through the IMF or other international efforts (Corsetti, Pesenti, and Roubini 1998b).

Capital inflows to the five Asian crisis economies totaled $211 billion between 1994 and 1996—$93 billion in 1996 alone—some three fourths of them short-term funds (Radelet and Sachs 1998). By 1996, the ratio of debt service plus short-term foreign debt to foreign reserves was 294 percent in Indonesia, 69 percent in Malaysia, 137 percent in the Philippines, and 123 percent in Thailand (Corsetti, Pesenti, and Roubini 1998b). The rapid growth of foreign short-term capital relative to foreign reserves left the currencies of the crisis economies highly vulnerable to speculative attack through a sudden reversal of capital inflows. In 1997, this reversal of short-term foreign capital inflows occurred, with net private inflows to the five crisis economies falling to –$12.1 billion, representing a swing of $105 billion, or

11 percent of the combined pre-crisis GDP. Fully $77 billion of the decline in inflows came from commercial bank lending (Radelet and Sachs 1998).

Overborrowing and Overlending in the Financial Sector

The high proportion of short-term capital in total inflows was an inherently risky basis for growth; this risk was multiplied by the investment of these funds in speculative and unproductive projects. With the exception of Indonesia, where foreign borrowing was mainly undertaken directly by private firms, foreign capital inflows were to a large degree channeled through the banking system. The average annual growth rate in bank lending to the private sector during 1990–96 was 17 percent in the Republic of Korea, 20 percent in Indonesia, 19 percent in Malaysia, 32 percent in the Philippines, and 22 percent in Thailand. Over this time period, bank lending as a percentage of GDP jumped from 71 percent to 93 percent in Malaysia, from 64 percent to 102 percent in Thailand, and from 19 percent to 49 percent in the Philippines (Corsetti, Pesenti, and Roubini 1998b).

The growth in bank lending was accompanied by a rapid decline in the quality of loans and investments, with a heavy emphasis on speculative real-estate investments (by themselves an indicator of the impending crisis) and other nonperforming investment projects. With the exception of Indonesia and the Philippines, the incremental capital output ratio declined sharply between 1987–92 and 1993–96. In the Republic of Korea, even prior to the onset of the crisis, seven of the 30 largest *chaebols* (conglomerates) were effectively bankrupt, with an average debt-equity ratio of 333 percent for the 30 *chaebols,* compared to about 100 percent for firms in the United States.

The pre-crisis (end of 1996) share of nonperforming loans as a proportion of total lending was 14 percent for the Philippines, 13 percent for Indonesia and Thailand, 10 percent for Malaysia, and 8 percent for Korea. With capital-asset ratios as low as 6–8 percent in these countries (somewhat higher in the Philippines), nonperforming loans were in many cases already above capital reserves prior to the crisis. The excessive reliance on investment in real estate is shown by the property exposure (property loans as a percentage of total loans) of 25–30 percent in Indonesia and 30–40 percent in Malaysia and Thailand, with a somewhat lower exposure in the Republic of Korea and the Philippines (Corsetti, Pesenti, and Roubini 1998b). Overinvestment in real estate fueled a classic bubble economy in Malaysia and Thailand, and to a lesser extent in the other crisis economies. Real-estate prices continued to rise well beyond levels justified by the productivity of the asset, but with continued price rises, existing investors were rewarded and collateral was created for new loans to finance further investment, until the inevitable crash. This boom was fueled by unrealistic expectations that in turn created the conditions for a crash (Warr 1998).

Why was the performance of these investments so poor? Much of the problem can be attributed to the structure of incentives governing the operation of the corporate and financial sectors in the crisis economies, in particular the pervasive moral hazard problem. On the corporate side, political pressures to maintain high rates of economic growth were sustained through public support for private projects, including direct subsidies and directed credit to favored firms and industries. Underlying profitability and riskiness were often ignored.

Industrial and financial policy was often conducted within networks of personal and political favoritism. With governments seemingly willing to intervene to help troubled firms, markets often appeared to operate under implicit guarantees against adverse shocks. Structural distortions were also pervasive in the financial sector, including weak prudential supervision; insufficient expertise in both banks and regulatory institutions; incentive-distorting deposit insurance schemes; inadequate capital adequacy ratios; nonmarket criteria for credit allocation emphasizing semimonopolistic relations between banks and firms; and corrupt lending practices (Corsetti, Pesenti, and Roubini 1998). These structural distortions were exacerbated by a lack of competition in the financial sector. In all of the crisis countries, weak and inward-looking domestic banks had been protected by law from more competitive, major international banks. In addition, the lack of global diversification of assets left the financial systems of the Asian crisis economies even more susceptible to economy-wide shocks (Fane 1998).

The Onset of the Crisis

The combination of factors described above made the five economies vulnerable to the withdrawal of foreign, short-term capital, while at the same time creating the conditions that induced a radical change in foreign investor expectations with respect to the performance of these economies. The increasing evidence of failures in the financial and corporate sectors, poor export performance, and rapid appreciation in real exchange rates in Thailand and Korea caused increasing concern among foreign and domestic investors. After years of rapid growth, the Thai stock market entered a rapid decline after January 1996. The Korean market followed in the second quarter of 1996. Thus the Thai currency crisis actually started in 1996, with rumors of devaluation of the baht beginning as early as May 1996. A year later, following continued deterioration in Thai macroeconomic fundamentals, speculative outflow overwhelmed monetary authorities and the baht fell sharply. Financial instability then spread through much of East and Southeast Asia during the third quarter of 1997 in what could be called a contagion of radically changed expectations. The collapse of the baht indicated that presumed government guarantees on currency parities, bank solvency, and corporate profitability were no longer secure. This event also forced many international investors to realize that they had a poor understanding of the functioning of the East and Southeast Asian financial markets, with a resulting large increase in the required risk premiums on investment throughout Asia. The increasingly diminished opportunities for trade and investment within the region constituted an additional source of contagion.

The severity of the crisis has varied across the affected countries. Indonesia was the most severely affected, followed by Thailand, Republic of Korea, Malaysia and the Philippines. Many of the differences could be seen above in the comparison of economic indicators leading into the crisis. At least two broad differences, however, are worth highlighting here. First, the quality of financial institutions and markets was a crucial determinant of the severity of the crisis. Financial institutions were weakest in Indonesia, quite weak in Thailand and Korea, and less so in Malaysia and the Philippines (Garnaut 1998a).

The Philippines had strengthened its financial sector significantly as a result of reforms carried out after an economic crisis in the early 1980s. It also had lower international exposure due to the more recent opening of its economy. The Philippine financial crisis in the early 1980s had been caused by many of the same factors that brought about the current economic crisis:

- supervision and monitoring of banks were poor;
- capital requirements and limitations on lending to related interests were weak;
- politically inspired lending was prevalent; and
- nonbank financial institutions were permitted "without-recourse" lending with essentially no oversight.

Financial reforms begun after the crisis and expanded in the late 1980s and early 1990s led to prudential regulations with respect to minimum capitalization, increasing compliance with capital asset ratios, limits on loans to related interests, and tightening of audit and reporting requirements. Restrictions on the entry of foreign banks were partially eased in 1994. Regulatory improvement and the easing of entry for foreign banks were important factors in the Philippine response to the East Asian crisis. The capital adequacy ratio for Philippine banks ranged from 16.9 percent to 20.2 percent between 1992 and June 1997, allowing the banking sector to weather the crisis comparatively well (Intal et al. 1998).

A second crucial difference can also be seen in the political and policy responses to the crisis. In Indonesia, for example, policy responses were mostly inconsistent and contradictory. McLeod (1998) cites a number of factors contributing to the mismanagement there:

- genuine confusion about managing a crisis not previously experienced;
- divisions within the government due to President Suharto's reliance on economic policymakers some of whom had become increasingly concerned about policy distortions benefiting his family and close friends, followed by concerns over the post-Suharto political transition;
- ethnic tensions focusing on the role of ethnic Chinese in the economy;
- economic nationalism among some ministers and their departments; and
- general lack of sympathy for the private sector and market processes within the bureaucracy.

The closed political system in Indonesia made it difficult to resolve these internal conflicts, which were therefore reflected in erratic policy responses to the crisis.

In Thailand, the government response to the crisis was better coordinated and managed, mainly due to the government's democratic structure (Warr 1998). The Chavalit government fell in a no-confidence vote in November 1997 as a result of the crisis. The new government of Prime Minister Chuan Leekpai did not have to defend its predecessor's performance during the period leading up to the crisis and instead could devote full attention to instituting a reform package.

In the Republic of Korea, a fortuitously timed presidential election ushered in a new government with a mandate for reform. The legitimizing force of a democratic transition was also apparent in the Philippines, where an orderly transition was accomplished while maintaining relatively stable policy responses to the crisis.

IMPACT OF THE CRISIS

Social Impact of the Crisis

The financial and economic crisis has had severe negative social impacts in the affected countries, including declining incomes and increasing layoffs, rising absolute poverty, increases in malnutrition, increased pressure on already underserved rural areas, declining social services, and threats to education and health status, in particular of children and women. Although all the crisis countries (and at least some other developing countries in Asia and elsewhere) are experiencing negative social effects from the crisis—rapidly transmitted by the strong trade, capital-flow, and migration linkages among the countries—the relative impacts have differed by country. For example, Indonesia's welfare undoubtedly has been hit hardest, while the social sector in the Philippines has fared relatively well (World Bank 1998e). More encouraging evidence emerging in late 1998 and early 1999 has also indicated that the worst early predictions of devastating social impacts throughout the affected countries have not been borne out and that negative social impacts have been to some extent ameliorated by both civil-society responses and government policy.

Declining Incomes

As the financial crisis hit the real economy in the form of near-zero or negative GDP growth, labor-market conditions deteriorated sharply. In addition to substantial retrenchments in the construction, financial services, and manufacturing sectors, there has been an abrupt decline in new hiring, sharply reducing the employment prospects of new entrants into the labor market and the re-employment prospects of displaced workers. The International Labor Organization (ILO 1998) projected that the combined impact of these factors would lead to at least a doubling of open unemployment rates in Indonesia and Thailand. Ranis and Stewart (1998) estimate that there was a reduction of 10–15 percent in formal-sector employment in Indonesia and Thailand in 1998, and that 1.5 million jobs may have been lost in Korea. By mid-1998, the open unemployment rate in Indonesia was estimated at 17 percent; the rate in the Republic of Korea of 6.7 percent in April 1998 was three times the rate of October 1997 (ADB 1998e). By the end of 1998, however, estimated unemployment in Indonesia had fallen to 7 percent, with labor-market adjustments resulting in falling wages rather than unemployment (World Bank 1999; Severino 1999).

The real wages and earnings of those still employed have fallen due to the decline in labor demand and the increase in inflation caused by the substantial currency depreciation. In Korea, the fall in real wages recorded in the fourth quarter of 1997, immediately after the financial crisis, was 2.3 percent and real wages fell significantly thereafter. Available evidence indicates that real wages in 1998 may have declined by 20–30 percent in Thailand and by 50 percent in Indonesia (ILO 1998; Ranis and Stewart 1998).

The impact has also been severe on the important overseas labor market. The deployment of migrant workers from the Philippines dropped by 23.4 percent in the first quarter of 1998, compared to the first quarter of 1997 (Pernia and Knowles, 1998). There has also been a significant rise in underemployment, both in terms of the hours of work and of wages, due to the increased influx of displaced workers and unsuccessful new job seekers into the rural and urban informal sectors. The larger number of workers seeking a livelihood in these sectors has reduced average incomes.

With unemployment and falling wages hitting the urban middle class the hardest and relatively robust agricultural incomes cushioning some of the impacts on the rural areas (see below), most of the income losses have been incurred by the relatively well-off in urban areas. Thus, in Indonesia, average real per capita household expenditure (a somewhat crude proxy for income, but the only measure available) fell by 30 percent in Jakarta and 42 percent in West Java between August/September 1997 and August/September 1998. But median expenditures fell by only two percent in Jakarta and by six percent in West Java. This differential decline in mean vs. median expenditures indicates that relatively well-off households suffered by far the largest declines in household expenditures. For all urban households in a nationwide sample in Indonesia, mean per capita household expenditures in urban areas fell by 34 percent and median expenditures by 5 percent, while mean rural expenditures fell by 13 percent and median rural expenditures by 1.6 percent (Poppele, Sumarto, and Pritchett 1998).

Rising Absolute Poverty

The increase in unemployment, declining nominal and real wages, and rapid inflation have increased the number of people with incomes below the poverty level. In Thailand, poverty incidence increased from 11.4 percent in 1997 to 12.9 percent in 1998, meaning that an additional one million people were pushed below the poverty line (Severino 1999). UNICEF (1998) estimates that, in Indonesia, the rural population below the poverty line will have increased from 17 million before the crisis to 24–26 million by the end of 1999. The World Bank (1998a) projects that the poverty rate in Indonesia has increased from 11.3 percent in January 1996 to 14.1 percent in March 1999 (an increase of 5.7 million in the number of the poor), with urban poverty increasing from 5.0 percent to 8.3 percent and rural poverty from 15.0 percent to 17.6 percent. These figures indicate that, although most poverty in Indonesia is expected to remain rural, the share of urban poverty is expected to increase significantly. Recent household survey results appear to confirm the lower estimates of increases in poverty in Indonesia, showing increases in total poverty of 2.8 to 3.4 percent (Poppele, Sumarto, and Pritchett 1998).

Increases in Malnutrition

Food prices were hit especially hard by inflation, with disproportionate impacts on the poor. Food items account for a large share in the consumption basket of the poor: in Indonesia, 71 percent and in Thailand, 55 percent. Moreover, large numbers of people are clustered around the poverty line, especially in Indonesia. Consequently, price increases have led directly to declines in household consumption and increases in poverty levels (Finance and Development 1998).

In Indonesia, the Consumer Price Index (CPI) for food increased by more than 50 percent between June 1997 and March 1998, compared with a 38-percent increase in the general CPI (World Bank 1998d). The overall inflation rate in Indonesia for the whole of 1998 rose to as much as 100 percent (ADB 1998e). In Thailand, the overall price increase was estimated at 10 percent in 1998, food prices increased by 7 percent in the first quarter of 1998, and energy and transportation prices increased by 11 percent. The dramatic increase in domestic food prices reduced both the amount and types of food available to poor households. This change is likely to have adversely affected child development and pregnancy outcomes, as well as to have increased the incidence of diseases like diarrhea (ADB 1998e).

Increased Pressure on Rural Areas

The drastic fall in employment in the urban sectors of the crisis countries has put further pressure on the rural nonfarm and agricultural sectors. A January 1998 Thai government survey of workers returning to rural areas illustrates the pressure placed on rural areas by the crisis. Nearly 200,000 people had returned to the countryside, with the highest proportion returning to the northeast. This migration pattern put pressure on the weakest parts of the Thai agricultural sector; adding further pressure, the north and northeast were hit particularly hard by drought at the end of 1997. Furthermore, remittances from urban areas had played a major role in sustaining living standards in these regions (ILO 1998). Indonesia has also had a series of relatively poor harvests that have made it difficult for the agricultural sector to support the increased population resulting from reverse migration from urban areas.

Declining Social Services—Threats to Education and Health

There is cause for concern that the severe immediate negative welfare effects may be intensified and prolonged, but there were also some signs of recovery in late 1998 and early 1999, due both to government policy and civil-society responses. The negative social effects of the crisis have been compounded by a lack of social safety nets for the newly unemployed and the newly poor. Traditional social systems that have supported the poor in the past, including the ability to return to subsistence agriculture, close family ties, and community support, have been weakened by the development of a dynamic urban economy (Ranis and Stewart 1998). New social safety-net systems have not been put in place to compensate for the traditional systems.

None of the crisis countries except the Republic of Korea has an unemployment benefit scheme or other social-welfare programs. Most countries do not have national pension schemes to provide protection for old age, invalidity, and survivors. In Indonesia, pensions are mainly provided to civil servants and military; in Thailand, only 10 percent of the labor force is covered by the pension system (Finance and Development 1998). The compulsory savings or provident fund schemes that exist in some of the countries provide only very modest benefits in the form of lump sum payments upon retirement. All of the countries have some kind of health insurance scheme, but coverage is often limited to the working population and the benefits tend to be limited.

Household demand for health services depends on the price of food, medical care, and time, as well as on the productivity of health investments and their perceived benefits.

In addition to increases in the prices of food and medicine, the cost of time of poorer parents may have risen due to the need to hold multiple jobs or carry on time-consuming informal activities (ADB 1998c). As a result of the crisis, total government expenditures on social services have been reduced; instead the share of government expenditure going to the rescue of the financial sector and to foreign debt service has increased. In Thailand, for example, the 1998 budget of the Ministry of Health fell by 10 percent compared to the previous year and the budget of the Ministry of Education fell by 6 percent (Pernia and Knowles 1998). Privately provided social services, which often operate with high foreign indebtedness and rely upon imported equipment, drugs, and other supplies, have also deteriorated; so have local donations. In Thailand, many private hospitals closed, at least temporarily, after the onset of the crisis. In the Philippines, the budget for vaccinations had to be cut (ADB 1998c). Nevertheless, adaptive responses have helped to dampen the negative impact of the crisis on health. In Thailand, recent evidence indicates no adverse impact of the crisis on health outcomes, even though households spent less on health: continued use of public facilities and an increase in use of public assistance and voluntary health-care schemes helped compensate for reduced household expenditures (Severino 1999).

Of particular concern are reductions in investment in human capital formation, like schools and universities, because the effects of such reductions on the long-run growth potential of these countries could well be irreversible. The costs of schooling are likely to increase, especially at the secondary level, because of budgetary constraints for public schools. In addition, child labor is likely to increase due to increased household demand for current income (ADB 1998e). In Indonesia, for example, gross enrollment rates fell from 62 percent to 52 percent at the junior secondary level during the economic shock of 1986–87 and took almost a decade to recover (Atinc and Michael [1998] as cited in ADB [1998e]). But again, government and civil-society responses have helped sustain enrollments in the current crisis. In Thailand, the government allowed tuition fees to be paid in installments, permitted schools to waive tuition fees when necessary, and introduced vouchers and loans. Neither school drop-out rates nor child-labor rates have increased. In Indonesia, increased resources for schools and scholarships have been effective in reaching the poor. Primary school attendance has been maintained, and only 5 percent of secondary-school children have left to work (Severino 1999).

Evidence indicates that women are disproportionately affected by all social dimensions of the crisis. The traditional gender difficulties faced by women have been exacerbated by the crisis. In addition, women tend to be the first ones laid off when companies feel the pressure of labor costs, girls are pulled out of schools before boys, and women face increasing difficulties in providing social services for their families. On the other hand, women tend to find work in the informal sector more easily, they tend to work in the less affected export-oriented sectors, and they are sometimes substituted for more expensive men's labor (World Bank 1998d; ADB 1998e).

Impact on the Agricultural Sector

The financial and economic crisis has had a mixed impact on the agricultural sector and evidence is still limited on the net effects of the crisis. The large currency depreciations increased the competitiveness of the agricultural export sectors, inducing higher production and income generation in those sectors. In the 14 months ending in September 1998 the Republic of Korea's real exchange rate depreciated by 24 percent, Malaysia's by 29 percent, the Philippines' by 27 percent, Thailand's by 22 percent, and Indonesia's by 58 percent (IMF 1998a). But even in the export sectors, financial-market problems have severely hampered operations. Expansion of exports is dependent on the availability of trade credit, which has been limited in the aftermath of the crisis. Other agricultural sectors, especially those that rely on the domestic market to sell their outputs and/or on imported production inputs, may have been badly harmed by the crisis. Indonesia's livestock industry was hit hard, with production reported to have contracted by

30 percent in the first six months after the crisis began. This contraction was due to dependence on imported feeds, which doubled or tripled in price as a result of the devaluation; the sharp drop in consumer demand; and the bankruptcy and closure of many livestock enterprises (UNICEF 1998).

The economic and financial crisis has had a large impact on the world prices of some agricultural commodities of which the Asian countries are large producers, like rubber. Thailand, a country less affected by the El Niño-induced drought than Indonesia, for example, experienced substantial increases in exports of agricultural commodities after domestic demand fell following the significant rises in consumer prices between the first quarter of 1997 and the first quarter of 1998. But average increases in agricultural export volumes of nearly 40 percent—more than twice the volume growth of exports as a whole—were partially offset by substantial declines in the dollar prices for most of these commodities (McKibbin and Martin 1998, cited in World Bank 1998d).

The long-term impact on agriculture could also be large, but to date forward-looking analyses show mixed results. Noland et al. (1998b) analyze the effects of the Asian crisis based on a computable general equilibrium (CGE) model consisting of 17 regional models, each with 14 sectors (three of them agriculture), and five primary factors of production. The authors find no large, long-term changes in agriculture, as this sector is more regulated than manufacturing, for example. The greater part of the external adjustment comes from declines in imports rather than increases in exports.

Stoeckel et al. (1998) also use a CGE model to assess the long-term impacts of the crisis on agriculture. They find that Asian currency devaluations provide no clear boost to agricultural production in the crisis countries. Agricultural output is projected to decline by 10 to 15 percent in the Republic of Korea, Thailand, and Indonesia, due to higher capital costs and other investment constraints on agriculture, declining agricultural productivity, and falling domestic consumption. Agricultural exports are projected to increase, with growth varying widely by country due to country differences in imports of agricultural inputs and the varying relative price changes between agricultural commodities. The agriculture sectors in the Philippines and Malaysia have been (and will continue to be) least affected, whereas Indonesia has experienced the largest negative effects on agriculture of the Asian developing economies. Negative impacts on agricultural production, consumption, and trade are likely to last until investor confidence is restored and capital flows and interest rates recover.

Rosegrant and Ringler (1998) explore the potential long-term global impact of the crisis on food supply and demand based on a comparative analysis of three alternative scenarios using IFPRI's International Model for Policy Analysis of Agricultural Commodities and Trade (IMPACT). The baseline scenario reflects the conditions prevailing before the onset of the crisis. In the "severe Asian crisis scenario," long-term real currency devaluation and sharp drops in long-term income growth rates in the region are postulated, whereas in the "moderate Asian crisis scenario" it is assumed that the long-term income growth rates will recover to closer to pre-crisis levels.

Global cereal demand in 2020 is expected to decline by 74 million mt (3.0 percent) in the severe crisis scenario and by 19 million mt (0.8 percent) in the moderate scenario, compared with the pre-crisis baseline scenario results. In Asia, the contraction will be slightly larger, 4 percent in the severe crisis scenario and 1 percent in the moderate. Repercussions will be larger for the livestock sector, which is more price- and income-sensitive. The most serious impact of a severe crisis will be on the food security of Asian developing countries. In the severe crisis scenario, the number of malnourished children in the group of developing countries is projected to increase by 15 million, from 143 million to 158 million by 2020. In the moderate scenario, the increase would still be 3 million. In the severe crisis scenario, the number of malnourished children will increase by 11 million in South Asia, by almost 3 million in the PRC, and by 2 million in Southeast Asia.

The long-run effects of the crisis on agriculture remain uncertain, in particular whether the countries can convert the ongoing agricultural export boom to long-term growth in production and trade. The long-term outcome will be determined by the balance between the impacts of the price effects due to real exchange-rate depreciation—which should boost agricultural production and trade—and the investment effects due to the sharp drop in income and contraction of the financial sector—which will tend to depress agricultural production and export growth. CGE models may well underestimate the dynamic impact of the restored competitiveness of agriculture due to real exchange-rate depreciation and structural reforms and overestimate the negative effects of the slowdown in investment on the agricultural sector, particularly if GDP growth continues to recover more quickly than had been feared at the onset of the crisis.

LESSONS FOR POLICY

In addition to causing severe short-term problems, the financial and economic crisis has revealed policy and institutional shortcomings that will challenge the East and Southeast Asian economies beyond the crisis period. Many of these problems are equally relevant to other regions in Asia. The new initiatives that evolve must help policymakers respond both to the crisis and to longer-term, revealed structural problems.

Social Services and Social Safety Nets

As noted above, the crisis countries do not have strong safety nets in place to sustain or enhance the welfare of poor or vulnerable groups in times of economic crisis; even basic social services in health and education are threatened. The immediate problem is to preserve existing economic and social services for the poor, while in the longer term building stronger social safety nets. Key social sectors include health, employment, and education.

In the health sector, short-term efforts should concentrate on maintaining real spending on health activities that have high public externalities, such as vaccinations and vector control; providing community-based delivery of essential health and nutrition care; spending on regional health centers and subcenters; and providing temporary subsidies for essential drugs during the transitional period of exchange-rate disequilibrium (World Bank 1998d; ADB 1998c). Increased priority in health budgets should be given to women and children, including maternal and child health-care programs, because of the high probability that short-term deprivation of children will cause long-term reduction in physical and intellectual capability and a drop in lifetime well-being and productivity.

In the longer term, health services must be improved in order to ensure quality and cost-effectiveness, while still benefiting the poor. Health-financing schemes in many of the countries are characterized by considerable overlap, inconsistency, inefficient targeting mechanisms, and inequitable distribution of government subsidies. Rationalization of fragmented public health-care programs could yield substantial benefits and better access to those most in need. Rural-urban health-care disparities should also be addressed through redeployment of health-care personnel to rural areas. In Thailand, for example, infant mortality rates in the northern, northeastern, and southern regions are about twice the rate in Bangkok, and the ratio of doctors to patients is only one third the level in the capital (ADB 1998b).

In the employment sector, labor-intensive public employment programs, such as construction of rural roads, sanitation and water facilities, and reforestation and other environmental projects, would stimulate demand and directly benefit the poor (see also Chapter 2). Direct support could be provided for laid-off workers and the unemployed; making poverty programs more effective would protect the poor in the informal sector. In Thailand, for example, social security for medical care, disability, and education benefits has been extended for a period of six months after loss of work.

As recovery takes place, sounder safety nets should be developed in order to help households improve the management of insecurity related to employment. Unemployment-insurance programs should be developed and extended to a significant share of the population. Funded pension systems or provident funds would create the institutional investors required to provide oversight of corporate governance and would provide greater stability for domestic bond and equity markets. The competitiveness of industry and the flexibility of workers would be enhanced through greater private investment in worker training, which could be encouraged through tax deductions for training (ADB 1998b; World Bank 1998d).

In the education sector, it is essential that real spending on primary schools—the level of education that most directly benefits the poor—be sustained; in addition, targeted subsidies linked to incomes—such as exist in Thailand and Indonesia, see above—should be provided to maintain participation rates in primary and secondary schools. In the longer run, improvements in the education system are essential in order to maintain and improve the quality and flexibility of the labor force and to foster international competitiveness. Educational bottlenecks have developed due to the relative neglect of secondary education in Thailand, upper secondary and college education in the PRC, and the relatively weak quality of education in much of the Philippines and Indonesia.

Institutional and policy reforms are required in education in order to provide the right skills for full participation in the increasingly knowledge- and information-intensive global economy. In addition to developing a sounder foundation of literacy, communication, numerical and analytical skills, curriculum reform at primary and secondary levels should emphasize team-building, flexibility, and adaptability (World Bank 1998d). At the same time that quality of education is improved, access to education must be made more equitable.

In Indonesia, the 1996 enrollment rate for junior secondary education was only 30 percent of the lowest-income quintile, compared to 90 percent for the highest quintile (ADB 1998c). Long-term targeted scholarship programs could ameliorate the disparity in access to education.

Good Governance

As noted above, the East Asian crisis countries performed well right up to the crisis in terms of many macroeconomic fundamentals. But the onset of the crisis underlined the necessity to broaden the concept of "fundamentals" to include the quality of the legal system, regulatory capacity, and intercountry cooperation (Soesastro 1998). Breakdowns in governance, including rapid growth in corruption and poor institutional performance, were fundamental causes of the crisis. Good governance implies that authority is based on the rule of law, that its policies are transparent, that it is accountable to society, and that it is based on institutions and not on individuals (Soesastro 1998). Although the most immediate governance problems in the crisis countries have involved financial and corporate oversight, the crisis exposed significant weaknesses in governance in most sectors of the economy. International cooperation, through forums such as APEC and ASEAN, is an important avenue for broadening the acceptance of good governance. In the end, however, good governance must result from efforts internal to each society, rather than external moral suasion or pressure.

Institutional reform to provide good governance is a complex and long-term process that requires both improvement in public administration and public-sector management and movement toward more diversified delivery of services that is responsive to stakeholders. Public-sector management should be improved to enhance transparency and accountability, improve efficiency and effectiveness, and reduce the opportunities for corruption. Management information systems, audit functions, and procurement systems should be upgraded to strengthen the capacity of governments to monitor expenditure; ensure control over disbursements; and reduce costs, fraud, and abuse. Pay increases and improvement in employment conditions for civil servants can reduce the incentives for illicit behavior. Improved procedures for recruitment and promotion would help avoid the abuses of patronage, nepotism, and favoritism and help foster the creation of an independent, meritocratic civil service (ADB 1998d).

Reform is also necessary in the relationship between the public sector and the recipients of public-sector services. A diversified approach to delivery of services that would involve government, civil society, and religious institutions can help reduce the risks of relying on only one delivery channel (World Bank 1998d). To diversify delivery successfully, it is important also to reform the "demand side" for services. Generation of effective demand for public services and monitoring of public-sector performance is enhanced by a pluralistic society with rights to associate and to organize interest groups that have access to information on government services and programs. Governments could reduce implementation problems and enhance public support for their programs by easing access to information and allowing affected communities the opportunity to voice their concerns. Decentralization of services to local or community-based institutions can be an important component of good services, but should not be seen as a panacea. Local elites may have weaker technical resources at their disposal and greater opportunities for corruption and lack of transparency.

Nongovernmental organizations (NGOs) can play an important role in good-governance reforms in both the supply and demand for services. Traditionally, NGO activity has concentrated on the supply side: delivering services or assisting the public sector in operating its programs. But NGOs are increasingly becoming involved in the demand side: helping communities articulate their concerns and preferences; negotiating with official bodies in order to amplify the community "voice"; and mixing technical operational skills with information-intensive communication, advocacy, and networking skills to enhance the influence of poor people (Clark 1993).

Effective NGOs can improve governance in several ways:

- by encouraging government ministries to adopt successful approaches developed within the voluntary sector;
- by educating and sensitizing the public about their rights and entitlements under public programs;
- by acting as a conduit to the government for public opinion and local experience;
- by collaborating with official bodies;
- by influencing local development policies of national and international institutions; and
- by helping government and donors fashion a more effective development strategy through strengthening institutions, staff training, and improving management capacity (Clark 1991).

NGO involvement is not always a positive force for good governance, however. Mutual distrust between NGOs and government is often deep-rooted. An NGO that operates according to predetermined principles coupled with preset plans of action might both overlook the real needs and desires of local communities and alienate local and national government agencies. NGOs that act in isolation as "missionaries" setting up "mini-kingdoms" can also hinder rather than encourage good governance (Clark 1993).

Financial and Corporate Management

In the short run, the hardest-hit crisis countries face the task of restructuring their devastated financial and corporate sectors. This necessitates the creation of an enabling environment that includes better accounting and disclosure standards, bankruptcy and foreclosure processes, and improved taxation and accounting laws. For the reconstruction of the banking sector, a combination of approaches, including bailouts, assisted mergers, recapitalization and sale, and liquidation and payoff of depositors has been used. The dimensions of the task of financial and corporate restructuring are immense. The resources required for recapitalization of banks have been estimated to be as high as 20 percent of GDP in Indonesia and Malaysia and 30 percent in the Republic of Korea and Thailand (World Bank 1998e). A fuller treatment of these issues is beyond the scope of this book, but can be found in World Bank (1998d, 1998e).

In the longer term, a critical component of good governance will be the prudential regulation of financial institutions to reduce the possibility of future financial crises. Garnaut (1998b) summarizes the objectives of prudential regulation: accuracy, honesty, and transparency in financial reporting; the avoidance of related-party and other noncommercial transactions; and the maintenance of relatively high capital/asset ratios—higher than the norm for developed countries. Many of the East Asian crisis countries, notably Indonesia, had in place elaborate regulatory structures that were simply too complex to enforce. Prudential regulation should focus on simple, enforceable targets, rather than attempting to regulate a wide range of indicators (Fane 1998). An essential reform component will be allowing the establishment of international banks, both to introduce competition and induce improved management in domestic banks and to diversify the investment portfolios of the banking sector beyond the national economy.

Corporate governance also needs long-term reform. Many of the necessary reforms are similar to those that are called for in the reform of state enterprises in the transition economies. They include monitoring of enterprises by commercial banks, for example. Enhanced disclosure and accounting practices and strengthened enforcement of corporate governance regulations, especially as they relate to capital markets, are also essential components of corporate-governance reform. Moreover, broader private oversight of management, for example through the representation of minority shareholders, should be enhanced. Institutions need to be strengthened so that the areas of analysis of corporate financing and monitoring of firm performance and behavior are comprehensively covered by a combination of private, semipublic, and public organizations (World Bank 1998e).

International Capital Flows

The East Asian economic crisis raised serious questions about the free convertibility of short-term foreign capital inflows. Stiglitz (1998) recommends temporary controls on capital inflows combined with domestic reforms and greater disclosure to help reduce the frequency and magnitude of shocks. He argues that the benefit of short-term capital is small or even negative, because, unlike foreign direct investment, it does not bring along technology and management innovations. When savings rates are already high and marginal investment is misallocated, short-term capital greatly increases the vulnerability of the economy. Available empirical evidence indicates that foreign direct investment and other long-term, relatively stable investments have significant positive impacts on economic growth, whereas the benefits of short-term capital are small and problematical (World Bank 1998e).

More judicious management of international capital flows should focus on the creation of an environment conducive to long-term investments (and discouraging to short-term capital inflows). Tax incentives and other distortions that favor short-term inflows over long-term investments should be eliminated. Moreover, both prudential regulations on currency positions of banks and strengthened supervision of these regulations and other risk-management procedures are required. Finally, short-term and unhedged borrowing by corporations should be disclosed to reduce the credit risk.

On the other hand, in those Asian developing economies that are plagued by weak institutional capacity and financial systems, controls on short-term capital may well be appropriate. Capital-account restrictions, if put in place, need to be explicit, transparent, and market-oriented (World Bank 1998e). The Chilean approach provides a possible model for market-oriented short-term capital controls in East and Southeast Asia: in Chile, first of all, 30 percent of all nonequity capital inflows must be deposited interest-free at the central bank for one year. Second, Chilean firms and banks can borrow from international capital markets only if they are rated as highly as Chile's government bonds. Third, any foreign inflows must remain in the country for at least one year (Soesastro 1998).

International Recovery Programs

The type of recovery package appropriate for the East Asian financial and economic crisis has also been widely debated. The IMF has come under severe criticism for the reform programs in Thailand, Indonesia and Korea (the Philippines in 1996 was already in an IMF-supported reform program, which was modified as a result of the crisis; and Malaysia did not seek support from the IMF). The main goals of the stabilization and reform programs were to restore macroeconomic stability and to address financial-sector and other structural distortions in the economy. Contractionary fiscal and monetary policies were central to the program, with a reduction in government budget deficits and contractions in money supply.

A contractionary fiscal policy was implemented to reduce inflationary pressures, but fiscal policy was quickly relaxed to provide funding for social programs and financial restructuring and to partially balance falling aggregate demand. Tight monetary policies were implemented to raise interest rates and reduce capital outflows in order to defend the exchange rate, restore investor confidence, and control inflation. Financial restructuring was implemented through aggressive bank closures and enforcement of capital-adequacy standards, both to send strong signals about the seriousness of reform and to restore the banking system to a solid footing as quickly as possible. Nonfinancial structural changes in the reform packages included cuts in tariffs, reductions in monopoly power, and opening of sectors to foreign investment (Radelet and Sachs 1998).

Critics of the stabilization and recovery programs have focused on the pace of structural reform, fiscal policy, and monetary policy. Garnaut (1998b) accepts that reform of the financial sector and removal of other structural weaknesses are essential in order to avoid future crises, but argues that it was a mistake to overload the policy agenda in the midst of the crisis with reform issues that were not crucial to immediate recovery objectives. An excessively broad reform agenda can overload political systems. In Indonesia, such an overload prompted the incoherence in the implementation of agreed-upon IMF reform packages and eventually contributed to the political instability that made the recovery process far more complicated. Overloading the political system resulted in a weakening of government credibility and probably contributed to the breakdown of the political system in Indonesia (Garnaut 1998b).

Early bank restructuring policies may also have caused unnecessary economic contraction. In Indonesia, aggressive bank closures appear to have created panic rather than signaling political resolve. A possibly excessive rate of bank recapitalization may have caused an overly strong contraction in lending as banks attempted to meet capitalization targets. The counterargument is that failure to address these structural problems would have been devastating to domestic and foreign confidence and would not have produced the basis for durable recovery (Nellor 1998). Particularly in the financial sector, keeping failing institutions alive through central bank liquidity injections would only have aggravated the banking problem, led to more runs on banks, and made control of inflation more difficult. Instead, a program of closures, mergers, and recapitalization, together with reform of bank supervision, was started immediately in Indonesia, the Republic of Korea, and Thailand (Neiss 1998).

Critics of the stabilization programs have also argued that monetary and fiscal policy targets in the crisis countries were too stringent, resulting in greater contraction of the economy and harsher impacts on the standard of living than were necessary and deepening the economic depression. In each of the four countries in which a recovery program was implemented, the initial IMF fiscal policies were impossible to meet and the reform programs were renegotiated to relax some of the targets. The changing targets increased uncertainty over policy responses in the countries, contributing to political instability. Even the revised fiscal targets were contractionary, at least through early 1998, contributing to the plunge in aggregate demand.

In 1998, however, central government fiscal balance targets were relaxed for each of the countries in successive revisions of the recovery/stabilization programs. In Indonesia, for example, an initial target of –1.0 percent (a one-percent deficit) was reduced to –8.5 in a third revision. In Thailand, the target evolved from 1.0 percent to –2.4 percent and in the Philippines from 0.0 percent to –3.0 percent (World Bank 1998e). A general consensus appears to be emerging that initial fiscal targets in the reforming countries were excessively contractionary; fiscal policies should be more flexible, in order to provide some counterbalance to declining effective demand, to strengthen the social safety net, and to accommodate the costs of financial-sector restructuring (Kochlar, Loungani, and Stone 1998).

The use of tight monetary policy to stabilize exchange rates poses even more difficult trade-offs between macroeconomic and financial-sector stabilization objectives. The orthodox contractionary tight-money policies followed under the recovery programs were supposed to stabilize exchange rates and to curb inflationary pressures. High interest rates were to strengthen the exchange rate by making investment more profitable. A stable exchange rate limits the damage to banks and corporations that have large foreign currency debts. But high interest rates can also damage highly leveraged and weak banks and corporations by increasing the cost of debt service, contributing further to excessive credit contraction, and triggering an additional decline in income. Tight monetary policy can therefore also have perverse effects on the exchange rate by reducing investor expectations of future output, demand for money, and interest rates. Given the uncertainty about the behavior of interest rates and exchange rates during crises and the known adverse impacts of high interest rates on economic activity, critics of a tight monetary policy argue for increased flexibility in a situation such as the East Asian financial and economic crisis, where the financial system is fragile, corporations are highly leveraged, and declines in aggregate demand are large (World Bank 1998e).

Nevertheless, the policies have resulted in exchange rates first stabilizing and then strengthening. Moreover, inflation has been largely contained, with the exception of Indonesia, and interest rates have declined dramatically from post-crisis highs. The exchange rate of the Indonesian rupiah reached a low against the US dollar of 14,900/1 in the first quarter of 1998 and has since strengthened to 7,550/1. The Philippine peso strengthened from 43.87/1 in August 1998 to 40.88/1 in October 1998 and has since continued a gradual downward trend. The Thai baht strengthened from 47.25/1 in the fourth quarter of 1997 to 37.78/1 in October 1998 (IMF 1998b).

Annualized inflation hit 60 percent in Indonesia in mid-1998, but has since declined to an annualized rate of about 45 percent in September 1998 and 30 percent in May 1999. Inflation increased only slowly in the other crisis countries, reaching 10 percent in Thailand and the Philippines and 8 percent in Korea in the second quarter of 1998. In Indonesia, short-term interest rates increased from less than 20 percent at the beginning of 1997 to about 40 percent in late summer of 1997. After a brief decline, they increased to about 60 percent by September 1998, but fell to 30 percent in May 1999. In Korea, the short-term interest rate reached about 20 percent at the end of 1997 and declined to less than 10 percent by September of 1998 and 5 percent in May 1999. In Thailand, the interest rate peaked at slightly above 20 percent at the end of 1997 and came down slowly to below 10 percent by September 1998 and 4 percent in May 1999. In the Philippines, the short-term interest rate dropped during 1998 and 1999, from almost 30 percent at the end of 1997 to below 20 percent by September 1998 and 9 percent in May 1999 (IMF 1998a; Severino 1999).

Moreover, and most importantly, with the improvement in balance of payments, exchange rates, inflation, and interest rates, the East and Southeast Asian crisis economies are moving toward recovery in income growth. The Republic of Korea is projected to have GDP growth of 5 percent in 1999, Malaysia and the Philippines 2 percent, and Thailand 1 percent, while Indonesia is projected to have a fall in GDP of 2.4 percent over the full year (though real GDP grew at an annualized rate of 1.3 percent during the first quarter of 1999). It is impossible to evaluate fully the tight monetary policy (as well as the full recovery program) in mid-1999, while the impact of the crisis is still being felt.

Tight monetary policy appears to have successfully stabilized exchange rates and interest rates, but additional analysis will be required to assess whether, for example, interest rates could have been reduced more quickly, once the exchange rate stabilized, in order to stimulate growth. It has to be acknowledged, however, that, by the time the IMF was called in to help each country, all the low-cost options had been foreclosed. Important decisions in several complex and painful areas had to be made quickly and without full information. As of mid-1999, the East and Southeast Asian crisis economies appear to have weathered the worst of the downturn, but still face the massive challenges described above, and in an uncertain external environment. Continued slow growth in Japan, a downturn in the economy of the United States, or a large currency devaluation by the PRC could be serious setbacks to recovery.

CONCLUSIONS

The financial and economic crisis in East and Southeast Asia was the culmination of real appreciation of currencies and exchange-rate misalignment that diminished the international competitiveness of these countries; of large and growing current-account deficits financed with short-term foreign capital inflows; of poorly supervised overborrowing and overlending in the financial sector, much of it to risky and low-profitability projects; and of excessive short-term lending by international investors in the 1990s. These were followed by a sharp reversal of the short-term capital flows in 1997, which set off a round of competitive devaluations. The crisis has had a severe negative economic and social impact, including sharply declining incomes and increasing unemployment, rising absolute poverty, increases in malnutrition, increased pressure on already underserved rural areas, and declining social services and threats to education and health status. The effects on agriculture have been mixed and in the longer term will be determined by the balance betweeen the impact of the price effects due to real exchange-rate depreciation, which should boost agricultural production and trade, and the investment effects due to the sharp drop in income and the contraction of the financial sector, which will tend to depress agricultural production and export growth.

In addition to causing severe short-term problems, the crisis has revealed policy and institutional shortcomings that will challenge the East and Southeast Asian economies well beyond the crisis period. In the short run, it is essential to preserve existing economic and social services for the poor, including health, education, and employment, while in the longer term, stronger social safety nets must be built. Improvement in key institutions is also essential to assure long-term recovery and economic growth. Institutional reform to provide good governance is a complex and long-term process that requires both improvement in public administration and public-sector management and movement toward more diversified delivery of services that is responsive to stakeholders. Good-governance reforms must seek greater transparency and accountability in public-sector activities. A critical component of good governance will be the prudential regulation of financial institutions and corporations to reduce the possibility of future financial crises.

X Environmental and Resource Challenges to Future Growth

INTRODUCTION

Finite quantities of land and water worldwide that are suitable for agriculture limit the scope for bringing new natural resources on line for food production. In addition, some contraction in resources for agriculture due to rising pressure to divert resources already in agriculture to nonagricultural uses may partially offset any expansion. This puts the pressure for future food supplies primarily on land already in agriculture. Yet environmental degradation in areas already in production can dampen projected growth in food supplies by eroding the productive capacity of the natural-resource base; any new areas brought under production may be even more susceptible to degradation than are current areas.

Advances in crop productivity may help mitigate degradation-induced slowdowns in the growth of the food supply as agriculture approaches the limits to its expansion. If environmental degradation should exceed expectations or efforts to improve productivity fall short, however, the impact on food supplies could be major. Asia, whose high population densities put more pressure on natural resources than is the case elsewhere, presents a compelling example of degradation trends and efforts to counterbalance productivity losses. More than this, because of Asia's growing importance to the world economy and food supplies, what happens can have worldwide impact.

This chapter examines possible environmental and resource constraints on agricultural production growth and future food supplies. What is the potential for bringing more land resources into agriculture, compared to the likely losses due to urbanization? To what extent will water scarcity limit crop production growth, especially when the trend is for water used in agriculture to be siphoned off to meet growing demands from other sectors? What is the probability that agriculture will bump up against limits by placing too heavy demands on the world's energy resources? Evidence for land degradation, including degradation due to misuse of water resources, is reviewed and estimates of the magnitude of the ensuing crop productivity losses presented. Strategies aimed at raising crop productivity are evaluated, including closing the gap separating yields in farmers' fields from current biophysical limits and raising farmers' yields by setting new, higher biophysical limits: fertilizer use; management of plant genetic resources; and biotechnology. Finally the evidence on potential agricultural effects of global warming is reviewed.

CROPLAND – WILL THE BASE FOR FOOD SUPPLY SHRINK OR EXPAND?

Scarcity of data and difficulties in interpretation pose a challenge for anyone hoping to account fully or accurately for the amount of land being added to, and taken out of, cultivation. While some country-level data or estimates on net changes in cropland are available, a look at broader estimates about the potential for bringing new land under production, as well as for losing arable land to nonagricultural uses, is needed to give an idea of the scope of the problem over the next couple of decades.

Potential for Cropland Expansion

In 1993, worldwide crop area was approximately 1,070 million hectares, 70 percent of it harvested for cereal and root crops. The food-crop portion alone of total crop area may go up an additional 50 million ha by 2010, virtually all in the developing world (see projections in the Appendix and Chapter XII). The crop area harvested in 1993 is about one third of the projected theoretical maximum area potentially suitable for crop production of 3,300 million ha out of a total of 12,400 million ha in land resources (consisting of arable land, permanent pasture, forest and woodland, and other land) (Buringh and Dudal 1987). Most of this potential cropland (2,600 million ha), is classified as having low or medium capability for crop production, with only 700 million ha in high-potential areas. The remaining 10,100 million ha are classified as having zero potential for growing crops

But is the theoretical maximum a practical one? Most current cultivation takes place on relatively good agricultural land; additional land converted to cropland would be expected to have lower productivity levels than does the existing land stock. Conversion to the maximum would also mean taking land out of uses—such as forest and rangeland—where they are fulfilling essential functions. According to Kendall and Pimentel (1994), the world's arable land might be expanded by at most 500 million ha, at productivity below present levels. Nearly 90 percent of this potential cropland is located in developing countries, but it falls mainly outside Asia, in Sub-Saharan Africa and Latin America. Asia has less room for expanding cropland, since farmers there already cultivate nearly 80 percent of the potentially arable land (Plucknett 1995). So the scope for greater food production coming from more agricultural land is limited, particularly for Asia. The land gained by expansion will probably produce less food than in existing crop areas, since it is likely to be of low potential, possibly better suited for uses other than growing crops.

Loss of Cropland to Urbanization

Juxtaposed against these limits to the expansion of cropland in Asia is the specter of inroads made on cropland by nonagricultural uses. While historically more potential cropland has been converted to agricultural activities and grazing than urbanization has taken away, it has been suggested that current, unprecedented increases in urban populations may constitute a potential threat to agricultural production, through the loss of prime agricultural land (Brown and Kane 1994).

In 1975, 38 percent of the world's population, or 1.5 billion people, lived in cities. By 1995, that number had grown to 2.6 billion, or 45 percent of the population. The developing and developed worlds, however, faced quite different levels of urbanization: for example, urban population accounted for 70 percent of the total in both North America and Europe, but only 34 and 35 percent in Africa and Asia, respectively. From 1995 levels, overall urban populations are projected to grow at 2.3 percent per annum until 2025, when they will number over 5 billion, and constitute a majority (61 percent) of the population. Nearly all urban population growth, about 90 percent, will occur in developing countries, where roughly 150,000 people are added to the urban population every day.

In Asia, a threshold of sorts will be met in 2020. In that year, one of every two Asians is projected to be an urban resident, almost double the 1980 figure (27 percent). Among East and Southeast Asian countries, only Thailand, Myanmar and the PRC are expected to still have large rural populations (at 33 percent, 40 percent, and a projected 49 percent, respectively). In South Asia, by contrast, only the most urbanized countries, such as Pakistan, will hit this 50 percent urban threshold by 2020; in Nepal, only 21 out of every 100 residents will live in cities (WRI 1998).

Does urbanization mean that cities will sprawl to take over vast amounts of agricultural land? For all developing countries, estimates put the annual loss of arable land transformed to urban uses due to expanding urban populations at 476,000 ha (USAID 1988). Between 1990 and 2020, this would mean a loss of nearly 10 million ha of land to urban uses. While for Asia (as elsewhere), data on urban absorption of land previously under cultivation are scarce, present-day urban densities for selected cities provide a clue as to how much land cities hosting populations the size of these 2020 projections might need.

In the PRC, for example, in order to support projected urban populations in 2020 at the existing urban density of about 150 people per hectare, urban area must expand beyond 1993 levels by approximately 2.4 million ha, or an average of 87,000 ha per year. In Southeast Asia, projections from 1993 to 2020 using population densities found in major cities of each country show Indonesian cities taking over 416,000 ha, Philippine cities occupying an additional 232,000 ha, and Thai urban areas increasing by roughly 74,000 ha. In South Asia, nearly 1.5 million ha in India and Pakistan combined (623,000 ha in India and 802,000 ha in Pakistan) will be converted to urban uses. While in order to chart actual cropland loss to urbanization, other factors would have to be considered, including the type of land converted into urban uses and the final urban per capita land area, this exercise does give some idea of the scale of possible land loss.

Net Expansion/Loss of Cropland

To determine actual changes in cropland requires balancing out many factors beyond expansion and contraction of crop area, for many of which data are scarce, easily misinterpreted, or both. The case of the PRC illustrates the challenges of finding and interpreting information. It also shows that despite the challenges, estimates have been possible, at least in this case and for some historical or short-term projections.

In the PRC, as elsewhere, figures of land leaving agriculture may be overestimated, in that they indicate gross rather than net conversion of land to nonagricultural use. In other words, while land converted from agricultural to nonagricultural use is reported, land reclaimed for agricultural use is not. In addition, the categories used to describe how land use is shifting may be misinterpreted as meaning an unequivocal loss to the food supply. Widely reported reductions in cereal area in the PRC between 1990 and 1994, for example, stem in significant part from conversion of cereal land to inland aquaculture and from shifts of land to other crops, like vegetables or fruits, and from temporary fallowing of land due to the declining profitability of cereals during this period (Alexandratos 1996; Lindert 1996). Moreover, for land- and water-scarce countries such as the PRC, conversion of cropland to higher-value uses often constitutes a win-win situation for both economic development and environmental sustainability (Paarlberg 1997). After 1994, as the profitability of cereals once again improved, harvested cereal area also increased, from 86 million ha in 1994 to 91 million ha in 1996.

Despite the difficulties in making precise projections, given the amount of cropland still available for cultivation, likely land conversion to urban and other uses does not present a serious threat to food supplies on an Asia-wide basis. In several Asian countries, however, withdrawal of land from agriculture due to urbanization will probably contribute to reductions in production growth, with only limited potential to compensate for this loss by expanding into new arable areas. Moreover, even where there is an ample margin to expand agricultural area, overall crop area is expected to grow only slowly (see also Chapter XI).

WATER RESOURCES – DOES SCARCITY ENDANGER FUTURE FOOD SUPPLY?

Despite legitimate concern about whether cropland will be sufficient to meet agricultural demand, the resource base that may pose the most serious threat to future global food supplies is water. In many regions of the developed and developing worlds, groundwater is being depleted as pumping rates exceed the rate of natural recharge. While mining of both renewable and nonrenewable water resources can be an optimal economic strategy, it is clear that groundwater overdrafting is excessive in many instances. In parts of the North China Plain, for example, groundwater levels are falling by as much as one meter per year; in portions of the southern Indian state of Tamil Nadu heavy pumping is estimated to have reduced water levels by as much as 25–30 meters in a decade (Postel 1993).

The potential for increasing food production by expanding irrigated area or tapping into new sources of water supply is limited by rising costs. Both South and Southeast Asia have felt the trend of rising costs in irrigation development over the last few decades. In India, for example, the real costs of new irrigation have doubled since the late 1960s and early 1970s; in Sri Lanka, they have tripled. Indonesia has seen irrigation development costs go up 200 percent, while in the Philippines costs have increased by more than 50 percent and in Thailand by 40 percent (Rosegrant and Svendsen 1993). The result of these increases in costs (combined with declining cereal prices) is low rates of return for new irrigation construction.

Coupled with rising environmental concerns about irrigation (see the section on land degradation below), reduced rates of return to new irrigation have greatly slowed the rate of expansion of irrigated areas. This is a worrisome trend for global food supplies: irrigated area accounts for nearly two thirds of world rice and wheat production and thus plays a much larger role than rainfed land in global (and Asian) food production, so that growth in irrigated output per unit of land and water is essential if growing populations are to be fed.

Turning to nontraditional sources of water may provide some relief under particular conditions, but this path does not hold the answer to Asia's need for new water supplies. Desalination offers an infinite supply of fresh water, but at a high price, and will not be a significant factor in most of Asia. The reuse of wastewater will similarly make an important contribution only in the most arid regions, where the cost of new water supplies is very high. Water harvesting (the capture and diversion of rainfall or floodwater to fields to irrigate crops) will be important in some local and regional ecosystems. Like methods tapping other nontraditional sources, results may be important locally and in certain low-potential areas, but will not have a significant impact on global food production and water scarcity; reliance on irrigated area remains (Rosegrant 1995; Rosegrant and Meinzen-Dick 1996).

To make matters worse, rapidly growing household and industrial demand for water will increasingly be seeking to draw on the same water as irrigated agriculture. In much of developing Asia, new urban water supplies cost three to four times more than already developed existing water sources (World Bank 1993b). Since irrigated agriculture generally accounts for around 80 percent of water diversion in Asia, transfer of some of this water to meet growing urban demands is often seen as a relatively low-cost alternative to developing new water supplies. This raises a particularly difficult challenge: to save enough water from agriculture without sacrificing crop yields and output growth, while at the same time allowing reallocation of water from agriculture to rapidly growing urban and industrial uses.

Meeting the challenge involves improving the efficiency of agricultural water use so as to generate physical savings of water as well as economic savings. In agriculture, more can be done to make sure water is used by crops before it can either escape into the air (increasing crop output per unit of evaporative water loss), or into the ground (improve water use before its loss to water sinks). Further gains could come from lessening degradation such as salinization and other water pollution that diminishes crop yield per unit of water (see below). It is unclear how large each of these potential water savings might be, but estimates of water use efficiency in irrigation give some idea. For the developing world, this falls typically in the range of 25 to 40 percent. Making urban supply systems more efficient may also help: for the developing world's major metropolitan areas, "unaccounted-for water," much of which is direct water loss to the oceans, is often 50 percent or more (Rosegrant and Shetty 1994; Rosegrant 1995).

These inefficiencies seem to imply that there is a potential for huge savings from existing uses of water and that the challenge can easily be met. The potential savings of water in many river basins, however, are neither as dramatic nor as easy to achieve as implied by these efficiency figures. Because much of the water "lost" from irrigation systems is reused elsewhere (Seckler 1996), improvements in irrigation-system efficiency do not necessarily mean overall efficiency gains. In fact, even though individual water users in these basins are inefficient, whole-basin water-use efficiencies are quite high due to reuse and recycling of drainage water. In sum, efficiency gains from existing systems may prove to be limited. This significantly narrows the scope for overcoming the threat of water scarcity to the food supply by reducing wastage in water use from existing sources.

The directions for water-policy reform to address these serious challenges were introduced in Chapter V. Dealing with water scarcity and quality will require both substantial new investments in urban water systems and irrigation systems and reform of water management. The reform of water-demand management for both groundwater and surface water will be essential in meeting new water demands by saving water in existing uses and in improving the quality of water and soils. The necessary reforms will involve changing the institutional and legal environment in which water is supplied and used to one that empowers water users to make their own decisions regarding use of the resource, while providing correct signals regarding the real scarcity value of water, including environmental externalities. Reforms will need to be tailored to the specific institutional context in any given region, but important elements include establishment of secure water rights for users; decentralization and privatization of water-management functions (to private sector or community-based organizations); and the use of incentives, including markets in tradable property rights, pricing reform and reduction in subsidies, and effluent or pollution charges.

ENERGY – PLENTIFUL ENOUGH TO SUPPORT A GROWING FOOD SUPPLY?

Direct forms of energy (farm machinery, animal and human labor) and indirect (manufacture of agricultural chemicals, farm machinery and irrigation) have been essential factors in bringing about increases in agricultural productivity. The green revolution raised the energy intensity of agricultural production a hundred-fold or more in some cases (but from a near-zero base), with plant breeding aimed at designing plants that could cope with high levels of fertilizer use (Kendall and Pimentel 1994). In developed countries, the manufacture and operation of farm machinery account for the largest, albeit declining, share of commercial energy uses in agricultural production (52 percent in 1982), followed by chemical fertilizers with an increasing share (44 percent in 1982). In developing countries, however, fertilizers take first place, accounting for 69 percent of energy share in 1982 (Bhatia and Malik 1995).

Despite increases in energy intensity in agriculture, agricultural uses of energy account for only a small fraction of total energy consumption. In 1990, production of fertilizer, the most energy-intensive agricultural input, required only about 2 percent of global energy consumption; this level is expected to decline to about 1.6 percent by 2020. Considerable energy efficiency improvements in fertilizer plants after the energy crisis of the 1970s help explain this low figure. Possibilities for further increasing energy efficiency lie on the horizon, in the globalization and privatization of fertilizer markets as well as the removal of energy subsidies and inefficient organizational structures (Bumb and Baanante 1996).

Furthermore, overall energy use in agriculture constitutes only a small part of agricultural production costs. During the last 20 years, direct farm expenses in the United States for fuels, oils, and electricity have varied between 3.5 and 7.4 percent of total farm production expenses. Together with expenditures for pesticides and fertilizers, the cost share was between 11.2 percent and 17.2 percent of total farm production expenses. A study of the effects of large energy price changes on the agricultural sectors of different regions concluded that even extremely large and sustained increases in energy prices lead only to small declines in agricultural output and land prices, even in the very energy-intensive United States (McDonald et al. 1991). The threat to Asian agriculture—which uses far less energy—from higher energy prices would be minimal.

Although overall energy use has been increasing during the last decades, there is some evidence that energy intensity has been decreasing in developed countries. Bonny (1993) showed a downward trend of direct energy use in French agriculture since the 1970s, citing a 30 percent drop in direct and indirect energy intensity in the production of one ton of wheat in a region in France between 1955–60 and 1990. Finally, energy prices are projected to decrease for the next decade: according to the World Bank (1996b), crude oil prices are expected to fall from US $51.22 per barrel in 1980 (constant 1990 dollars) to US $13.23 per barrel by 2005.

Energy use clearly was an essential factor in bringing about the green revolution in the 1960s and will remain essential for achieving food security in the coming decades. However, with the prospects of increasing energy efficiency, with lower energy prices, and with only a small proportion of overall energy devoted to agriculture, energy availability may not be a serious resource constraint to long-term agricultural growth.

LAND DEGRADATION – WHAT EXTENT AND PRODUCTIVITY EFFECTS?

Land degradation can lead to lower crop yields, through several avenues. It may require farmers to increase input levels just to maintain yields, reducing total factor productivity; it may lead them to convert the land to lower-valued uses; or it may cause them to temporarily or permanently abandon their plots. Within the last decade, efforts to examine the extent of degradation and its productivity impact, worldwide and regional, have been stepped up, allowing the beginnings of an empirical look at the threat environmental degradation poses to future food production.

Extent of Land Degradation

Oldeman, Hakkeling, and Sombroek (1990) provide the most comprehensive assessment of global land degradation to date; further work on regional degradation and productivity effects build on it. It classifies the main types of land degradation as soil erosion from wind and water, chemical degradation (loss of nutrients, soil salinization, urban-industrial pollution, and acidification), and physical degradation (compaction, waterlogging, and subsidence of organic soils). Out of the total land resource base, Oldeman, Hakkeling, and Sombroek estimated that 1,964 million ha, roughly 16 percent, suffered from some degree of degradation. By far the greatest cause of this degradation, roughly 84 percent, is erosion, with water erosion accounting for a majority, 56 percent, and wind erosion responsible for a further 28 percent). Water, then, plays an important role in land degradation: in addition to water erosion's large impact, water-related effects (for example, from salinization and waterlogging) are incorporated within overall figures for chemical degradation (12 percent of degraded land), and physical degradation (4 percent of degraded land).

In South and Southeast Asia, degradation levels are proportionally even higher: the Assessment of Human-Induced Soil Degradation in South and Southeast Asia (ASSOD) found that agricultural activity had led to the degradation of 27 percent of all land and to the deforestation of 11 percent (Scherr 1999). Out of the 1,843 million hectares surveyed under ASSOD, loss of topsoil from erosion by water affected 19 percent, while erosion by wind affected 5 percent.

Productivity Effects of Land Degradation

Not all this degradation affects crop productivity. Of all degraded land worldwide, an estimated 562 million ha, or just under 30 percent, is degraded agricultural land. Here, chemical degradation plays a much larger role than it does in degradation overall, accounting for 40 percent of degraded land. Similarly, in Asia, the degradation figures given above refer to all land, not just arable land, which constituted only part, roughly 21 percent, of the ASSOD survey area. Of the arable area surveyed, the ASSOD study found 30 percent suffered from a decline in soil fertility, and a further 17 percent was degraded because of water-related problems—10 percent from salinization and 7 percent from waterlogging (Van Lynden and Oldeman 1997 as cited in Scherr 1999).

Are these trends alarming for future food supplies? As prevalent as degradation seems from these figures, ultimately the threat to food production must be gauged in relation to degradation's effects on productivity. Building on work that measures degradation's extent, more studies that measure the productivity effects of degradation have emerged, for Asia as elsewhere.

For Asia, all human-induced soil degradation in the period since World War II has resulted in an average cumulative productivity loss of 12.8 percent for cropland, according to estimates by Oldeman (1998) as cited in Scherr (1999). Over roughly the same period (1945 to 1990), South Asia has suffered a slightly higher accumulated loss of yield of approximately 16.5 percent, according to Crosson (1998), using productivity loss figures for the region from the FAO (1994). Both figures from Asia fall well above a global estimate of 5 percent for the same period (as cited in Crosson 1998).

In terms of percentage of productivity lost, Asian rainfed areas have been hit only slightly harder than those under irrigation, but only because the extent of the area hit is somewhat larger and not because the productivity effects of degradation were substantially greater. Dregne and Chou (1992) estimate that 10 percent of rainfed areas in Asia, as opposed to a slightly smaller 8 percent of irrigated land, have experienced a loss in productive potential of at least 25 percent. Furthermore, over half of all rainfed lands in area, and more than one third of irrigated land in Asia, experienced a loss of 10 percent of productive potential. Some information is also available on productivity loss due to specific types of degradation: soil erosion, soil fertility decline, and the water-related land degradation problems of salinization and waterlogging.

Soil losses have led to yield declines in crops in rainfed upland areas, as well as in perennial crops in plantation systems, according to separate case studies undertaken in different parts of Asia. In the upland rainfed areas of Java, Indonesia, erosion has caused annual yields for several crops (corn, soybeans, and grounduts considered jointly, and cassava considered separately) to decline an estimated 4 to 7 percent (Magrath and Arens 1989 as cited in Crosson 1998). Mingsarn and Benjavan (1999) cite two studies on yield decline for perennial tree crops due to loss of soils in Sri Lanka. In one, loss of topsoil on a Sri Lankan tea plantation resulted in yield declines of only 0.7 kg per hectare from an average yield of 1,634 kg per hectare. In the other (Samarappuli et al. 1997), rubber yields fell by 175 kg per hectare per year for every 1-cm loss of topsoil. That productivity effects of soil loss have been shown to vary under different conditions is in keeping with expectation: soil depth, structure, and slope in a given area, combined with known crop needs, would all help shape the point at which soil loss would be expected to hurt productivity substantially. To the extent that knowledge of the resource base in Asia and elsewhere improves, therefore, hypotheses about the extent of the productivity effects of soil losses in Asia can be refined and tested.

Losses in soil fertility can also mean yield declines. In the Asian environment, this effect is commonly observed in intensive rice monoculture systems as well as of rice-wheat systems, through a decline in the partial factor productivity of nitrogen fertilizer (Hobbs and Morris 1996; Pingali, Hossain, and Gerpacio 1997). When fertilizer levels are held constant, intensive rice culture can, over time, reduce the soils' ability to meet the plants' nitrogen requirements, especially in the later period of crop growth, thereby affecting crop yields and resulting in a declining partial factor productivity of nitrogen (Cassman et al. 1994). Declining soil nitrogen supply is caused not by the prolonged use of chemical fertilizers, but rather by intensive rice monoculture systems themselves (Cassman et al. 1994, Pingali and Rosegrant 1998). Cassman and Pingali (1993) estimate the magnitude of yields foregone due to declining soil nitrogen supply to be around 30 percent over a 20-year period, at all nitrogen levels, based on long-term experiment data from the IRRI farm. Farmers have been increasing the amount of chemical fertilizers applied in order to maintain their yield levels. Because of a lack of nutrient balance in fertilizers applied, increased application has not stemmed an observed trend of increased incidence of phosphorus, potassium and micronutrient deficiency (De Datta, Gomez, and Descalsota 1988). Because of increased cropping intensities and the predominance of year-round irrigated production systems, which have depleted naturally occurring soil nutrients, phosphorus and potassium deficiencies in particular are becoming widespread across Asia in areas not previously considered to be deficient (Pingali and Rosegrant 1998).

Still, at least for the PRC and according to one study, the impact of soil fertility declines on crop yields has not as yet been too severe. Lindert (1996) shows mixed trends in soil quality between the 1930s and the 1980s and concludes that soil degradation is neither nationwide nor concentrated in the more erosion-prone areas. In order to evaluate these conflicting trends in economic terms, Lindert estimates agricultural production functions for the period of 1981–86 with agricultural outputs and inputs in 1985. His results indicate that acidity, alkalinity, and potassium, which have not deteriorated since the 1930s, are also the most significant soil-chemistry parameters for agricultural yields in the PRC. In addition, shifts away from the main staple crops (grains, oils, and cotton) towards vegetables and animal products show positive effects on soil quality. Other trends may also brighten prospects: improved security of property rights, for one, by encouraging on-farm conservation investments, might actually advance soil conservation efforts in the future. Not all accounts for the PRC are as encouraging, however. Other experts find strong signs of degradation other than soil fertility losses hitting the agriculturally most favorable areas of the PRC. The degradation stems from deforestation, overgrazing, and salinization (see, for example, Huang and Rozelle 1995).

That the water-related degradation problems of salinization and waterlogging are particularly serious in the PRC and elsewhere in Asia should come as no surprise, given the importance of water management and the extent of irrigated agriculture in the region. Intensive use of irrigation water in areas with uneven toposequence and poor drainage can lead to a rise in the water table due to the continual recharge of groundwater. In semiarid and arid zones this leads to salinity buildup, while in humid zones it leads to waterlogging.

Salinization can occur when evapotranspiration exceeds rainfall, causing a net upward movement of water through capillary action, causing a concentration of salts on the soil surface. High water tables prevent the flushing of salts from the surface soil. The overall effect can be serious even when the groundwater itself is not saline; over the long term, salt can build up with evaporation of continuously recharged water of even low salt content (Moorman and van Breemen 1978). In the short term, salinity buildup leads to reduced yields; in the long term, it can lead to abandonment of cropland (Samad et al. 1992; Postel 1989; Mustafa 1991). Excessive irrigation and poor drainage (especially seepage from unlined canals) can induce salinity problems, with poor irrigation system design and management often driving factors. Postel (1989) estimated that 24 percent of irrigated land worldwide suffers from salinity problems, with several of the most affected countries—the PRC, India, and Pakistan, in a group that also included the United States and the then-Soviet Union—located in Asia.

The PRC's salinity problems date back centuries. By 1992, 7.7 million ha (more than 15 percent of total irrigated area), mostly in the North China Plain, was said to suffer from the combined effects of salinity and alkalinity (Huang, Rosegrant, and Rozelle 1996). The North China Plain and far-west regions, where large-scale irrigation has expanded rapidly in the past several decades, bear the brunt of the country's salinization problems; more recently, though, these areas have started to face serious water shortages (Huang and Rozelle 1995).

In Pakistan, salinity affected large areas in Sind Province after the introduction of extensive irrigation, which led to a rise in the water table from a depth of 20–30 m to 1–2 m within 20 years (Moorman and Van Breeman 1978). In India, Dogra (1986) estimates that salinity problems afflict nearly 4.5 million ha. Other examples from South Asia can be found in Chambers (1988), Abrol (1987), Dogra (1986) and Harrington et al. (1992).

In higher-rainfall areas, such as East India, rain flushes out accumulated salts, relieving salinity buildup. Still, in these zones similar causes—excessive water use and poor drainage—lead to another type of degradation that also lowers productivity, waterlogging. Waterlogged fields have lower productivity levels because of lower decomposition rates of organic matter, lower nitrogen availability, and accumulation of soil toxins. In the case of wheat, low plant populations in some areas can be attributed to waterlogging, especially when it occurs early in the growing season, during the germination and emergence stages. Hobbs et al. (1996) report for the Nepal Tarai that waterlogging reduced yields by half a ton per hectare. Throughout India, Dogra (1986) estimates area affected by waterlogging at 6 million ha (slightly exceeding area affected by salinization).

In some geographic regions, then, land degradation may be of overriding importance, leading to yield declines (or greater use of inputs to counteract degradation's effects), lower-value uses of the area, or even abandonment of plots. Still, the aggregate effect on productivity has not yet reached the stage where it poses a serious threat to aggregate food supply in Asia. If rates of degradation accelerate dramatically, particularly in high-potential areas (due to continued intensive use—see Chapter VI) but also in low-potential areas (because of pressure to expand agriculture there), the story could change dramatically, placing future food supply at risk.

IMPROVED CROP PRODUCTIVITY – THE KEY TO A SECURE FOOD SUPPLY?

Asian food production can rise through expansion of cropping area and increases in cropping intensity or through improvements in agricultural productivity. Given the projected slow growth in crop area in the future (see Chapter XII), higher agricultural productivity will have to bear most of the burden for achieving the necessary production rates to meet global food demand. Yet crop-yield growth has already slowed significantly in Asia (see Chapter V) and land degradation has also started to take a toll on yields (see above). In addition, new sources of water to meet agricultural needs are increasingly expensive to tap, while room for efficiency gains from existing sources may not be sufficient (see above). Will agricultural productivity, as the main engine of agricultural production growth, be able to keep up with global food needs, or are the biophysical yield limits already in sight?

Along with a theoretical maximum of arable area across the globe comes a biophysical limit of food production for the earth. It is reached when all land suitable for agriculture is cropped and irrigated, the potential yield on each field attained, and the remaining suitable grazing land grazed. The potential crop yield on any given piece of land has its own specific upper limit, determined by soil type, climate, crop properties, and available irrigation water. It is reached when the farmer selects the optimal combination of crop species and management practices (Penning de Vries et al. 1995).

Maximum theoretical yields, the highest limit of biological potential for specific crops at given locations, are generated on the basis of photosynthetic potential, land quality, length of the growing season, and water availability. They have been calculated in grain equivalents (with rice in milled form) by Linneman et al. (1979). Biophysical limits vary from one region to another due to different underlying conditions in the agricultural sectors.

Asia has a comparatively low biophysical limit, with an estimated maximum production of 13 tons per hectare per year. In South America, on the other hand, the potential is 18 tons per hectare; in Africa, 14 tons per hectare; in North and Central America, 11 tons per hectare, and in Europe and Australia, 10 tons per hectare. As measured in 1990–92, a wide margin separates actual yields, which vary from 0.7 to 3.8 tons worldwide per ha per season, on average, from these theoretical maximum figures. Thus, despite the slowdown in yield growth over the past 15 years, overall yield trends by country and region indicate ample room for yield improvement for most crops and regions against the theoretical maximum (Plucknett 1995).

That ample room exists for realizing gains in crop yields, however, is no cause for complacency, as the data on slowdown in yield growth underlines. Agricultural research at several levels will be essential. Some of it must be aimed at sustaining existing crop yields—productivity maintenance research to counter threats, for example, of evolving pest populations. Another portion must target ways of closing the gap between farm and research-station yields (which stand reasonably close to the theoretical ceilings), by reviewing the role fertilizer can play in future crop productivity and by investigating new techniques, including improved and extended resistance to biotic and abiotic plant stresses using existing plant genetic resources. Finally, strategic research aimed at raising the theoretical maximum yield ceilings, primarily through advances in genetic manipulation and biotechnology, is also needed (Plucknett 1995). Such gains are expected to translate into some yield improvements on farmers' fields (and not merely to widen the gap between realized and theoretical yields). Still, most strategies to improve crop productivity do contain some downside risk—some deleterious effects to the environment and/or future crop productivity—that must be kept in mind.

Fertilizer—Raising Yields or Damaging the Environment?

The role of fertilizer in raising yields is well documented, but excessive use may pose risks to the environment. Can continued expansion of fertilizer use support the yield gains necessary to meet effective food demand in the next decades without damaging the environment? Global fertilizer use increased from 64 million tons in 1966 to 158 million tons in 1988, then declined to 134 million tons by 1993 before increasing again to 158 million tons by 1996 (FAO 1998). The dip in global fertilizer use at the beginning of the 1990s resulted primarily from steep declines in fertilizer application in the reforming economies of Eastern Europe and the former Soviet Union (Bumb and Baanante 1996).

In Asian developing countries, fertilizer consumption increased rapidly during the last three decades, from 6 million tons in 1966 to 75 million tons in 1996. Thus the share of developing Asian countries in total fertilizer consumption rose dramatically from 9 percent in 1966 to 48 percent by 1996. As could be seen in Chapter V, fertilizer application rates are extremely high in some Asian developing countries, in particular the PRC, the Republic of Korea, and Viet Nam, and relatively high in most other countries of the region. At the same time, there is substantial potential for increased application in Myanmar, in most of South Asia other than high-intensity agricultural regions such as the Indian Punjab, and to a more moderate extent, in the Southeast Asian countries of Indonesia (particularly outside of Java), the Philippines, and Thailand.

With high long-term growth rates in fertilizer use, declining growth rates in yield, and very high fertilizer levels in relatively favorable areas of Asia, increasing amounts of fertilizer are being used to maintain current yield levels (although the yield declines are due not to fertilizer per se, but to the intensive systems themselves; see above). In areas like West Java in Indonesia, the Indian Punjab, and parts of the PRC, fertilizers are used at or above economically optimum levels at the border prices that would prevail in the absence of subsidies. In much of East Asia, further increases in fertilizer application will be small, but there is considerable room for improvement in fertilizer-use efficiency, nutrient uptake rates, and nutrient balance (Rosegrant and Pingali 1994).

Bumb and Baanante (1996) estimated effective demand growth for fertilizer, taking into account foreign-exchange availability, exchange rate, crop and fertilizer prices, the development of irrigation and other infrastructure, and the impact of policy reforms on fertilizer demand. During the 1990-2020 period, global fertilizer demand is projected to increase

1.2 percent per year and by a rapid 2.1 percent per year in Asia. In absolute terms, this means the world would use about 208 million tons in 2020, up from 144 million tons in 1990. Asia would virtually double its fertilizer consumption, from 53 million tons in 1990 to 101 million tons in 2020.

Can the production of fertilizer keep up with its projected effective demand? The projections of supply potential developed by the World Bank/FAO/UNIDO Industry Fertilizer Working Group (1994) and the International Fertilizer Development Center (Bumb 1995) suggest that the world will have the capacity to produce between 147 and 163 million tons of fertilizer nutrients in the year 2000. In order to meet the projected effective demand in 2020, an additional 55 to 71 million tons of nutrients will have to be produced. Assuming the lower capacity figure for 2000, fertilizer production should be increased at an annual rate of 1.4 percent during the 2000–2020 period to satisfy the projected effective fertilizer demand. Given the 5.7 percent annual growth in fertilizer production during the 1960–90 period and continued low energy prices, reaching this required growth should not be difficult. Bumb and Baanante (1996) also show that raw materials are not likely to be a constraint in meeting future global fertilizer demand.

The one constraint that could slow the expansion of fertilizer capacity is continued low fertilizer prices. In 1993 the real price of urea dropped to only one third of its 1980 price before beginning to recover and in 1995 was still only 60 percent of the 1980 value. The 1995 prices of diammonium phosphate, phosphate rock, potassium chloride, and TSP were also in the range of 50 to 60 percent of their 1980 values. World Bank (1996b) projections indicate that fertilizer prices will be stable or slightly lower through 2005. If these price levels constrain future investment in fertilizer production capacity, fertilizer prices could increase in later years. Although this would improve efficiency of fertilizer use, it would also induce a reduction in the growth of its use; the net result might be negative effects on crop yield growth.

Are the levels of fertilizer used in Asia a threat to the environment? The two major direct environmental effects of high levels of fertilizer use are nitrate leaching or runoff and eutrophication. Nitrates can leach from the soil or run off in drainage water when the supply of nitrogen from fertilizer and other sources exceeds nitrogen uptake by plants. Nitrates that leach into drinking water supplies can cause severe health problems. Eutrophication occurs when fertilizer is carried by soil erosion and water runoff into lakes, rivers, or other water bodies, potentially causing excess growth of algae, oxygen depletion, and fish mortality. These side effects are of considerable concern in Western Europe and parts of North America and policies are being put in place to selectively reduce fertilizer use there (Leuck et al. 1995). There is little evidence to date of nitrate leaching and eutrophication from fertilizer use in Asia, however. With the possible exception of intensively cultivated areas of East and Southeast Asia and pockets of high fertilizer use elsewhere, fertilizer use in Asia is low enough so that nitrate leaching and eutrophication do not pose a significant problem.

In developing Asia, as elsewhere, then, some danger from fertilizer to the environment and human health could develop. With fertilizer use in Asian developing countries likely to increase greatly over the next decades, emphasis must be placed on improving fertilizer use efficiency to limit additional surface and groundwater pollution. Much of the danger can be mitigated through more appropriate use of fertilizer, particularly removal of policies, such as fertilizer price subsidies, that encourage its overuse. In this way, the important yield gains from proper fertilizer use need not be negated in order to preserve the environment.

Plant Genetic Resources—A Narrowed Base?

Concerns about long-term effective use of fertilizer have combined with fears about lost genetic diversity through agricultural modernization to focus attention on the need to define how best to use the plant genetic base to help boost crop productivity and meet future food supply. Can plant genetic resources sustain further growth in food-crop yields, so that farmers' results approach the promise held by physical limits to crop productivity? Deriving the answer involves maintaining the resources in a way that heighten their chance of contributing to future yield gains.

Genetic resources can be conserved *ex situ* (not in the original or natural environment), or *in situ* (where naturally recurring). *Ex situ* strategies preserve plant seeds and propagating parts in gene banks, preventing the loss of species and subspecies. *In situ* conservation allows observation of the evolution of species as they interact with pests and pathogens (Smale and McBride 1996). *In situ* conservation of genetic resources may be an important complement to *ex situ* conservation because it allows adaptive and evolutionary processes to continue and may provide as yet unknown genetic characteristics for future breeding (Wright 1996; Smale and McBride 1996). Since these processes are longer-term in nature, though, for the foreseeable future crop-yield increases will rely on germplasm drawn from breeding lines stored *ex situ* (Wright 1996; Evenson and Gollin 1994).

Global *ex situ* storage of germplasm is substantial for the major food crops. Approximately 75 to 90 percent of the estimated genetic variation in the major crops and about 50 percent for minor crops is found in gene banks (Wilkes 1992). Concerns have been expressed, however, about the availability of information on sources, propagation techniques, basic characteristics, and the quality of some of the germplasm held in gene banks (McNeely et al. 1990). Even to maintain adequately, evaluate thoroughly, and document properly the system of germplasm banks already in place—without trying to expand its coverage of genetic diversity—will require sustained funding.

Although the available germplasm is characterized by wide genetic variation, the number of varieties actually tapped and utilized to develop new varieties is relatively small at any given time. This practice has led to the criticism that the development of modern rice and wheat varieties has narrowed the genetic base in farmers' fields, thereby increasing the threat of disastrous yield declines if, for example, genetic resistance to an insect or disease breaks down.

This criticism is based on a narrow understanding of genetic diversity in terms of just a couple of its many dimensions. Criticism usually focuses on spatial diversity and ignores trends even in that measure, at least for some crops. For rice, a decline in spatial diversity may well have followed the introduction of modern varieties in the 1960s. For wheat, on the other hand, spatial diversity (measured as the concentration of leading varieties in farmers' fields at a given time) is on the rise and is greater now than in the early 20th century (Smale 1996; Smale and McBride 1996). And for both rice and wheat, components of genetic diversity other than spatial diversity have improved over time. They include temporal diversity (average age and rate of replacement of cultivars); polygenic diversity (the pyramiding of multiple genes for resistance to provide longer-lasting protection from pathogens); and pedigree complexity (the number of landraces, pureline selections, and mutants that are ancestors of a released variety) (Evenson and Gollin 1994; Smale 1996).

In short, genetic diversity is multi-dimensional and extraordinarily complex, as well as difficult and expensive to measure. Trends in genetic diversity of cereal crops are mainly positive. This diversity, moreover, can do more than just protect against large downside risk for yields: it has in fact been generated primarily as a byproduct to breeding for yield and quality improvement and provides a pool of genetic resources for future yield growth. The threat of unforeseen, widespread, and catastrophic yield declines striking as the result of a narrower genetic base must be gauged against this reality.

Biotechnology—Higher Yield Ceilings?

The key to tapping the potential represented by the available genetic resources (and to increasing genetic diversity) will increasingly be the application of biotechnology techniques in tandem with conventional plant breeding. Biotechnology for agriculture includes

- agricultural microbiology;
- cell and tissue culture to propagate plant species more rapidly and facilitate "wide crosses" (crosses between different species);
- new diagnostic methods using monoclonal antibodies or nucleic acid probes to identify diseases and viruses;
- genetic mapping techniques for faster identification of useful genetic material to make conventional plant breeding more efficient; and
- genetic engineering to incorporate "alien" or novel genes into plant species (Persley 1994; Leisinger 1995).

The benefits from biotechnology include the introduction of higher plant resistance to pests and diseases; the development of tolerance to adverse weather conditions; the improvement in nutritional value of some foods; and ultimately the increase in the genetic yield potential of plants. While conventional breeding can have similar aims, genetic engineering can create "transgenic" crops that include genetic material that would otherwise never or only in extremely rare cases belong to a certain species (Kathen 1996), with a potential for greater gains made in each category.

The main successes of biotechnology thus far have been in improved pest and disease resistance, increasing yields through reduction in yield losses and extension of potential areas for production of high-yielding crops, rather than direct increases in crop-yield potential. A recent survey of releases of transgenic plants in developing countries identified 159 releases, nearly one half of which conveyed herbicide resistance, one third provided insect resistance, and the remainder virus-resistance, product-quality and other improvements (Kathen 1996).

Biotechnology research is currently dominated by the private sector in developed countries: it is estimated that some US $900 million was spent on agricultural biotechnology research and development in 1985, of which US $800 million was spent in developed countries and US $550 million by the private sector (Livernash 1996). The International Agricultural

Research Centers (IARCs), after a relatively slow start, have been increasing their research in crop-related modern biotechnology. Over the 1985–95 period, about US $260 million have been provided for international agricultural biotechnology programs, including US $206 million for 25 international agricultural research programs and about US $7 million for four international biotechnology networks (Cohen 1994).

The small share of developing countries in biotechnology research is partly due to time lags caused by the development of a complex and expensive technology that originated in the developed world. But it is also a function of what appears to have been a conscious decision on the part of developing-country research centers and the IARCs to "go slow" on biotechnology. This approach stemmed from the perception that (a) biotechnology research had not yet reached the state of "tool development" where large expenditures would be justified; (b) biotechnology research in the modern era of intellectual property rights is inherently a private-sector activity; and (c) the support system for the IARCs and National Agricultural Research Institutes (NARs) is oriented towards the development of technology, not upstream science (Evenson and Rosegrant 1993). Although all three justifications have some validity, since most current agricultural biotechnology research undertaken in developed countries is aimed at plants suitable for temperate climates, it will be crucial to increase biotechnology research aimed at the situations prevalent in developing countries in order to give these countries access to the next-generation yield potentials (Livernash 1996). In addition, more active biotechnology research in developing countries may be needed to help compensate for deleterious effects should new biotechnology products from the developed world replace exports from the developing world, as well as to guarantee proper compensation of biotechnology development based on developing-world genetic resources (Leisinger 1995). It will also be important for IARCs to react quickly to new developments in this area, given the fast pace of research and, in particular, decisions on property rights.

Fortunately, new institutional arrangements between developed and developing countries have been put in place recently and some developing countries, like the PRC and India, have increased their annual budgets for their research institutes. The IARCs could play an essential role in developing local biotechnology capacity, sharing information across countries, and collaborating with private-sector partners (Livernash 1996). This process would be greatly facilitated by the removal of unnecessary barriers to the free movement of plant materials, clarification of biosafety regulations, and provision of improved property rights protection for new products (Yudelman 1996). If funding and collaboration efforts between international centers continue to grow, biotechnology will provide a significant boost to crop production in the next century, proving particularly beneficial to developing countries with strong technological potential (Commandeur and von Roozendaal 1993 as cited in Leisinger 1995).

At the same time, given the reliance of biotechnology on new combinations of genetic elements, there are some uncertainties about potential negative effects, as underscored by the reluctance among some consumers to accept products generated by biotechnology. Will new creations crowd out native species, to the ultimate detriment of the gene pool? What are the hazards of a dangerous substance being created, then unleashed (or evolving into something dangerous after release)? Significant research into the extent of the danger is needed; policies that ensure caution as regards release are already a step in this direction.

CLIMATE CHANGE—DIFFERENT CROP PRODUCTIVITY EFFECTS FOR DEVELOPING VS. DEVELOPED WORLD

According to many studies, in the coming decades, global agriculture faces the prospect of a changing climate, which might adversely affect the goal of meeting global food needs. The prospective climate change consists of global warming and associated changes in hydrological regimes and other climatic variables, such as higher temperatures, shorter growing seasons, changing moisture regimes, and extreme weather patterns. It also includes secondary effects on social and economic systems induced by increasing concentrations of greenhouse gases from human activities, especially carbon dioxide (CO_2), which is projected to double by the year 2100, producing an expected temperature rise in the range of 1.5°–4.5°C (Wolfe 1996; Downing 1993; Kendall and Pimentel 1994).

Global warming could have both negative and positive effects on agriculture. A 1°C increase in mean annual temperature may advance the thermal limits of cereal cropping in the mid-latitude Northern Hemisphere by 150–200 km (Schimmelpfennig et al. 1996). At higher latitudes, increased temperatures can lengthen the growing season and ameliorate cold-temperature effects on growth. In warmer mid-latitude environments, adverse effects of climate change include increased pests and disease on crops and livestock, soil erosion, and desertification due to more intense rainfall and prolonged dry periods, as well as reduced water resources for irrigation (Downing 1993). Despite the many studies on global warming since the 1980s, however, there is no consensus on the impact of three major variables on agriculture: the magnitude of regional changes in temperature and precipitation, the magnitude of the beneficial effects of higher CO_2 on crop yields, and the ability of farmers to adapt to climate changes (Wolfe 1996).

Sensitivity studies of world agriculture to potential climate changes have indicated that global warming may have only a small overall impact on world food production, because reduced production and yields in some areas are offset by increases in others. Tropical regions, however, including many Asian regions, may suffer negative impacts from droughts, due to the nonlinear relationship between temperature and evapotranspiration, even though climate changes in these regions are expected to be less. These regions will also face greater difficulties in shifting planting dates, as they are limited more by rainfall than by temperature (Reilly 1995). Although results vary by climate-change scenario and by study, regions critically vulnerable in terms of resources to support their populations and projected decreases in water available to plants include parts of the semi-arid tropics and sub-tropics, such as western Arabia, southern Africa, or eastern Brazil, and some humid tropical and equatorial regions, such as Southeast Asia and Central America (Downing 1993). Most studies also conclude that changes will benefit Japan and the PRC.

Moderate global warming can have positive impacts on crop yields. Most plants growing in experimental environments with enhanced CO_2 levels exhibit a "CO_2 fertilization" effect that increases crop yields. Under experimental conditions, for rice, wheat, and more than 90 percent of the world's plant species, the estimated effect from a doubling of CO_2 is a 30-percent yield increase. For maize, millet, sorghum, and sugar cane, the effect is a much lower 7-percent yield increase (Schimmelpfennig et al. 1996). Under field conditions, with CO_2-stimulated weeds, potential lack of water and other nutrients, yield increases are estimated to be only one quarter to one third of the effect under experimental conditions (Kendall and Pimentel 1994).

In order to assess the potential impact of climate change on agriculture and food supply, complex climate, crop growth, and economic/food trade models have been linked. Between 1989 and 1992, a comprehensive study of alternative scenarios for the direct effects of greenhouse gas-induced climate changes on crop yields (wheat, rice, maize, and soybean) was conducted at 112 sites in 18 countries with the help of crop-growth models. The study concluded that with a continuation of current trends in economic growth rates, partial trade liberalization, and medium population growth rates; with assumed modest farm-level adaptations to climate change; and without the CO_2 fertilization effect; the estimated net impact of climate change would be a reduction in global cereal production of up to five percent by 2060. This global reduction could be largely overcome by major forms of adaptation such as installation of irrigation. The climate change would increase the disparities between developing and developed countries: production in the developed world could benefit from climate change and production in developing nations could decline. In scenarios that simulate more aggressive economic and farm-level adaptations to changing climate and that include the CO_2 fertilization effect, negative global cereal-yield impacts are nearly eliminated (with estimated yield changes in the range of +1.0 percent to –2.5 percent) and persist only in developing countries (Rosenzweig et al. 1993).

Similar scenarios focusing on results in Asia showed large ranges of crop-yield swings with doubling of CO_2. Matthews et al. (1995, as cited in Rosenzweig and Hillel 1998) largely bear out the result of lower yields in more tropical areas (up to a 30-percent drop), and higher yields at higher latitudes (up to a 38-percent rise), at least for rice and given current varietal vulnerabilities. One study for the southern PRC suggested higher temperatures would extend northward by as much as 10 degrees of latitude the area suitable for double- or triple-cropping rice (Jin et al. [1995] as cited in Rosenzweig and Hillel [1998]). Rising sea levels could also encroach on agricultural lands, a particular threat for Bangladesh and also for Malaysia (Karim et al. [1994] and Parry et al. [1992] for the two countries, respectively, both as cited in Rosenzweig and Hillel [1998]).

More recent studies concur that the negative effects of climate change on agriculture are likely to have been overestimated by studies that do not take into account broader economic and environmental implications or account for economic adjustments. Utilizing a modeling approach that captures some of these adjustment processes, Darwin et al. (1995) conclude that

- global changes in temperature and precipitation patterns are not likely to endanger food production for the world as a whole;
- farmer adaptations are the main mechanisms for keeping up world food production under global climate change;
- costs and benefits of global climate change will not be equally distributed around the world; and

- although water supplies are likely to increase as a whole under climate change, regional and local water shortages could occur.

The impact on crop yields is generally more positive: world cereal production increases by between 0.9 percent and 1.2 percent, even without the CO_2 fertilization effect (Darwin et al. 1995).

Prospective global temperature increases will occur gradually and not until far into the next century; crop yield reductions and economic losses due to global warming appear manageable (perhaps even positive over the next few decades). Thus, global warming will have little or no impact on global food production through the year 2010, but could increasingly disrupt Asian agriculture in the more distant future.

CONCLUSIONS

Asia is fast approaching its limits in terms of area available for cropland, and Asian agriculture may literally lose ground to urbanization. Likewise, development of new irrigation is likely to be slow for Asia, principally due to its high costs. Nontraditional water-development options are not realistic alternatives in enough areas to solve the problem or are also too expensive. Here too, the nonagricultural sector's growing demands threaten to siphon water from agriculture and reduce productivity. While some gains in efficiency of water use may mitigate this effect, complex tradeoffs are involved. Environmental degradation is widespread and its productivity effects are beginning to be felt. While these are not yet at a level that threatens the food supply, impacts are being felt primarily in high-potential areas and are direct byproducts of precisely the technologies used to date to boost yields, for example, overuse of fertilizer and improper management of irrigation systems. If not countered, this trend could spell danger in the future, especially given the aforementioned limits to expansion.

Other avenues for yield growth being built into research strategies are undertaken to close the gap between the yields farmers achieve and theoretical maxima and to push up those maxima. At the same time, buffers to catastrophic yield declines now or in the future are starting to be more explicitly factored into strategies for preserving genetic diversity on site and in germplasm banks.

The broad conclusion is that environmental and resource constraints are not intrinsically limiting to the necessary growth in crop production to meet Asian food demand in the coming decades. But continued weakness in environmental policies could lead to significantly increased resource degradation; any negative changes of unexpected magnitude or in a shorter period of time than expected could mean a slowdown in projected agricultural production growth significant enough to pose a real danger to food security.

For land, policies to counteract degradation should be targeted toward high-risk zones. In these zones, significant public investments in research, technology development, extension services, and rural infrastructure may be necessary to stabilize or reverse degradation. In addition, high risk should be defined not just by the extent of degradation, but also by crop-productivity effect, taking into account the area's importance to overall food supply. More broadly, land degradation can also be mitigated through policy reforms, such as the establishment of property rights to land, market and price reforms, and the elimination of subsidies to agricultural inputs.

For water, many questions remain unanswered. Can significant real water savings be achieved through improved water-management policies? What will be the impact on food production and food security of transfers of saved water out of agriculture? Understanding the contributions of water-management and -investment policies to future food security would provide important guidance to national and international policymakers and could generate large benefits for food producers and consumers in developing countries. But it is clear that both significant new investments in water-supply and sanitation systems and irrigation systems and reform of water-demand management will be necessary. Water management will need to move to market-oriented allocation methods as well as community-based management.

In any scenario, improvements in crop yields must occur for future food security to be guaranteed. The threat of land and water degradation and scarcity only up the ante in this regard. And awareness has been raised through experience with the environmental consequences of techniques that aimed to, and did, substantially improve crop yields, staving off a danger to world food supplies in the past: more care must be taken in the design and implementation of continued efforts to further boost yields, lest they inadvertently have the opposite effect. That said, raised awareness should not prevent careful evaluation of risks and benefits. While efforts to preserve and improve genetic diversity need to be stepped up and made more sophisticated, explicitly taking into account evolutionary pressures on plants and pathogens, a strong foundation for this exists. And, while the unknowns in biotechnology or the potential risk from a misstep may be great, there are also potentially great rewards from steps already known to carry less risk; these must be pursued.

While it may seem unrealistic to expect simultaneous improvement in all these areas, especially given the complexity and amount in each area that is still unknown, the fact that the set of policies necessary to encourage food-production growth while protecting the environment is quite consistent across the resource issues reviewed here should heighten chances that they can be conveyed and implemented. In the broadest sense, these are the same policies that have been discussed in several other chapters and several other contexts above, to improve the flexibility of resource allocation in agriculture: removal of subsidies and taxes that distort incentives; establishment of secure property rights; investments in research, education and training, and public infrastructure; better integration of international commodity markets; and a greater inclusion of populations in Asian economies into these markets.

XI Challenges for Less Favored Areas

INTRODUCTION

Past agricultural development strategies in Asia, including the green revolution, have concentrated on irrigated agriculture and "high-potential" rainfed lands in an attempt to increase food production and stimulate economic growth. This strategy has been spectacularly successful in many countries and was responsible for the transformation of rural Asia, and the resulting general economic development, described heretofore in this book.

At the same time, however, large areas of agricultural land have been neglected and lag behind in their economic development. These lands are characterized by lower agricultural potential, often because of poorer soils, shorter growing seasons, and lower and uncertain rainfall, but also because past neglect has left them with limited infrastructure and poor access to markets. Despite some out-migration to more rapidly growing areas, the population of these areas continues to grow and this growth has not been matched by increases in yields. The result is often worsening poverty and food insecurity problems, as well as widespread degradation of natural resources (for example, mining of soil fertility, soil erosion, deforestation, and loss of biodiversity) as people seek to expand cropped area.

It has become increasingly clear that, if the general goals of poverty alleviation and environmental sustainability through development are to be met, the less favored areas cannot continue to be bypassed by the revolution that has transformed so much of rural Asia. More attention will have to be given to these lands in setting priorities for policy and public investments. Two key policy questions need to be addresssed and answered: first, what level of investment can be justified in less favored areas? Second, how should the resources allocated to less favored areas be used to promote development that is beneficial to the poor and environmentally sustainable?

EXTENT OF POVERTY AND ENVIRONMENTAL DEGRADATION IN LESS FAVORED AREAS

Less favored areas (LFAs) are extensive in Asia. According to a report prepared by the Technical Advisory Committee of the Consultative Group for International Agricultural Research (CGIAR) (TAC 1996), 550 million ha of land currently used for agricultural purposes in Asia can be classified as "marginal," and another 340 million ha are sparsely populated arid lands. In contrast, there are only 305 million ha of high-potential or "favored" environments currently in agricultural use.

The LFAs are also home to a significant share of Asia's rural poor. Using FAO and other data sources, Hazell and Garrett (1996) estimated that 263 million rural poor live in less favored lands in Asia. Given that there were 633 million rural poor living in Asia in 1988 (IFAD 1995), this implies that about 40 percent of Asia's rural poor live in LFAs. While a very rough estimate, this figure is supported by more precise estimates for India and the PRC.

For India, Fan and Hazell (1999) used regional data available for 47 out of 65 agroclimatic zones to arrive at the results in Table XI.1. In 1993, of the 184 million rural poor living in the 47 agroclimatic zones analyzed, 16 percent, or 30 million, lived in irrigated areas (defined as areas with more than 40 percent of the cropped area irrigated), while the vast majority— 84 percent, or 154 million—lived in rainfed areas. Fan and Hazell also classified the rainfed areas as high- or low-potential based on climate and soil data for each agroclimatic zone. On this basis, 76 million people, or 41 percent of all of India's rural poor,

Table XI.1: Changes in Rural Poverty by Type of Regions in India 1972, 1987, and 1993

	Irrigated Areas	Rainfed Areas		
		Total	High Potential	Low Potential
Number of Poor (Millions)				
1993	30	154	78	76
1987	35	167	79	88
1972	37	155	80	75
Percentage of Poor in Total Population (%)				
1993	28	39	44	36
1987	32	46	48	44
1972	39	52	59	47

Source: Fan and Hazell (1999).

Note: Only 47 agroclimatic zones (of total 65) are included in the calculation due to data unavailability. An agroclimatic zone is defined as rainfed if less than 40 percent of the total cropped area is irrigated, and as irrigated if irrigated share is over 40 percent.

lived in low-potential rainfed areas in 1993. That same year, the incidence of rural poverty as a percentage of the total population was also higher in the rainfed areas; (39 percent in rainfed as compared to 28 percent in irrigated areas), though it was higher in the high-potential rainfed areas (44 percent) than in the low-potential rainfed areas (36 percent). Table XI.1 also shows the corresponding poverty data for earlier years of 1987 and 1972. Despite the fact that India's total rural population increased by about 50 percent between 1972 and 1993, the total number of rural poor declined a little (from 192 million in 1972 to 184 million in 1993). Virtually all the reduction occurred in irrigated areas, and there was very little change in the number of poor in either type of rainfed area. But because of population growth, the poor represented a declining proportion of the total rural population in irrigated and rainfed areas.

Similar poverty data are currently being compiled for the PRC by Shenggen Fan at IFPRI. Tentative estimates are that 43 percent of the total rural poor live in regions that can best be characterized as low-potential agricultural areas (Fan, personal communication).

Poverty and Productivity in the LFAs: Evidence of Links

Further confirmation of the high concentration of poverty in LFAs comes from the analysis by Sharma et al. (1996) of the relationship between the incidence of malnourished children and the CGIAR's agro-ecological zones (AEZs). For Asia, the highest incidence of malnourished children (62 percent) occurs in the warm semi-arid tropics, where food production per hectare and per person are the lowest. The incidence of malnutrition is lowest (20 to 25 percent) in the cooler sub-tropical AEZs, where food production per person is among the highest of all AEZs. In most of the semi-arid and subhumid AEZs (with productivity levels between these extremes), the incidence of childhood malnutrition runs at about 45–50 percent. More evidence from India comes from Hossain (1995), who shows a strong and negative correlation between foodgrain yield and the incidence of poverty across states. Poverty is especially acute in regions where foodgrain yields have remained low, such as Bihar, Orissa, and Madhya Pradesh.

Many LFAs with lower agricultural production have also seen that production grows more slowly than in irrigated and high-potential rainfed areas. Hossain (1995), for example, reports that most of Asia's growth in rice production in recent decades has arisen from yield growth within irrigated areas and from expansion of irrigated areas. There has been relatively little growth, by contrast, in rainfed, particularly upland, rice-growing areas. From 1964–93, yields grew by 2.1 percent per year on average in intensively irrigated areas (more than 50 percent irrigated), but in largely rainfed areas (less than 50 percent irrigated) they grew by only 1.2 percent per year. Yields actually grew more slowly than population in the rainfed areas, and to compensate, farmers expanded their cultivated rice area at a faster rate in the rainfed than in the irrigated areas (0.5 percent compared to 0.3 percent). Even so, production grew at only 1.7 percent per year in the largely rainfed areas compared to 2.4 percent in the mostly irrigated areas. Current rice yields exemplify the widening disparities among regions: irrigated areas yield about 5.0 t/ha of rice, compared to 2.3 t/

ha for rainfed lowlands, 1.5 t/ha for flood-prone areas, and 1.1 t/ha for upland areas. Since the low-yielding areas (flood-prone and rainfed uplands) account for about one third of Asia's total rice area (Crosson 1996), or 31 million ha, the gap has important implications for large numbers of rural people.

In India during 1992–94, total agricultural production per hectare (including crop and livestock production) in the low-potential rainfed areas was only 40 percent of that in the irrigated areas and 46 percent of that in the high-potential rainfed areas (Fan and Hazell 1999). The gap between irrigated and low-potential rainfed areas has changed little since 1970 because agricultural production grew at similar rates in both types of areas (2.26 and 2.68 percent per year, respectively). But the gap between low- and high-potential rainfed areas has widened; in 1970, total agricultural production per hectare in the low-potential rainfed areas was 66 percent of that in high-potential rainfed areas, but in 1992–94 it was only 46 percent as large. These gaps seem destined to widen; since 1990, low-potential rainfed areas have undergone an apparent slowdown to 1.9 percent growth per year in agricultural production, while both irrigated and high-potential rainfed areas have seen slightly higher agricultural production growth rates (2.6 and 3.6 percent per year, respectively).

In another worrying trend, labor productivity has grown much more slowly in the low-potential rainfed areas than elsewhere in India, averaging only 0.25 percent per year since 1970 (Fan and Hazell 1999). Since wage rates and per capita incomes are highly correlated with labor productivity, this trend does not bode well for the future welfare of the rural poor in these regions.

Environmental Degradation in the LFAs: Outcome of Low Productivity and Expansion of Cropped Area

The low baseline productivity and slower growth rates of the LFAs lie behind a good part of the continuing and sometimes worsening poverty and food insecurity problems there. With continuing population growth and a scarcity of good land, cropped area can only be expanded by encroaching into forested and woodland areas and onto steeper slopes, with increasing soil erosion (see, for example, Zeigler, Hossain, and Teng 1994; Penning de Vries et al. 1998; Scherr and Hazell 1994). Soil erosion contributes not only to lower yields on site but also to siltation problems downstream, reducing the capacity and productivity of reservoir and irrigation schemes and thereby affecting an even broader area. Likewise, deforestation in upper watershed regions can also have broader effects, for example by contributing to flooding problems in lowland areas.

These problems are already serious in many "hot spot" areas in Asia such as the foothills of the Himalayas, sloping areas in southern PRC and Southeast Asia, and the forest margins of Indonesia, Malaysia, Viet Nam, Cambodia and the Lao PDR (Scherr and Yadav 1995). This kind of degradation can result in high economic costs. One study of soil erosion in Java, for example, estimated the economic costs at between US$340 million and US$406 million per year, of which about 90 percent were on-site costs and 10 percent were downstream damages from siltation (Magrath and Arens 1989). Unfortunately, few reliable studies of this kind have been undertaken, and while there are good reasons to be skeptical about some of the available estimates of the extent and costs of land degradation in Asia (see Chapter VIII and Crosson and Anderson 1992), there are, nevertheless, good reasons to expect that the problems are most severe within the LFAs themselves.

RETURNS ON PUBLIC INVESTMENTS IN THE LFAS

The amount of public investment that can be justified in any region should depend on the net social returns realized through productivity growth, poverty reduction, and the containment of environmental degradation. While investments that contribute to all three of these goals ("win-win-win" outcomes) are generally better, an investment that involves some tradeoff among these social goals may also be attractive, providing any sacrifice of one goal is adequately compensated by gains in others.

Returns on Investment in High-Potential Areas: New Realities Challenge Conventional Wisdom

Conventional wisdom suggests that the productivity returns on investment are highest in irrigated and high-potential rainfed areas, and that growth in these areas also has substantial trickle-down benefits for the poor, including those residing in the LFAs. Even though investing in LFAs might have a greater direct impact on the poor living in those areas, it is argued that investments in high-potential areas give higher social returns for a nation than investments in low-potential areas and a more favorable impact on poverty in the long run. The logic behind this position is as follows: investment in high-potential areas generates more agricultural output and higher economic growth at lower cost than in less favored areas. Faster economic growth leads to more employment and higher wages nationally and greater agricultural output leads to lower food prices, both of which are beneficial to the poor. LFAs will benefit from cheaper food, from increased market opportunities for growth, and from new opportunities for workers to migrate to more productive jobs in the high-potential areas and in towns. Fewer people will try to live in less favored lands; this will help reduce environmental degradation and increase per capita earnings. Migrants may also send remittances back to LFAs, further increasing per capita incomes there, especially for the poor.

Many of the expected benefits arising from rapid agricultural growth in high-potential areas have been confirmed in Asia (Pinstrup-Andersen and Hazell 1985; Hazell and Ramasamy 1991; David and Otsuka 1994). Nevertheless, the rationale for neglecting the LFAs is being increasingly challenged by a) the failure of past patterns of agricultural growth to resolve growing poverty, food insecurity, and environmental problems in many LFAs; b) increasing evidence of stagnating levels of productivity growth and worsening environmental problems in many high-potential areas that make further investments less attractive (see Chapter V and Pingali and Rosegrant 1994); and c) emerging evidence that the right kinds of investments can increase agricultural productivity in many LFAs to higher levels than previously thought.

It now seems plausible that increased public investments in many LFAs may have the potential to generate competitive if not greater agricultural growth on the margin than comparable investments in many high-potential areas, at the same time having a greater impact on the poverty and environmental problems of the LFAs to which they are targeted. If so, then additional investments in less favored areas may actually give higher aggregate social returns to a nation than additional investments in high-potential areas. In fact, they might even offer "win-win-win" possibilities.

Productivity and Poverty Impact of Investment in LFAs: Emerging Evidence

There have been few rigorous attempts to test these competing hypotheses, hence a recent study completed by IFPRI is of relevance (Fan and Hazell 1999). This study analyzed the growth and poverty-alleviation impact of alternative types of investments in high- and low-potential areas in India over recent decades (environmental data were not available). India provides a good example because past public investments have been biased towards high-potential areas, and the remarkable productivity gains achieved in those areas (which have led to national food surpluses) can now be juxtaposed with the lagging productivity, widespread poverty, food insecurity, and environmental degradation that exist in many less favored rainfed areas.

The study combined data on cross-sections (districts for production and agroclimatic zones for poverty) and time series (annually from 1970 to 1993 for production, and 1972, 1987 and

1993 for poverty) to estimate an econometric model that included production and poverty functions. These functions included, among other things, public-investment variables on the right-hand side. This enabled calculation of the marginal contribution of an additional unit of investment in each type of infrastructure to agricultural production and poverty alleviation. The model was estimated for three different types of lands: irrigated, high-potential rainfed, and low-potential rainfed. Table XI.2 summarizes the key results.

For every investment, the highest marginal impact on production and poverty alleviation occurs in one of the two rainfed lands, while irrigated areas rank second or last. Moreover, all types of investments in low-potential rainfed lands, except regulated markets, give some of the highest production returns and have some of the most favorable impacts on poverty. These results provide strong support to the conjecture that more investment should now be channeled to LFAs in India.

More specifically, the marginal impact of high-yield varieties (HYVs) on production is much larger in high- and low-potential rainfed areas (243 and 688 rupees per hectare of HYVs adopted, respectively) than in irrigated areas (63 rupees per hectare). HYVs also contribute more to poverty alleviation in rainfed areas; another hectare of HYVs raises 0.02 and 0.05 persons above the poverty line in high- and low-potential rainfed areas, respectively, compared to no measurable poverty impact in irrigated areas. Roads have sizeable productivity impacts in all three types of areas, but a much larger impact on poverty alleviation in rainfed areas, particularly the low-potential rainfed lands. Rural electrification and education have their biggest productivity impacts in rainfed areas; they also impact favorably on the poor in these areas. Their impacts in irrigated areas are very small. Canal irrigation has its biggest productivity and poverty impacts in high-potential rainfed areas, while private irrigation has its biggest impacts in low-potential rainfed areas. These results seem consistent with the relative water endowments of the two types of rainfed lands; there is little surface or groundwater to capture through private

Table XI.2: Marginal Returns to Infrastructure and Technology Inputs by Type of Region, India, 1994

		Irrigated	High-potential Rainfed	Low-potential Rainfed
Returns to Production (1990 prices)				
HYV	Rps/Ha	63	243	688
Roads	Rps/Km	100,598	6,451	136,173
Markets	Rps/Number	(276,745)	7,808,112	(4,794,073)
Canal Irrigation	Rps /Ha	938	3,310	1,434
Private Irrigation	Rps/Ha	1,000	(2,213)	4,559
Electrification	Rps/Ha	(546)	96	1,274
Education	Rps/Labor	(360)	571	102
Returns to Poverty Reduction				
HYV	Persons/Ha	0.00	0.02	0.05
Roads	Persons/Km	1.57	3.50	9.51
Markets	Persons/Number	(2.62)	537.79	(312.72)
Canal Irrigation	Persons /Ha	0.01	0.23	0.09
Private Irrigation	Persons/Ha	0.01	(0.15)	0.30
Electrification	Persons/Ha	0.01	0.07	0.10
Education	Persons/Labor	0.01	0.23	0.01

Source: Fan and Hazell (1999).

Notes: Numbers in parentheses are negative; in most cases they are not statistically significant.

investment in low-potential areas because of their low rainfall. Market development has a huge marginal impact on production and poverty alleviation in the high-potential rainfed areas, but not in irrigated and low-potential rainfed areas.

It should be noted that the marginal impact of different investments is reported gross of their costs. It could be argued that some investments are more expensive to undertake in the LFAs, because of the diverse and generally less favorable agro-ecological conditions. The development of HYVs for less favored areas, for instance, may be more difficult, and widespread adoption more constrained by the diversity of growing conditions, than past experience with high-potential areas, if extrapolated to low-potential areas, might suggest. Investments in roads and other infrastructure may also be more costly per kilometer in many less favored lands because of difficult topographical conditions or remoteness from major population centers or markets. But data obtained at the state level for India (Fan, Hazell, and Thorat, 1998) suggest that the unit costs of key investments are not all that different across states, despite considerable diversity in the proportions of their irrigated and rainfed areas.

STRATEGIES FOR DEVELOPING LFAS: RECOGNIZING COMMONALITY AMID DIVERSITY

Many past attempts to develop LFAs in Asia (for example, integrated rural development projects and watershed-development projects), including a good deal of agricultural research undertaken by national and international agricultural research centers, were not very successful. The reasons for this are complex, but include

- inappropriate macro, trade and sector policies that penalized agriculture;
- insufficient levels of investment in agricultural research in LFAs and research of the wrong type;
- insufficient investments in rural infrastructure and human capital targeted on LFAs;
- inappropriate development strategies that were too top-down; and
- poor performance by, and coordination among, many public-sector institutions working in LFAs.

Structural-adjustment programs and market-liberalization policies have created a more enabling economic environment for the development of many LFAs, although recent reductions in government budgets as a result of macro-economic policy reforms and the economic crisis are also constraining the needed expansion of public investment in these areas. But future investments in LFAs need to be based on new or improved paradigms for sustainable development.

LFAs are very diverse in their agroclimatic conditions and hence in their potential for agricultural growth. In some areas agricultural development may not be an economically viable alternative, and solutions will have to be sought through development of the rural nonfarm economy and through accelerated out-migration. Possibilities for achieving these alternatives are most promising when the national economy is growing rapidly and when agriculture has become a relatively minor share of national income and employment (as in many of the fast-growing East and Southeast Asian economies). Prospects are much less promising in slow-growing and predominantly agrarian economies, since the rural nonfarm economy is then constrained by local demand for its output, which is in turn constrained by the level of per capita incomes. Without agricultural growth, incomes in these areas remain low and the demand for rural nonfarm goods and services remains stagnant (Hazell and Reardon 1998).

Migration and nonfarm diversification will also have to play an important role in the long run for most types of LFAs if their per capita incomes are to keep pace with rising national living standards. This longer-term view needs to be kept in mind when strategies for the short to medium term are being developed, particularly when those strategies are focused on alleviating poverty and environmental problems. Policymakers need to avoid inadvertently locking too many people into marginal areas where their long-term prospects are limited.

But for many less favored lands, agricultural intensification must be a key component of their development strategy, particularly over the next few decades while the number of people living in these areas continues to grow. But because of poor infrastructure, low to moderate yield potential, fragile soils, and high climate risk, the strategy will typically need to be different from the green-revolution approach adopted in irrigated and high-potential rainfed areas.

Agricultural development strategies need to be tailored to local agroclimatic conditions and to the social and economic conditions that determine the type of development pathway that local communities are best suited to follow (Pender, Place, and Ehui 1998). There are no "one-size-fits-all" approaches. Nevertheless, some common elements of appropriate development strategies can be identified.

Promote Broad-Based Agricultural Development

Broad-based agricultural development that reaches small and medium-sized family farms as well as larger commercial farms should be promoted. There are few economies of scale in agricultural production in developing countries (unlike many agricultural processing and marketing activities), hence targeting family farms is attractive on both equity and efficiency grounds. Broad-based development strategies require that small and medium-sized farms receive priority in publicly funded agricultural research and extension and that they obtain adequate access to markets and credit and input supplies. These requirements demand special attention at a time when markets and agricultural services are being privatized, since the high transport costs and thin markets of many LFAs do not make them attractive to private agents. Special attention must also be given to women farmers, who have traditionally been discriminated against in their access to resources and improved technologies, credit and farm inputs.

Improve Technologies and Farming Systems

Because of the poor infrastructure, low yield response and high climate risk in many LFAs, the intensive use of modern inputs like fertilizers is unlikely to be economical. Monocrop farming systems, moreover, can be environmentally destructive as well as very risky. Agricultural researchers and farmers need to step back from narrow commodity approaches and take a more holistic approach to improving resource-management practices at the farm and landscape levels. These may need to include a) management at the watershed level of water catchment and use and soil erosion control; b) improved soil moisture and fertility management, including improved crop rotations and intercropping and better integration of farm trees and livestock into cropping systems to generate and recycle plant nutrients; and c) more rational exploitation of favorable niches in the landscape for production of high-value crops and trees (Scherr and Hazell 1994). This will require that research be more multidisciplinary, more site-specific, and more responsive to farmer (both men and women) and community needs.

Ensure Equitable and Secure Access to Natural Resources

The distribution of land is often quite inequitable in the LFAs, land leases tend to be short term and insecure, and the incidence of landlessness is high. The poor are also dependent on access to common property resources (for example, woodlots, grazing areas, and wetlands) to supplement their incomes, yet these resources are increasingly being degraded as more people use them and the ability of local organizations to regulate and manage their use erodes. These problems affect poverty and economic development in a number of ways. For example, uneven access to land can prevent the most efficient allocation of land, labor and other inputs, with too much land tied up in the hands of larger, less efficient producers and away from more efficient smallholders. Insecure tenure inhibits land-improving investments and may encourage unsustainable farming practices. Land access is highly correlated with poverty and households with even the smallest holdings face a much lower risk of absolute poverty than landless households (Mearns 1999).

Land-reform programs are politically difficult to implement and have not been very successful in redistributing land to the poor. Market-assisted land redistribution offers a new and potentially promising way of avoiding the usual political barriers to achieving such change (Van Zyl, Kirsten, and Binswanger 1996). Efficient land-lease markets can also help offset many of the worst effects of an inequitable distribution of land ownership. Unfortunately, land leasing is often discouraged by government policy; in India, for example, many states discourage land leasing and some even outlaw it altogether, an outgrowth of "land-to-the-tiller" movements that have sought to protect farmers' rights to their land. While such laws rarely succeed in eliminating all land leasing, they do constrain its volume and, by forcing its concealment from the authorities, lead to short-term leases with little protection for the tenant. There is a need to reform these policies. Mearns and Sinha (1999) suggest opening up land-lease markets in combination with the credible enforcement of land-ceiling laws and clearly defined, enforceable lease contracts.

At the same time, land fragmentation is often cited as a constraint to agricultural development (for example, Singh 1990), but evidence shows that it performs a useful role in helping farmers spread risk in rainfed areas with variable microclimates (Ballabh and Walker 1992; Blarel et al. 1992). Consolidation programs have not made any headway in rainfed areas and are not likely to be a useful investment.

Farmers also need assured long-term access to land if they are to pursue sustainable farming practices and to make long-term investments in improving and conserving resources (such as tree planting, continuous manuring, and terracing and contouring for soil and moisture conservation); see, for example, recent work by Otsuka, Suyanto, and Tomich (1997) and Pender and Kerr (1996). Feder et al. (1988) have demonstrated the value of land titles in areas where property rights are insecure, such as in newly settled areas.

Many resources are owned and managed as common properties in less favored lands because this provides a more effective way to share risks and to ensure equitable access to resources by all members of the community. If these resources are to remain common properties and are not to be privatized or overexploited, effective local organizations are needed to manage them. Often, governments have undermined indigenous institutions by nationalizing important common-property resources, such as forests and rangelands. Public institutions have then failed to manage these resources effectively and they have degenerated into open-access areas. The most successful institutions for managing common properties are local organizations dominated by the resource users themselves.

Conserving or improving natural resources often requires collective action by groups of users even when the resources are not commonly owned. Examples include organizing adjacent farmers in a landscape to invest labor in land terracing, bunding or water catchment, or for biological pest control. Organizing farmers into effective and stable groups for collective action is difficult and success is conditioned by a range of physical, social, and institutional factors (Uphoff 1986; Ostrom 1994; Rasmussen and Meinzen-Dick 1995). Collective action is facilitated if there is a smaller number of users, if there is homogeneity of members in terms of shared values and economic dependence on the activities of the group, and if the net benefits from group membership are substantial and equitably distributed. Institutional design is also important. Ostrom (1994) has identified seven design principles for effective local organizations:

- a clear definition of the members and the boundaries of any resource to be managed or improved;
- a clear set of rules and obligations adapted to local conditions;
- the ability of members to modify those rules collectively in response to changing circumstances;
- adequate monitoring systems; with
- enforceable sanctions, preferably graduated to match the seriousness and context of the offense;
- effective mechanisms for conflict resolution; and
- the protection of the organization, if not its empowerment or recognition by government authorities, against being challenged or undermined by those authorities.

Policymakers can facilitate more effective community management by

- legitimizing the ownership rights of the group;
- providing institutional options for resolving disputes, particularly with outsiders; and
- recognizing the role of some NGOs in helping local organizations to manage common-property resources.

Ensure That Risks Are Managed Effectively

Risk management is important in all rainfed environments, but the problems are most severe in low- and high-rainfall areas that are susceptible to catastrophic droughts and floods. The economic problems resulting from climate risks are the most severe in areas where poverty is widespread and the regional economy is heavily dependent on agriculture. Risk of crop or livestock losses due to bad weather, pests, or diseases can discourage investments by farmers in land improvements and adoption by them of productivity-enhancing technologies.

Agricultural research can help reduce risk, for example, by improving drought or pest resistance in crops or developing better ways to conserve soil moisture. Additionally, governments may need to assist farmers in coping with catastrophic losses, particularly losses arising from risks that affect most farmers in a region at the same time (drought, for instance), and to provide effective safety-net programs and credit and insurance markets. Care should be taken in designing such interventions, however, for if heavily subsidized, they can all too easily lead to changes in farming practices that increase the dependence of the beneficiaries on subsidized assistance in the future. Subsidized drought insurance, for example, increases the profitability of risky farming practices beyond their true economic value and encourages their adoption, even though this may lead to greater financial exposure in future droughts, as well as to resource degradation.

Agricultural insurance has often appealed to policymakers as an instrument of choice for helping farmers and agricultural banks manage climate risks like drought; many countries in Asia spend large sums of public money each year on such insurance. But the experience has generally not been favorable (Hazell, Pomareda, and Valdés 1986; Hazell 1992). Publicly provided crop insurance has without exception depended on massive subsidies from the government and even then, its performance has been plagued by moral-hazard problems associated with many sources of yield loss, by high administration costs, by political interference, and by the difficulties of maintaining the managerial and financial integrity of the insurer when government underwrites all losses (Hazell 1992). Nor has crop insurance been able to reach the poorer farmers or to assist nonfarm members of rural communities who also suffer in catastrophic agricultural years (among them landless laborers, agricultural traders, and shopkeepers). Area-based yield insurance may offer a better alternative, and has recently been tried in India with some success (Mishra 1996). Unfortunately, it remains very costly to the government, because the premium rates are set far too low in relation to costs. It is also unnecessarily restricted to farmers who grow the insured crops. A more promising approach would be area-based insurance based on rainfall rather than yield. This could be a useful risk-management aid to all kinds of rural households and could be simpler and cheaper to operate than area-yield insurance schemes (Hazell 1992).

Invest in Rural Infrastructure and People

LFAs are often poorly placed to compete in liberalized markets because of their restricted access to markets and high transport and marketing costs. The public sector has an important role to play in building and maintaining roads in these areas and in promoting expansion of private transport, marketing, input supply, and financial services that are competitively priced. Investments in electricity and telecommunications are also needed if the private sector is to grow. Investments in clean water and the education and health of local people not only increase their productivity in agriculture, but enhance their opportunities to diversify into nonfarm activities, including out-migration to better-paying jobs. The results from the Fan and Hazell (1996) India study summarized above show that these kinds of public investments in less favored lands can yield favorable growth as well as poverty alleviation payoffs. As such, these investments do not have to be a net drain on the national economy.

Investment in rural infrastructure should be closely linked to other agricultural policies (such as development strategies and provision of credit) as well as to other sectoral programs in education, health, and communications. Priorities for targeting investment by geographic area and type of infrastructure should be guided by at least three criteria: population density, agricultural development, and potential market integration (Wood 1998; Pender, Place, and Ehui 1998).

Provide the Right Policy Environment

Market reforms, including price and trade liberalization, are necessary to ensure that prices provide the right production signals to farmers and that production and input markets can be competitive and work well. Available evidence suggests that, prior to recent reforms, LFAs were typically penalized along with the rest of the agricultural sector by distorted macro, trade and sectoral policies (see Chapter VII). As a result, many of the ongoing policy reforms have improved the terms of trade for less favored areas and have increased their market opportunities.

In order to take advantage of these new opportunities, however, adequate investments in rural infrastructure are needed to improve market access and to reduce transport and marketing costs. If market reforms are not matched by appropriate levels of investment in local infrastructure, they can actually be quite destructive for many rainfed areas. For example, market reforms have reduced the availability of inorganic fertilizers and increased their costs in many backward areas; the resulting reduction in their use (often from modest levels to begin with) is now contributing to worsening soil fertility problems. The associated reduction in food production also adds to the food insecurity problems of the poor. Transitional policies, sometimes including targeted subsidies, may be necessary in some LFAs to manage some of the negative impacts of market reforms, at least until such time as the required infrastructure investments have been made.

Strengthen Public Institutions

Many of the public institutions that service agriculture and rural areas have tended to neglect LFAs; moreover, they are often poorly positioned to address the unique problems of these regions. Agricultural research and extension systems, for example, have been structured to serve the needs of irrigated and high-potential rainfed areas, and while reasonably efficient at promoting green-revolution technologies in these areas, they are much less able to deliver the kinds of multidisciplinary or farmer-oriented natural-resource management approaches needed in most LFAs. Similar biases have existed in rural credit and insurance institutions. Public agencies with resource mandates (such as forestry and rangeland departments), on the other hand, have often been very active in LFAs, but have taken top-down approaches to the management of these resources. Not surprisingly, by excluding local users from any real stake in the ownership and management of these resources, these approaches have resulted in resources being exploited and degraded, while the relevant public departments are hamstrung by their inability to regulate resource use effectively on the ground.

The development of LFAs will require significant changes in the objectives and operational modalities of many public agencies. More participatory approaches that build on the interests and abilities of local people to manage resources are needed; this calls for very different incentive structures within public institutions, with greater accountability to intended beneficiaries.

Another problem that has plagued the effectiveness of public institutions has been their seeming inability to coordinate relevant activities in rural areas. Key functions are compartmentalized within different ministries and at different levels (local, regional and national) of government; only rarely is there an effective institutional mechanism for coordinating their plans and activities. Integrated development projects attempted to overcome this problem, but with few exceptions, they failed to move the coordination beyond the planning stage. Coordination units at the highest level of government (as in the prime minister's office) have rarely worked in practice; more effective solutions probably require greater devolution of authority to local governments.

CONCLUSIONS

In order to promote economic growth and redress poverty and environmental problems, Asian policymakers will need to pursue appropriate and sustainable methods of agricultural intensification for both high- and low-potential regions. This dual strategy will be particularly challenging if government budgets for investment in agriculture and rural areas continue to remain tight; striking the right investment balance between irrigated and rainfed regions, and between high- and low-potential rainfed areas, will be particularly important. Investments in irrigated and high-potential rainfed areas cannot be neglected, because these areas still provide much of the food needed to keep prices low and to feed growing urban and livestock populations.

On the other hand, the poverty, food-security, and environmental problems of many less favored areas are likely to remain serious in the decades ahead as populations continue to grow. While out-migration and economic diversification should become increasingly important in the development of areas with low agricultural potential, agricultural intensification will often offer the only viable way of raising incomes and creating employment on the scale required in the near future. Even when the investments needed to achieve this growth yield lower economic returns than investments in high-potential areas, they might still be justified on the basis of their significant social benefits in the form of poverty alleviation and improved environmental management. Moreover, with worsening income disparities between many favored environments and LFAs, policymakers are likely to come under increasing pressure to invest more in low-potential areas.

The size of the potential tradeoffs between investing in favored environments and LFAs has yet to be widely quantified in Asia and it is possible that it may be changing. Productivity levels in many high-potential areas have reached a plateau, while at the same time recent agricultural research in some low-potential rainfed areas is suggesting new avenues for increasing these regions' productivity (Scherr and Hazell 1994). Results from an IFPRI study of the returns on public investment in India raise the tantalizing possibility that greater public investment in some LFAs could actually offer a "win-win" strategy for addressing productivity and poverty problems.

The successful development of less favored lands will require new and improved approaches, particularly for agricultural intensification. These will require stronger partnerships than needed in high-potential areas, between agricultural researchers and other agents of change, including local organizations, farmers, community leaders, NGOs, national policymakers and donors. It will also require time and innovation; new approaches will need to be developed and tried on a small scale before they are scaled up and their testing will take time to assess and evaluate, particularly given the noise introduced by climatic variability. All this will require patience and perseverance on the part of policymakers and donors, perhaps more than the current aid culture allows.

XII ALTERNATIVE FUTURES FOR ASIAN AGRICULTURE AND FOOD SECURITY

INTRODUCTION

The preceding chapters have explored the rapid economic growth and massive transformation in agriculture and the rural economy in Asia and have analyzed many of the challenges and uncertainties that continue to face Asian policymakers. In this chapter we explore how alternative policies can influence food supply, demand, and trade and food security in Asia. This chapter starts out by highlighting these and other emerging trends for developing Asia in the global economy to 2010, based on the best available assessment of likely policies and population growth scenarios as captured by IFPRI's International Model for Policy Analysis of Agricultural Commodities and Trade (IMPACT). For more detail on the model, plus a description of best-assessment "baseline" results for critical supply and demand outcomes in 2010 compared to 1993, broken down by region, and for much of Asia by country, please see the Appendix.

How robust, though, are best-assessment baseline outcomes in the face of a range of realistic, but different, policy strategies? Investment in agriculture, water resources, and social sectors may fall off in Asia, with policies focusing on other sectors. Slow progress on economic policy reform may result in a more prolonged economic crisis dampening income growth in Asia. Alternatively, policymakers may take a more aggressive stance towards improving agriculture and other rural economic sectors, boosting investment and accelerating the pace of policy reform in agriculture and the general economy. Would either of these policy directions substantively change projections of developing Asia's contribution to cereal and livestock markets, and, if so, with what implications for the people of this region, particularly its children? Strikingly, policy packages that moderately disfavor agriculture, natural resources, and social sectors and moderately slow economic reform lead to a much worsened overall impact vis-à-vis the baseline, while those moving more aggressively to strengthen agriculture and economic policy reform yield an outcome that, while far from utopian, is much improved. Together, the alternative scenarios plus the baseline establish a range of possible outcomes for Asia in 2010 that are vastly different in terms of human suffering and are directly dependent on policy decisions.

This chapter lays out and contrasts results from IMPACT simulations of two alternative policy scenarios for Asia in 2010: one containing a series of plausible, moderate declines in investment with slower policy reform, the other consisting of renewed, but again moderate and plausible, policy efforts in the agricultural sector, irrigation and water, and social sectors. After the summary of projected Asian trends, the two scenarios are described in more detail. This is followed by a look at selected supply, demand, and trade figures for Asian regions under the alternative scenarios, as against 2010 baseline results and 1993 figures. The implications of these alternatives for food availability and for malnutrition in children are presented in a separate subsection. Finally, to explore the potential for more aggressive policies and investments to eliminate malnutrition in Asia, we assess an alternative scenario designed to launch a major attack on childhood malnutrition through income growth, agricultural growth, and social investment.

GLOBAL TRENDS—A PROMINENT ROLE FOR ASIAN DEVELOPING ECONOMIES

As described in considerable detail in the Appendix, the long-term prospects for food supply, demand, and trade indicate a strengthening of world cereal markets. Real world prices of cereals will be virtually constant through 2010, indicating a much stronger market for these commodities than in the past few decades, when real prices declined rapidly. The stronger price picture is the result of the continued gradual slowing in the rate of growth in both production and demand. Even taking account of the likely effects of the current economic crisis, developing Asia is projected to play a major role in global cereal and livestock markets.

On the production side, there will be virtually no growth in crop area in Asia. Crop yield growth will therefore account for nearly all production growth. In Asia and most other countries and regions, however, the gradual slowdown in crop yields that began in much of the world in the early 1980s will continue. Nevertheless, yield growth will outstrip the decline in prices, presenting the potential for improved long-term profitability in cereal production. Livestock production in Asia will grow considerably faster than crop production, but will also slow down relative to its growth in the past decade.

Countering the continued gradual slowing of production will be a matching decline in the growth rate in food demand. Fundamental changes are occurring in the global structure of food demand, driven in large part by economic growth, rising incomes, and rapid urbanization in the developing economies. Population growth rates in Asia (and most of the world) will be declining throughout the projection period. Rising incomes and rapid urbanization—particularly in Asia, and even with the slowdown due to the Asian financial crisis—will change the composition of demand. Per capita food consumption of maize and coarse grains will decline as consumers shift to wheat and rice with increasing incomes. As Asian incomes rise further and lifestyles change with urbanization, there will be also be a shift from rice consumption to wheat. Growth in incomes in developing countries will stimulate strong growth in per capita and total meat consumption, which will in turn induce strong growth in feed consumption of cereals, particularly maize. These trends will lead to an extraordinary increase in the importance of Asia in global food markets. A full 84 percent of the increase in global cereal and meat demand between 1993 and 2010 will come from the developing countries. By 2010, developing countries will account for 63 percent of global cereal demand and 58 percent of global meat demand. Even more strikingly, the share of developing Asia in the increase in global cereal demand to 2010 will be 52 percent and the share in the increase in global livestock demand will be even higher at 57 percent.

Asia will not be able fully to meet its rapidly growing food demand through growth in its own production. Food import demand from Asia (and other developing countries) will grow rapidly, particularly for cereals. Although regions outside Asia will be important growth centers for imports of some of these commodities, such as West Asia and North Africa (WANA) for wheat, it is East and South Asia that will drive the boom in cereal import demands.

In developing countries worldwide by 2010, rates of malnutrition among children 0–5 years old will have fallen to 25 percent from 1993 levels of 32 percent. Still, absolute numbers will remain unacceptably high—165 million children malnourished. Of these, 68 percent, or 113 million, will reside in developing Asian countries, particularly South Asia. While rates in the rest of Asia and other developing countires will have rates of child malnutrition falling to 3 or 4 percentage points from a base of 25 percent or less, in South Asia rates of child malnutrition started out at 57 percent of all children in 1993 and will drop to 46 percent in 2010.

ALTERNATIVE SCENARIOS FOR AGRICULTURE IN DEVELOPING ASIA: WEAK VS. STRONG INVESTMENT/POLICY REFORM STRATEGIES

The baseline results briefly summarized above and described in more detail in the Appendix represent our best assessment of future directions in the world food situation. These results, however, may be sensitive to rates of agricultural productivity growth stemming from assumptions about research investment levels or to underlying assumptions about population and income growth. There are continuing debates in developed nations over possible large cuts in foreign aid that would slash public investment in international agricultural research. What would be the impact on food prices and malnutrition if, instead of maintaining investments, national governments and international institutions were to continue to reduce their investments in agricultural research, irrigation and water, and social-sector expenditures? What if, at the same time, slow economic reform in response to the ongoing economic crisis resulted in slower- than-projected income growth? What if, in addition, water-policy reform failed to respond to growing demand for nonagricultural water uses, resulting in more rapid loss of water from agricultural uses; and if population growth increased relative to the "medium" UN population projections? Would these steps, jointly taken, make a difference for malnutrition and food prices for developing Asia in 2010? Alternatively, can increased investment and improved policy reforms in each of these areas make a significant positive impact on food security and malnutrition?

Low Investment and Weak Policy Reform

This scenario incorporates the following changes to the baseline scenario:

- A 25-percent reduction in annual nonagricultural income growth in Asian countries (i.e., a growth rate of 5.0 percent per year is reduced to 3.75 percent per year), beginning in 1998. This assumption results in projected income growth consistent with relatively low income projections that assume that relatively slow policy reform will lead to prolongation of the crisis in East and Southeast Asia and relatively slow economic liberalization in South Asia.

- A 10-percent cut in the growth rate in public investment in Asian national and international agricultural research systems relative to the base case.
- A 10-percent reduction in investment in health, education, and sanitation compared to baseline 2010 projections, leading to a worsening of the projected percentage of women with access to secondary education, the projected percentage of the population with access to clean water, and projected social expenditures.
- An increase in population growth in Asia and the world relative to the baseline projections, to the 1996 United Nations "high" population growth rates (UN 1996). Under this scenario, population growth still slows relative to past rates of growth.
- An increase in soil degradation that reduces crop area and both yield growth and animal numbers growth in Asia by 0.05 percent per year.
- Increased transfer of water away from agriculture, assuming weak reforms in institutions, policies, and technologies to achieve water savings and mitigate the impact of the transfer, resulting in

    - no increase in irrigated area to the year 2010 (with any current pipeline investment balanced by investment cutbacks and loss of existing irrigated area due to degradation and urban encroachment), leading in 2010 to a reduction of 15.8 million ha (11 percent) in irrigated area in Asia compared to the 2010 baseline projection;
    - phased reductions in agricultural water use over the projection period, averaging 10 percent by the end of the period (2010) across countries and regions, which assume slow improvements in water use efficiencies for agricultural as well as for domestic and industrial uses (see also the description of the baseline scenario in the Appendix);

- declines in crop area growth in proportion to the reduction in agricultural water use; and
- reduction in crop yield growth, in proportion to changes in relative water supply, based on the relative water supply/crop yield function approach.

The combined effect of the full scenario is a reduction in projected constant-price cereal yield growth (i.e., the yield growth in mt/ha that would obtain if commodity prices were constant) for the Asian developing countries, 1995–2010 of 0.31 percentage points, from 1.31 percent per year under the baseline to 1.00 percent per year under the low-investment/weak policy-reform scenario. Constant-price area growth is projected as assumed to fall from an average across Asian countries of 0.12 percent per year to –0.20 percent per year. Livestock yield growth (in kg carcass weight) is projected to decline from 0.80 percent per year to 0.68 percent per year, and growth in animal herd numbers from 1.71 percent per year to 1.19 percent per year.

High Investment and Strong Policy Reform

This scenario incorporates the following changes to the baseline scenario:

- A 25-percent increase in annual nonagricultural income growth in Asian countries, 1999–2010, with projected growth rates consistent with relatively rapid recovery from the financial and economic crisis in East and Southeast Asia and relatively fast economic liberalization in South Asia.
- A 10-percent increase in the growth rate in public investment in Asian national and international agricultural research systems relative to the base case.
- A 10-percent increase in investment in health, education, and water and sanitation compared to baseline 2010 projections, leading to an improvement

in the projected percentage of women with access to secondary education, the projected percentage of the population with access to clean water, and projected social expenditures.
- A reduction in the rate of increase in population growth in Asia and the rest of the world relative to the baseline population growth rate, to the 1996 UN "low" population growth scenario (UN 1996).
- A decrease in environmental degradation that increases Asian crop area and yield growth and animal numbers growth each by 0.05 percent per year.
- Effective water policy reform and increased irrigation investment that results in

  - a five percent increase in irrigated area relative to the baseline 2010 projections;
  - phased increases in agricultural water use compared to the baseline, amounting to 10 percent by the end of the period (2010);
  - increases in crop area growth, in proportion to the increase in agricultural water use; and
  - increases in crop yield growth, in proportion to changes in relative water supply, based on the relative water-supply/crop-yield function approach.

The combined effect of the full scenario is an increase in projected constant-price cereal yield growth for the Asian developing countries, 1995–2010, from 1.31 percent per year under the baseline to 1.58 percent per year under the high-investment/strong policy-reform scenario. Constant-price area growth is projected to increase from an average across Asian countries of 0.12 percent per year to 0.33 percent per year. Livestock yield growth (in kg carcass weight) is projected to increase from 0.80 percent per year to 0.93 percent per year, and growth in animal herd numbers from 1.71 percent per year to 1.81 percent per year.

ALTERNATIVE SCENARIOS: IMPACT ON ASIAN FOOD MARKETS AND FOOD SECURITY IN 2010

If the policy steps in the low-investment/weak-reform package are taken, the profile of developing Asia by 2010 is projected to look substantially worse for the poor than it would under the baseline: higher food prices, reduced levels of production, and higher dependence on imports, with concomitant sharp falls in per capita food consumption and increased prevalence of malnutrition among children. With the policy steps described in the high-investment/strong-reform package, on the other hand, the picture projected for the region in 2010 brightens considerably across all these measures when compared with the baseline, and even more so when compared with the bleak low-investment scenario.

The sections below quantify the extent of these different outcomes for the two scenarios simulated by using the IMPACT model. They concentrate first on market factors of supply, demand, net trade and international commodity prices for cereals, roots and tubers, and livestock products (plus milk), then more specifically on one component of demand—demand for food per capita—along with welfare indicators of per capita food availability and child malnutrition. Results (other than international prices) are presented in the aggregate for all developing Asia, then disaggregated for the major subregions of South and Southeast Asia, as well as the PRC.

Scenario Effects on Supply, Demand, and Net Trade

Alternative Scenarios for Crops

Both policy scenarios, as laid out in Table XII.1, show cereals production up in 2010 for developing Asian countries as a group. In the low-investment scenario, Asian cereals production is 46 million mt (5.4 percent) below the baseline result in 2010 of 852 million mt. Under the high-investment scenario, production increases an additional 44 million mt (5.2 percent), to 897 million mt. Both scenarios project changes in the same direction as production and of roughly similar magnitudes in component area and yield outcomes. The changes in area and yield over time are lower than might be expected given the projected constant-price yield and area growth assumptions described above. The initial assumed reduction in underlying yield and area growth causes a reduction in supply, which drives up commodity prices, generating a positive area and yield response and ultimately resulting in equilibrium area, yield, and supply at higher levels than predicted from the initial constant-price assumptions alone.

Downside and upside cereals production figures for these scenarios do not straddle the baseline so neatly everywhere in the region, however. In the PRC, for example, projected cereals production under the low-investment strategy drops 2 percent from the baseline (primarily due to yield declines, as area increases slightly), but rises nearly 6 percent from the baseline under the more favorable strategy, with increase in both area and yield contributing. In both South and Southeast Asia, by contrast, cereals production falls more under the low-investment package than it rises under the high-investment one (down 10 percent vs. up 6 percent for South Asia, and down 6 percent vs. up 1 percent for Southeast Asia). The striking drop in South Asia under the low-investment strategy can be traced to substantial shrinkage in area under production and yields (both down roughly 5 percent from the baseline).

The production picture for roots and tubers shows relatively little change across the region as a whole for either of the two scenarios as compared to the baseline in 2010. In the PRC, the main Asian producer for this group of crops (across all scenarios), however, yield and production for this group of crops actually improves under the low-investment strategy due to large increases in world prices because of population increases in Sub-Saharan Africa. On the other hand, South Asian root and tuber production drops considerably (roughly 12 percent) under the low-investment strategy scenario, albeit compared to a baseline already at a relatively low level of production (45 million mt).

Table XII.1: Area, Yield, and Production for all Cereals and Roots & Tubers, Developing Asia, 1993 and Projected 2010, Baseline and Alternative Policy Scenarios (area in M ha; yield in mt/ha; production in M mt)

	1993	2010 Base	2010 Low Inv/ Weak Reform	2010 High Inv/ Strong Reform
All Cereals				
Asia (developing)				
Area	265.6	271.7	264.9	277.8
Yield	2.5	3.1	3.0	3.2
Production	667.7	852.2	806.2	896.6
PRC				
Area	88.6	89.5	91.2	92.1
Yield	3.9	4.7	4.5	4.8
Production	343.3	418.1	409.9	441.9
South Asia				
Area	126.1	129.6	122.4	132.8
Yield	1.7	2.2	2.1	2.3
Production	214.9	289.8	259.6	308.4
Southeast Asia				
Area	47.2	48.9	47.5	49.3
Yield	2.1	2.7	2.6	2.7
Production	97.5	130.8	122.9	132.6
All Roots & Tubers				
Asia (developing)				
Area	16.1	16.4	16.0	16.6
Yield	14.7	18.1	18.5	18.0
Production	235.7	296.6	296.4	297.6
PRC				
Area	9.6	9.4	9.4	9.5
Yield	16.2	20.2	20.7	20.1
Production	155.8	191.2	195.3	191.8
South Asia				
Area	1.9	2.2	2.0	2.3
Yield	14.8	20.3	20.3	20.4
Production	28.1	45.1	39.7	46.7
Southeast Asia				
Area	4.4	4.5	4.4	4.5
Yield	11.2	12.6	13.1	12.2
Production	48.6	56.5	57.4	55.3

Source: IFPRI IMPACT.

While production for cereals in most parts of developing Asia decreases under the low-investment/weak-reform strategy, total cereal demand changes relatively little. This is because the offsetting impact from higher population growth boosts total demand, while higher prices and lower per capita incomes reduce total demand (Table XII.2). The net effect from significant changes in production and modest changes in total demand: dramatic shifts in imports—from 89 million mt under the baseline to 150 million mt under this scenario (a jump of nearly 70 percent). For the PRC, responsible for nearly half of total cereal demand for the region, consumption grows by 4 percent over the baseline; with drops in production, this means an 80 percent rise in cereal imports compared with the 35 million mt needed in the baseline scenario. For South and Southeast Asia, with the more dramatic drops in supply already described coupled with only slight dips in demand, import needs in 2010 more than double from what they would under the baseline, to 46 million mt (from 19 in the baseline), and 11 million mt (from 5 in the baseline), respectively.

Cereal imports decline significantly for the region under the high-investment scenario. Consumption still increases over the baseline scenario (by about 2 percent for the region as a whole and in each of the subregions), but the substantial gains in supply more than compensate in most areas, leaving a narrower trade gap for cereal imports, amounting to 30 percent less than the baseline regionwide (and nearly 88 million mt less in cereal imports than the unfavorable scenario). The trend is most pronounced where supply improves the most from favorable policy: in the PRC and South Asia, with import levels at 45 and 64 percent of the baseline, respectively (and only a third and a little more than a sixth, again respectively, of the low-investment scenario import result). The exception to the trend is Southeast Asia, where cereal imports increase slightly from the baseline under the positive policy scenario, as income-driven demand growth outpaces supply gains, but import levels remain considerably below those under the negative policy scenario.

Table XII.2: Total Supply, Demand, Net Trade (in Mmt) and Per Capita Food Demand (in kg/cap) for All Cereals and Roots & Tubers, Developing Asia, 1993 and Projected 2010, Baseline and Alternative Policy Scenarios

	1993	2010 Baseline	2010 Low Inv/ Weak Reform	2010 High Inv/ Strong Reform
All Cereals				
Asia (developing)				
Total Supply	667.7	852.2	806.2	896.6
Total Demand	695.1	941.3	955.7	958.6
Food (per capita)	183.9	187.7	183.1	193.2
Net Trade	-27.4	-89.1	-149.5	-61.9
PRC				
Total Supply	343.3	418.1	409.9	441.9
Total Demand	344.3	453.4	473.5	461.4
Food (per capita)	213.7	217.5	214.2	220.8
Net Trade	-0.9	-35.3	-63.6	-19.4
South Asia				
Total Supply	214.9	289.8	259.6	308.4
Total Demand	218.1	308.8	305.1	315.1
Food (per capita)	162.2	170.9	165.2	178.3
Net Trade	-3.3	-18.9	-45.5	-6.7
Southeast Asia				
Total Supply	97.5	130.8	122.9	132.6
Total Demand	100.7	136.0	134.1	138.9
Food (per capita)	169.1	171.1	165.8	178.3
Net Trade	-3.2	-5.2	-11.3	-6.3
All Roots & Tubers				
Asia (developing)				
Total Supply	235.7	296.6	296.4	297.6
Total Demand	214.9	278.0	290.9	267.5
Food (per capita)	40.0	39.4	42.8	37.0
Net Trade	20.8	18.7	5.5	30.2
PRC				
Total Supply	155.8	191.2	195.3	191.8
Total Demand	155.9	192.1	199.1	184.7
Food (per capita)	61.0	57.2	63.9	51.9
Net Trade	-0.1	-0.8	-3.9	7.1
South Asia				
Total Supply	28.1	45.1	39.7	46.7
Total Demand	28.0	44.5	43.6	47.1
Food (per capita)	18.6	21.9	21.2	23.6
Net Trade	0.1	0.7	-3.9	-0.3
Southeast Asia				
Total Supply	48.6	56.5	57.4	55.3
Total Demand	25.6	34.8	40.9	29.5
Food (per capita)	44.9	47.8	54.9	41.5
Net Trade	23.0	21.7	16.6	25.8

Source: IFPRI IMPACT.

For roots and tubers, the baseline scenario projects developing Asia to remain an exporter, but for exports from the region to decline slightly from 1993 levels of 21 million mt to 19 million mt in 2010. The alternative policy scenarios both show Asia remaining a net exporter of this category of crops in 2010, but the export levels are much lower under the low-investment than under the high-investment scenario. Indeed, under the high-investment policy package, regional exports would be expected to rise from their 1993 levels to around 30 million mt, whereas the low-investment scenario would bring export levels in 2010 down to around 5.5 million mt. As noted above, the two scenarios make no substantial difference to projections of supply in 2010, so the difference in imports across scenarios must lie in demand. Regional demand does differ considerably across scenarios, rising by about 5 percent vis-à-vis the baseline for the poor policy scenario and declining by close to 4 percent under the positive investment/reform policy, again using the baseline as comparator. Behind these regional figures lies substantial variation in root and tuber markets in each subregion for the alternative scenarios.

The PRC once again plays a paramount role, responsible for close to 65 percent of total regional supply and almost 70 percent of total regional demand for roots and tubers. The baseline scenario shows the PRC moving from a slightly negative net trade position to that of a net importer, but of less than one million mt in 2010. The low-investment scenario would exacerbate this trend, so that the PRC would need to import 4 million mt of roots and tubers by 2010. Under the high-investment scenario, on the other hand, PRC exports of roots and tubers would reach 7 million mt in 2010. Southeast Asia in 2010 is expected to account for nearly 19 percent of regional supply and between 11 and 15 percent (depending on the scenario) of total consumption. As in the PRC, in Southeast Asia demand for roots and tubers rises under the low-investment scenario (by about 6 million mt from the baseline figure of 35 million mt—17 percent of the baseline figure), and drops under the high-investment package (by about 5 million mt, or 15 percent of the baseline, to 30 million mt). Unlike the PRC, though, Southeast Asia starts out in 1993 and finishes in 2010 as a net exporter (with virtually all exports from Thailand), regardless of the scenario examined, with the level of exports considerably lower in the low-investment scenario (17 million mt) than in the high (26 million mt). These figures establish a range of nearly 24 percent below, and 19 percent above, the Southeast Asian baseline export figure of 22 million mt.

Alternative Scenarios for Livestock and Dairy Products

Developing countries in Asia will sharply increase production as well as consumption of meat from 1993 levels to 2010 under the baseline scenario (see the Appendix for more detail on growth rates and types of meat projected to increase the most by location in Asia). As was the case for crops, depending on what type of policy package is adopted in the region, developing Asia is projected to face two substantially different potential futures for livestock product markets in 2010. Again, as Table XII.3 shows, the general trend holds of production levels lowered from the baseline due to a low-investment strategy, and boosted from the baseline due to a high-investment strategy. Here, though, the low-investment strategy depresses demand for livestock products compared to the baseline. The high-investment strategy, conversely, induces even greater meat consumption than was the case in the baseline. So, for developing Asia, meat production and consumption move up in tandem in response to the high-investment scenario and down in response to the low-investment scenario.

In the baseline, meat consumption in the region roughly keeps pace with production, leaving net trade for all developing Asia similar in 2010—a net exporting position on the order of 0.5 million mt—to 1993 levels (exporting 0.4 million mt). The downside shifts in supply and demand for meat from the low-investment strategy are of similar magnitude (roughly 7 percent), but demand falls slightly more, leaving the region with higher net exports than in the baseline (but still under one million mt, at 0.9), but at significantly lower production levels (91 million mt, compared to 98 in the baseline). Under the high-investment scenario, demand again responds slightly more than

Table XII.3: Total Supply, Demand, Net Trade (Mmt), and Per Capita Food Demand (kg/cap) for All Meats and Milk, Developing Asia, 1993 and Projected 2010, Baseline and Alternative Policy Scenarios

	1993	2010 Baseline	2010 Low Inv/ Weak Reform	2010 High Inv/ Strong Reform
All Meats				
Asia (developing)				
Total Supply	56.1	98.3	91.4	104.7
Total Demand	55.8	97.7	90.5	104.7
Food (per capita)	19.1	26.9	24.4	29.7
Net Trade	0.4	0.5	0.9	0.0
PRC				
Total Supply	39.4	69.1	63.8	73.5
Total Demand	38.6	68.6	63.2	72.7
Food (per capita)	32.8	51.1	46.0	55.5
Net Trade	0.9	0.5	0.6	0.8
South Asia				
Total Supply	6.1	10.5	9.7	11.7
Total Demand	6.0	10.6	10.2	11.3
Food (per capita)	5.0	6.7	6.3	7.4
Net Trade	0.1	-0.1	-0.4	0.4
Southeast Asia				
Total Supply	7.1	13.0	12.4	13.7
Total Demand	7.0	12.2	11.4	13.4
Food (per capita)	15.1	20.5	18.7	23.4
Net Trade	0.1	0.8	1.0	0.3
Milk				
Asia (developing)				
Total Supply	93.2	173.6	162.3	189.1
Total Demand	99.2	181.5	169.4	198.0
Food (per capita)	28.5	44.6	40.5	50.3
Net Trade	-6.0	-7.9	-7.1	-8.9
PRC				
Total Supply	8.3	14.3	13.0	14.8
Total Demand	9.2	15.0	13.9	15.6
Food (per capita)	6.8	9.9	9.0	10.6
Net Trade	-0.9	-0.7	-0.8	-0.8
South Asia				
Total Supply	81.1	153.6	143.9	168.3
Total Demand	81.6	154.4	144.0	169.3
Food (per capita)	57.7	87.5	79.4	99.0
Net Trade	-0.6	-0.7	0.0	-1.0
Southeast Asia				
Total Supply	1.6	2.6	2.4	2.8
Total Demand	5.5	8.8	8.3	9.4
Food (per capita)	11.4	14.3	13.1	15.8
Net Trade	-4.0	-6.2	-5.9	-6.6

Source: IFPRI IMPACT.

supply, but this time on the upside, so that all that is produced is consumed in the region at a level of 105 million mt (as compared to the baseline, this is 6.6 percent higher on the supply side, and 7.2 percent higher on the demand side). Note that the supply-and-demand profile of each alternative creates a band of possibility for levels of production and consumption of about 7 percent around the baseline for the region.

Net export and import positives change little under the high-and low-investment scenarios. The PRC, responsible for about 70 percent of regional production and consumption (across all 2010 scenarios), could end up exporting between 0.6 and 0.8 million mt of meat (the extremes projected under low and high-investment scenarios, respectively). South Asia is the exception: under the low-investment strategy, the subcontinent needs to import 0.4 million mt of meat to meet demand. Note, however, that the picture for East Asia other than the PRC, not reported in Table XII.3, is also an exception (under the high-investment scenario): taken as a whole, East Asian developing countries under the favorable policy package would be net importers of 0.7 million mt of meat in 2010. This explains why, with all regions reported in Table XII.3 as net exporters under the high-investment scenario, net meat trade for developing Asia nevertheless stands at 0.

Under the baseline, milk production for all developing countries in Asia is projected to rise 3.7 percent annually from a 1993 level of 93 million mt, to 174 million mt in 2010 (Table XII.3). With the unfavorable policy package, though, in 2010 production levels will only climb to 162 million mt (6.4 percent less than in the baseline). The high-investment/strong-reform scenario is projected to result in 2010 levels of milk production of 189 million mt, 9 percent above the baseline level. Milk consumption will also rise under the baseline scenario, at 3.6 percent annually from 1993 to 2010, to 182 million mt. Developing Asia, then, will continue to be a net importer of milk, with imports increasing from 6 million mt in 1993 to 8 million mt under the 2010 baseline. Southeast Asia continues its role as the major net importer of milk among the subregions throughout this period.

As was the case for meat, under the low-investment scenario demand for milk in the region is reduced significantly relative to the baseline level in 2010 (to 7 percent less), and under the high-investment scenario substantially exceeds the baseline level (this time by 9 percent). Under the low-investment scenario, import needs in 2010 will still be up from 1993 levels to 7 million mt, but at lower levels than the baseline. More imports will be needed to satisfy milk consumption under the high-investment scenario—nearly 9 million mt. As in the baseline, Southeast Asia dominates other subregions within Asia as regards milk imports.

South Asia, however, will continue to account for the dominant share of production and consumption of milk for developing Asia, ranging from 85 to 90 percent of total regional supply and demand in 2010 (amounting to between 144 and 170 million mt, depending upon the scenario). Under the baseline, South Asian imports of milk are projected to increase slightly (from 0.6 to 0.7 million mt) from 1993 to 2010 (when supply and demand are projected at close to 154 million mt). Under the low-investment strategy, South Asia supply and demand of milk balance out at roughly 144 million mt. Under the high-investment strategy, imports of 1 million mt are needed to meet milk consumption of 169 million mt (production projected at 168 million mt).

Even taking into account the variability across subregions and commodities, the overall difference that either of these policy scenarios makes vis-à-vis the baseline for developing Asia is clear: for a broad range of crop and livestock/dairy products and across the major subregions within Asia out to 2010, the low-investment/weak-reform policy package means lower production and weakening trade positions, while the high-investment/strong-reform policy package means improved production scenarios and stronger trade positions.

Alternative Scenarios: Effects on International Commodity Prices

As shown in Table XII.4, international commodity prices will be affected strongly and in different directions by the policy scenarios for developing Asia (low-investment/weak-reform and high-investment/strong-reform) underlining the importance of the region in global markets.

More specifically, under the low-investment scenario, cereals prices would all turn upward, not just from the baseline but from 1993 levels, hurting the food security of the poor. The international price of wheat, for example, is projected to go up 21 percent from its 1993 levels, to US$179/mt under this scenario, instead of dropping 1 percent under the baseline. Similarly, international maize prices would rise to US$151/mt (up about 20 percent from 1993 levels). Rice, one of few commodities expected to increase in price under the baseline (by 5 percent from 1993 levels), would have an even higher price of US$387/mt (up 35 percent from 1993) under this low-investment scenario. Figures for specific roots and tubers tell a similar story:

Table XII.4: Commodity Prices, Developing Asia, 1993 and Projected 2010, Alternative Scenarios

	1993	2010 Baseline	2010 Low Inv/ Weak Reform	2010 High Inv/ Strong Reform
	(US dollars per metric ton)			
Wheat	148	146	179	120
Maize	126	127	151	111
Rice	286	303	387	222
Other Grains	122	116	138	97
Potatoes	160	159	208	119
Sweet Potatoes	91	90	143	55
Cassava & other R&T	54	56	87	38
Beef	2,023	1,835	1,956	1,732
Pigmeat	1,366	1,260	1,330	1,202
Sheep&Goat	2,032	1,915	2,067	1,773
Poultry	1,300	1,175	1,238	1,124
Milk	234	217	233	209

Source: IFPRI IMPACT.

higher prices and less food security for the poor (international potato prices, for example, are projected to climb 30 percent under the low-investment strategy/weak policy-reform scenario.

With the low-investment strategy in place in Asia, international prices for most meats and milk are projected to decline less from 1993 to 2010 than they would under the baseline but retain a slightly downward trend (or remain practically stagnant, as is the case for pork prices, down just US$30/mt over the period under the low-investment strategy, to US$1,330/mt; or milk, remaining close to US$230/mt from 1993 to 2010 under this scenario). With the high-investment strategy in place, on the other hand, international crop prices decline more sharply than under the baseline and downward price trends in livestock and dairy product prices (although still slight) become more firmly discernible.

Alternative Scenarios: Effects on Per Capita Food Demand

The changes in per capita income growth and commodity prices under the alternative scenarios have large impacts on per capita food demand, which is a significant determinant of food security. As can be seen in Tables XII.2 and XII.3, the impacts include shifts in the composition as well as the level of food demand. Per capita food consumption of roots and tubers in 2010 actually increases by 8.6 percent in the low-investment scenario relative to the baseline. Although white potatoes have strongly positive income elasticities of demand in Asia, they account for just over one fourth of direct consumption of roots and tubers in the base year. Sweet potatoes, cassava, and other roots and tubers, which account for the bulk of roots and tubers food consumption, are inferior goods characterized by negative income elasticities. The depressed per capita income growth under the low-investment/weak-reform scenario therefore results in a shift in consumption away from preferred staples such as wheat to inferior staples such as cassava and sweet potatoes. For the same reasons, the high-investment/strong-reform scenario results in a decline in per capita consumption of roots and tubers in 2010 compared to the baseline projection.

The changes in per capita consumption of cereals under the scenarios are also slightly less dramatic than might be expected because, for many of the East and Southeast Asian countries, rice is an inferior good. Under the low-investment/weak-reform scenario, there is thus a shift out of wheat and into rice that somewhat compensates for the decline in consumption of wheat. Nevertheless, under the low-investment scenario, per capita food consumption of all cereals in Asia declines by 4.6 kg/capita (2.5 percent) relative to the baseline, and in fact is projected to be nearly one kilogram lower than in the base year of 1993 (Table XII.2). The sharpest fall is in South Asia, a decline of 5.7 kg/capita compared to the baseline projection. Under the high-investment scenario, cereal consumption increases by 5.5 kg/capita (2.9 percent) compared to the baseline 2010 projection, with South Asia showing the most dramatic increases of 7.4 kg/capita (4.3 percent) relative to the baseline, and 16.2 kg/capita compared to 1993 (Table XII.2).

The relative impact of the alternative policy scenarios on meat and milk consumption per capita is even more dramatic. Projected meat consumption per capita declines by 2.5 kg/capita (9.3 percent) under the low-investment/weak-reform scenario (Table XII.3). The biggest impact is for the PRC, which would see a decline in meat consumption relative to the baseline of 5.1 kg/capita, or 10 percent. Food consumption of milk would also be hard hit, with a decline of 4.1 kg/capita (9.2 percent) for developing Asia. In South Asia, which relies much more heavily on milk than on meat, the projected decline in milk consumption is 8.1 kg/capita (9 percent) compared to the baseline level in 2010. The improvements in meat and milk consumption under the high-investment/strong-reform scenario are comparably large. For developing Asia, projected meat consumption would increase by 2.8 kg/capita (10.4 percent) and milk consumption by 5.7 kg/capita (12.8 percent). The PRC is projected to have the biggest absolute gain in per capita meat consumption, 4.4 kg/capita, but the most dramatic relative gain would be in Southeast Asia, where the 2.9 kg/capita increase represents a 14.4 percent increase in meat consumption compared to the baseline projection for 2010. South Asia is the largest gainer in milk consumption, with projected consumption in 2010 reaching 99 kg/capita compared to 87.5 kg/capita in the baseline, an increase of 13.1 percent.

Scenario Effects on Food Security: Childhood Malnutrition

The alternative scenarios have profound effects on food security as measured by childhood malnutrition. For the low-investment/weak-reform scenario, reductions in per capita income and increased food prices reduce per capita food consumption as described above. The reduction in food consumption reduces per capita calorie consumption, which in turn increases the percentage of the childhood population that is malnourished. Reductions in public investment in the social sector, including total social expenditures, education and water, and sanitation, result in further increases in the percentage of malnourished children. The total childhood population also increases due to the higher fertility rates incorporated in the high population-growth scenario. The combined result of these effects is an increase in the number of malnourished children. The high-investment/strong policy-reform scenario has the opposite effects on the number of malnourished children.

The projected levels of per capita food availability, as measured by kilocalories per day, are shown in Figure XII.1 for the alternative scenarios. Under the low-investment/weak-reform scenario, projected calorie consumption for developing Asia declines in comparison to the baseline, from 2,734 kilocalories to 2,646 kilocalories, a drop of 3.2 percent, eliminating nearly 45 percent of the gains in calorie consumption projected under the baseline scenario. The order of magnitude in the decline in projected food availability is similar across the countries and subregions shown in Table XII.5. The high-investment scenario boosts daily calorie consumption to a projected 2,842 kilocalories in 2010, an increase of 4 percent compared to the baseline projection. The biggest increase in calorie consumption is in Southeast Asia, an increase of 132 kilocalories per day, or 4.8 percent (Table XII.5).

Figure XII.1: Per Capita Food Availability, Developing Asia, 1993 and Projected 2010, Alternative Scenarios

■ 1993 ▫ Baseline ▨ 2010 Low Inv/Weak Reform ▧ 2010 High Inv/Strong Reform

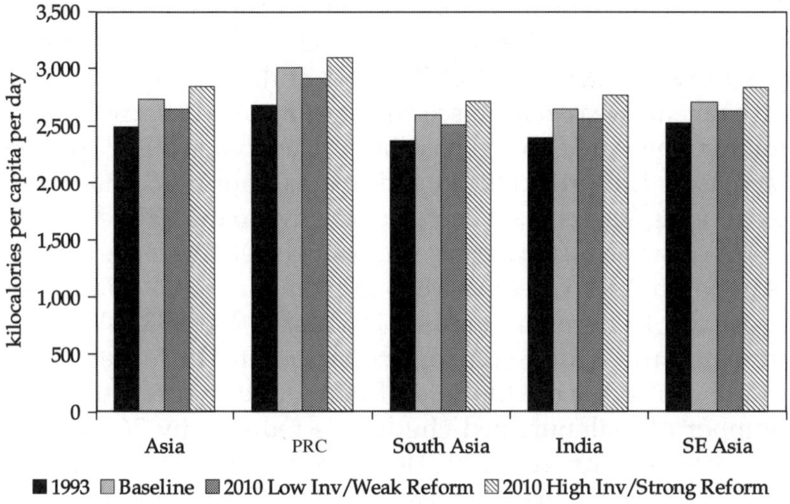

Source: IFPRI IMPACT.

Table XII.5: Per Capita Food Availability, Developing Asia, 1993 and Projected 2010, Alternative Scenarios

	1993	2010 Baseline	2010 Low Inv/ Weak Reform	2010 High Inv/ Strong Reform
		(kilocalories per day)		
Asia	2,488	2,734	2,646	2,842
PRC	2,680	3,008	2,913	3,096
South Asia	2,370	2,599	2,510	2,719
India	2,397	2,644	2,559	2,764
Southeast Asia	2,525	2,707	2,626	2,838

Source: IFPRI IMPACT.

The decline in calorie consumption under the low-investment scenario, combined with reduced social expenditures, results in significant increases in the percentage of malnourished children (aged 0 to 5 years) in the Asian developing countries. The proportion of children malnourished is projected to increase by 2.4 percentage points in the PRC, by 4.3 percentage points in South Asia, and by 2.5 percentage points in Southeast Asia (Figure XII.2). The increase in the proportion of children malnourished, together with the higher population of children, causes a dramatic absolute worsening of childhood malnutrition. The low-investment/weak-reform scenario results in an increase of 27.9 million compared to the baseline projection, a full 25 percent increase. The projected number of malnourished children in developing Asia actually increases relative to the 1993 level. Two thirds of the projected increase in numbers relative to the baseline is in South Asia, where nearly 3 million additional children are projected to be malnourished compared to 1993. The high-investment/strong-reform scenario, on the other hand, dramatically lowers the rate of childhood malnutrition. The number of malnourished children is reduced by 36.8 million compared to the baseline 2010 level, a reduction of fully one third and of almost one half compared to 1993 (Figure XII.3).

The results of this scenario analysis indicate that a series of moderate and plausible policy differences can generate dramatic changes in food security and welfare in developing Asia within the relatively brief span of years to 2010. Under the low-investment/weak-reform scenario, Asian cereal imports of 150 million mt are projected to be five times the 1993 level and more than double the 2010 level under the high-investment/strong-reform scenario. Cereal prices will be nearly 50 percent higher under the former scenario than under the latter. The difference between the low-investment/weak-reform and high-investment/strong-reform scenarios generates a swing of 10 kg/capita in consumption of both cereals and milk, and more than 5 kg/capita in consumption of meat. Most importantly, the high-investment/strong-reform scenario would reduce the number of malnourished children in developing Asia in 2010 by nearly 65 million compared to the low-investment/weak-reform scenario, from 141 million to 76 million children.

This high-investment/strong-reform scenario combines three broad courses for reducing the projected levels of malnutrition: the first way is through broad-based and rapid agricultural productivity growth and economic growth to

Alternative Futures for Asian Agriculture and Food Security 369

Figure XII.2: Percentage of Malnourished Children in Developing Countries, 1993 and Projected 2010, Alternative Scenarios

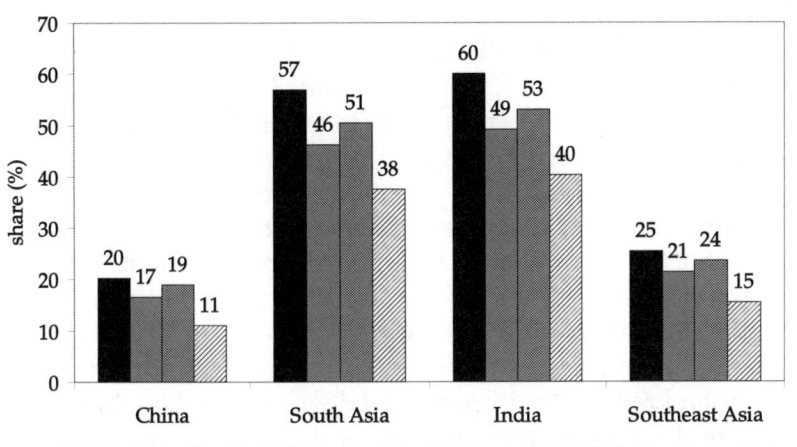

Region	1993	Baseline	2010 Low Inv/Weak Reform	2010 High Inv/Strong Reform
China	20	17	19	11
South Asia	57	46	51	38
India	60	49	53	40
Southeast Asia	25	21	24	15

Source: IFPRI IMPACT.

Figure XII.3: Number of Malnourished Children in Developing Countries, 1993 and Projected 2010, Alternative Scenarios

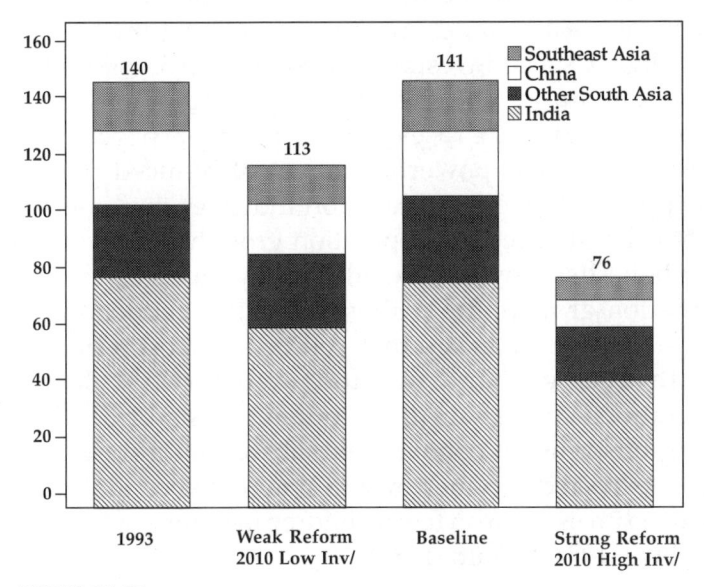

Stacked bar chart (Southeast Asia, China, Other South Asia, India):
- 1993: 140
- Weak Reform 2010 Low Inv/: 113
- Baseline: 141
- Strong Reform 2010 High Inv/: 76

Source: IFPRI IMPACT.

increase effective incomes, effective food demand, and food availability; the second way is through a reduction in population growth rates; and the third is through investments in education, social services, and health (the latter proxied by access to clean water in the model). The relative contribution of these three courses to reducing malnutrition can be estimated by undertaking a series of simulations embodying specific interventions such as lower population growth or higher social investments. These disaggregated simulations indicate that 31 percent of the improvement in childhood malnutrition from the high-investment/strong-reform scenario can be attributed to higher income growth and higher agricultural production growth due to increased research investment, improved water policy, and reduced environmental degradation. The combined contribution of improved income and productivity cannot be disaggregated further because a significant portion of the contribution is an interaction affect: the improved agricultural production growth reduces prices and increases the effective demand created by the increased incomes. Conversely, in the absence of improved income growth, the impact of production growth would be reduced, because agricultural prices would be driven down, dampening the production growth.

Higher social investment accounts for 32 percent of the improvement in childhood malnutrition and reduced population growth rates account for 37 percent of the improvement. The powerful impact of reduced population growth on childhood malnutrition indicates that policies and investments that reduce population growth rates would have large benefits. However, while policy changes can reduce population growth, the pathways for this reduction are not always direct, neither are they felt in the short term. Income growth does go hand in hand with fertility decline as expectations shift about the economic costs of bearing and rearing offspring and about their subsequent income benefits. Higher expected incomes overall may enhance incentives for parents to invest more in fewer children (see, for example, Becker and Lewis 1974; Schultz 1981).

Broader social development also has important effects on fertility. For example, more education for women may reduce fertility rates via delayed onset of marriage and childbearing as well as being an increased incentive for women to enter the formal workforce, thereby cutting into incentives for childbearing and -rearing. But it takes time for this effect on fertility decisions to play out in terms of lower population growth rates (since this effect is balanced against declining mortality and since not all of the childbearing cohort is affected, at least initially). Indeed, analysis of the timing of fertility decline in relation to economic development indicators has revealed wide variation from country to country (see Bongaarts and Cotts Watkins 1996; they attribute much intercountry difference to sociocultural factors). Given the uncertainty about short-term payoffs from population-oriented policies, it is important to examine whether significant inroads can be made on child malnutrition without influencing population growth rates.

Can Childhood Malnutrition in Asia Be Eliminated by 2010?

The high-investment/strong policy-reform scenario shows that relatively moderate improvement in policies in key strategic areas can significantly improve food security in Asia, as measured by childhood malnutrition. But even in this scenario, 76 million children in Asia remain malnourished, including 59 million in South Asia, hardly a desirable outcome. Instead of moderate improvements that leave millions of children malnourished, would it be possible to launch an attack on food insecurity that would eliminate childhood malnutrition in Asia by 2010? Given the difficulty and uncertainty in directly reducing population growth rates in the relatively short time between now and 2010, we examined whether plausible combinations of income growth, agricultural productivity growth, and social investments could attain this goal while maintaining the baseline medium population-growth assumptions.

The results reveal that a combination of high income growth, high productivity growth for crops and rapid expansion in numbers of livestock, and large social investments can dramatically reduce (but not entirely eliminate) childhood malnutrition by 2010 and virtually eliminate it by 2020. In order to achieve these food security gains, the improvements in projected performance must be much higher for South Asian countries than for the rest of Asia, because of the far greater initial severity of childhood malnutrition. The "rapid reduction of malnutrition" scenario combines the following elements: annual income growth rates are increased by 50 percent compared to the baseline for the East and Southeast Asian countries and 75 percent for South Asian countries. The resulting aggregate income growth rates are 9.6 percent for India, 8.8 percent for Pakistan, and 7.9 percent for Bangladesh. While these rates are high, they are within the bounds of growth rates achieved in East and Southeast Asia from 1975 to 1995. For the latter regions, the growth rates of, for example, 9 percent for the PRC and 9.75 percent for Indonesia and Malaysia would imply a return to the peak growth rates of the 1980s and early 1990s, while for the Philippines, the growth rate of 7.5 percent would imply a take-off into sustained growth similar to the East Asian success stories.

Crop productivity, expressed in terms of annual yield growth per hectare for cereals, roots and tubers, and soybeans, is assumed to increase by 50 percent for East and Southeast Asia and by 75 percent for South Asia. The projected realized cereal yield growth rates to 2010 under these assumptions would be 1.45 percent per year for East Asia, 1.90 percent per year for Southeast Asia and 2.44 percent for South Asia. These growth rates are higher than the 1989-95 yield-growth trends but lower than the yield-growth rates achieved during the peak green-revolution growth (see Chapter VI). Higher growth rates in production of meat and livestock products are assumed to be generated by more rapid expansion in animal numbers, with a 50 percent increase in annual growth rates in animal numbers for East and Southeast Asia and a 75 percent increase for South Asia.

The third pillar for rapid reduction in childhood malnutrition is increased social investments. All Asian countries are assumed to increase female access to education to 90 percent in 2010 and to 95 percent in 2020. This compares to average values in South Asia of 42 percent and in East and Southeast Asia of 48 percent in 1993. Access to clean water is assumed to increase to 95 percent in 2010 and 98 percent in 2020, compared to 57 percent in South Asia and 55 percent in East and Southeast Asia. Finally, government social expenditures are assumed to reach levels that are 50 percent higher than baseline projected levels in 2010 and 2020 for East and Southeast Asia, and 75 percent higher in South Asia. This implies average annual growth rates in public social expenditures of 6.8 percent per year in South Asia and 5.0 percent per year in East and Southeast Asia.

What would be the impact on malnutrition of these dramatic improvements in economic, agricultural, and social performance? In Southeast Asia, it is projected that childhood malnutrition would decline from 16.1 million persons to 4 million in 2010, with complete elimination of malnourished children in Thailand and less than 0.2 million remaining in Malaysia and Indonesia. By 2020, childhood malnutrition would be eliminated in Southeast Asia. The PRC would achieve dramatic reductions in childhood malnutrition, from 24.4 million in 1993 to only 0.4 million in 2010, and would have completely eliminated malnutrition by the following year.

In South Asia, extraordinary improvements would also be made, but the high initial levels of malnutrition mean that significant numbers of malnourished children would persist in 2010. It is projected that numbers of malnourished children in South Asia would decline from 99.8 million in 1993 to 40.1 million in 2010. In India, the numbers would fall from 76 million to 30.9 million, in Pakistan from 9.9 million to 5.7 million, in Bangladesh from 11.4 million to 3.6 million, and in the rest of South Asia from 2.4 million to 0.7 million. By 2020, the number of malnourished children would further decline to 7.7 million in India and 0.3 million in Pakistan, and childhood malnutrition would be eliminated in Bangladesh and the rest of South Asia.

CONCLUSIONS

Under the baseline scenario, the Asian and global food supply-and-demand picture in the most aggregate sense is relatively positive: if governments and the international community maintain their commitment to agricultural growth through sustained, cost-effective investment in agricultural research, extension, irrigation and water development, human capital, and rural infrastructure, there will not be overwhelming pressure on aggregate Asian and world food supplies from rising populations and incomes. Projected per capita availability of food in Asia will increase and real world food prices will be steady or declining slowly for the main food commodities. However, these positive aggregate outcomes hide the massive human suffering that would continue under the business-as-usual baseline scenario. Despite gains from trade and the overall ability of the world's productive capacity to meet effective demand for food, there will be only slow improvement in food security in Asia as measured by the number of malnourished children, with 113 million children suffering from malnutrition in 2010, including 83 million in South Asia. Moreover, food security is vulnerable to relatively small declines in policy efforts relative to business as usual. Policies that moderately disfavor agriculture and natural resources, reduce social investment, and slow economic reform lead to a much worsened food security impact compared to the baseline. Complacency toward agricultural and social investments risks a severe negative food security impact.

Conversely, the results shown here indicate that a significant assault on childhood malnutrition in Asia could be mounted within the boundaries of plausible long-term performance of the Asian economies. Moderate but important reductions in childhood malnutrition can be achieved through relatively small improvements in income growth, investment in agricultural research and irrigation, improved water policy, reductions in environmental degradation, slower population growth, and increases in social investments. But the near-elimination of childhood malnutrition will require policy reform and public investment that produce dramatic long-term gains in income growth, agricultural productivity, and social indicators. Although the precise set of policy reforms and the priorities and magnitudes of increases in investments required to eliminate childhood malnutrition would need to be determined in detail for each country, the results here confirm that the three foundations for success are broad-based economic growth, growth in agricultural production, and investment in social services including education and health. Failure in any of these three areas will severely hamper efforts to eliminate childhood malnutrition.

XIII Lessons Learned

INTRODUCTION

The ongoing, revolutionary transformation of the Asian countryside underlies the apparent miracle of economic growth in the region. This revolution, first visible in dramatic gains in agricultural output, was built on farm-based technology and the increased use of basic inputs, particularly labor, fertilizer, and water. It then spread to rural towns and nonfarm activities. The success of the revolution released labor and capital for broad industrial and economic growth, provided the mass demand for the products of economic growth, and generated much of the income that initiated a rapid decrease in poverty in Asia.

Like any revolution, this one has also produced unintended consequences, leaving some lands environmentally degraded and leaving some regions, even within dynamic countries, behind. Like most revolutionary transformations, this one played out differently under different circumstances, shaped by the type, timing, and sequence of actions taken by those in power. Perhaps most importantly, though, this revolution is as yet unfinished: large parts of Asia—some of which have only recently opened up to the possibility of reform as they emerge from an era of central planning—lag behind its most dynamic economies.

In this concluding chapter are summarized the lessons learned as the first wave of this transformation swept over parts of rural Asia, what those lessons mean, and how this unfinished revolution can play itself out in areas hitherto largely or partly left out of the transformation. In the wake of an economic shock that has threatened to slow the dynamism of even those parts of rural Asia already transformed, we take another look at the unresolved challenges that left these economies vulnerable to just such a shock. The pivotal role of changing styles and structures of governance in completing the revolution highlights what might otherwise be masked: a revolution that at first glance may seem to hinge solely on its market orientation in fact gained much of its impetus from, and will in significant part owe its succesful completion to, the public sector's effectively playing its role in supporting and supplementing the functioning of the market.

AGRICULTURAL GROWTH: AN ENGINE FOR ASIA'S ECONOMIC DEVELOPMENT

Stable macroeconomic policies, market-friendly policies, relatively open trade policies, and aggressive public investments in education and infrastructure have driven the accumulation of capital and technological change that produced rapid growth in East Asia. Economic reforms to apply this policy package in South Asia and in several Asian transition economies show promise of accelerated growth in these regions. But successes with this policy agenda were in most cases built upon rapid agricultural and rural economic growth during the early stages of the transformation. This agricultural growth was driven by the green revolution, a cost-reducing technological package that led to significant improvements in productivity on small and large farms alike. Because of the size of the agricultural sector, the productivity gains that were achieved had economy-wide significance and the benefits of growth were distributed widely across income groups in rural areas. Rapid growth in agriculture freed up labor and capital for the nonfarm economy, maintained a downward pressure on the prices of food and key primary inputs for agro-industry, contributed to foreign exchange earnings (through reduced food imports and increased agricultural exports), and provided a buoyant domestic demand for nonfarm goods and services. These results of agricultural growth not only led to rapid growth in the rural nonfarm economy, but also contributed importantly to the transformation of the urban-based economy.

Agricultural growth and the move to more open, market-oriented policies were synergistic. Economies with massive public intervention (for example, the centrally planned countries and India), weak infrastructure, and inward-looking rather than export-oriented economic direction were least successful in using the agricultural revolution to stimulate a broader economic transformation. For example, the PRC inititiated rapid agricultural growth through green-revolution technologies relatively early, but failed to capitalize on this for successful national economic growth until the process of economic liberalization was begun. National economic growth in India and the Philippines was also relatively slow despite successful green revolutions, but their growth performance is improving as they reform and liberalize their economies.

As economic growth proceeds and agriculture declines in relative size, economy-wide policies that support factor accumulation and productivity growth, including fiscal discipline, market-oriented policies, open trade policies, investment in education, and institutional quality, are increasingly important in determining the pace of economic transformation. But recent experiences in the transition economies and with the Asian financial and economic crisis have shown the continuing importance of agriculture in Asia. The agricultural sector has played a pivotal role in determining the pace of reform and constitutes an important backbone for the acceleration of economic growth in transition economies. The rigid, centralized, and collectivized agriculture structure in the Central Asian economies fell apart with the collapse of the Soviet Union and paralyzed growth prospects in the initial reform years, despite the gradual adoption of market reforms.

The central control of agriculture was much weaker in the East Asian transition economies at the onset of reform. The higher prevalence of smallholder agriculture contributed to a smoother transition in the agriculture sector and thus in the overall economy. The existence of surplus labor that could be released to the industrial sector was also favorable for economic growth during the transition process. As economic transition proceeded, the agriculture sector helped cushion the adverse impacts of initial dislocations due to reform in the nonagricultural sectors, stimulating economic sectors in several ways:

- through its linkages with related industries, like food-processing;
- through the provision of social stability while a large segment of the population remained employed in the agriculture sector;
- through the provision of food security and savings in foreign exchange on food imports; and
- through the creation of foreign exchange through exports of cash crops.

The agricultural sector has been an important factor in countering some of the negative effects of the financial and economic crisis. The urban economy has been the hardest hit by the crisis, and sharp increases in unemployment in urban areas have been compounded by the lack of social safety nets for the newly unemployed and the newly poor. However, the agricultural and rural regions in Thailand and Indonesia have absorbed large numbers of persons returning from urban areas, relieving pressures on overburdened urban social services. Moreover, the agricultural sectors in these countries have responded strongly to the real exchange-rate depreciation that has improved the competitiveness of the sector. Increased agricultural production and exports have helped compensate for the negative income effects of the crisis.

MAKING GROWTH PRO-POOR: A KEY TO POVERTY REDUCTION IN ASIA

Rapid agricultural and economic growth has benefited the poor. The incidence of poverty in East and Southeast Asia was reduced by two thirds between 1975 and 1995, while in South Asia, which grew more slowly and had more rapid population growth, the incidence of poverty nevertheless declined by one third. Comparable progress has been made in reducing poverty in rural areas, but poverty in Asia remains overwhelmingly a rural problem. More than three fourths of all the people living below the poverty line in Asia are in rural areas and tend to be illiterate, to depend on subsistence agriculture—often in resource-poor areas—and to depend on agricultural or low-skill labor for their livelihoods.

Since poverty is largely a rural phenomenon and since many of the poor depend, directly or indirectly, on the farm sector for their incomes, growth that raises agricultural productivity and the incomes of small-scale farmers and landless laborers is particularly important in reducing poverty. In India, growth in agricultural output per hectare is an important factor in explaining cross-state differences in rural poverty reduction between 1958 and 1994. Moreover, the initial investments in physical infrastructure and human resources played a major role in explaining the trends in rural poverty reduction: higher initial irrigation intensity, higher literacy, and lower initial infant mortality all contributed to higher long-term rates of poverty reduction in rural areas. Progress in reducing rural poverty is thus directly related to public investments in agriculture, education, and health.

Growth alone is therefore not sufficient to reduce poverty rapidly. Policies must also reach out directly to the poor, particularly through investments in their human capital. Investments in health, nutrition, and education not only directly address the worst consequences of poverty, but also attack some of its most important causes (see also below). Moreover, even with rapid economic growth, some of the poor will be reached slowly if at all and many of them will remain vulnerable to economic reversals. These groups can be reached through income transfers or through safety nets that help them through short-term stresses or disasters. In the agricultural sector, the poor are benefited most when

- land is distributed relatively equitably;
- agricultural research focuses on the problems of small farmers as well as large;
- new technologies are scale-neutral and can be profitably adopted by farms of all sizes;
- efficient input, credit, and product markets ensure that farms of all sizes have access to needed modern farm inputs and receive similar prices for their products;
- the labor force can migrate or diversify into the rural nonfarm economy; and
- policies do not discriminate against agriculture in general, and small farms in particular, (for example, no subsidies for mechanization).

The green revolution met most of these requirements in the more successful countries, particulalrly when preceded by a land redistribution, as in the cases of Taipei,China and the Republic of Korea, and by massive public investments in rural infrastructure.

AGRICULTURAL AND RURAL ECONOMIC GROWTH: MARKETS ENHANCED BY PUBLIC POLICIES

The past three decades have been remarkable for the failure of central planning and the subsequent turn toward market-oriented economic development in Asia. However, the experience of these decades has also demonstrated the paramount importance of government in implementing an enabling environment for market-based development and in particular for investing in the critical public goods of agricultural research, rural infrastructure, and education.

Agricultural and rural growth have been driven fundamentally by public investment in agricultural research and extension to generate productivity and income-enhancing technologies, by public investment in rural infrastructure, and by the existence (or introduction) of secure property rights to land. As agricultural growth took off, other factors were increasingly important to sustaining and enhancing this growth, including economic liberalization, especially trade and macroeconomic reform and deregulation of agriculture; development and liberalization of rural financial markets; and investment in the social sectors, particularly education, health, and nutrition.

Agricultural Research and Extension

Strong agricultural growth in most Asian economies has been based both on rapid growth in input use and on productivity growth. The main sources of productivity growth have been public agricultural research and extension, expansion of irrigated area and rural infrastructure, and improvement in human capital. The rates of return to public research are high, showing the continued profitability of public investment in agricultural research and strongly indicating that governments are underinvesting in the sector.

In addition to driving overall agricultural growth, research has also facilitated the process of commercialization and diversification of agriculture and the rural economy by generating new technologies that increase productivity and farmer incomes. Improved technologies provide farmers with the flexibility to make crop choice decisions and move relatively freely between crops and to increase the linkages to the nonfarm rural economy. Crop-specific research includes increases in yield potential, shorter-duration cultivars, improved quality characteristics, and greater tolerance to pest stresses. System-level research includes land management and tillage systems that allow for shifts of cropping patterns in response to changing incentives and farm-level water-management systems that can accommodate a variety of crops within a season.

The role and structure of agricultural research are changing over time. The relative importance of productivity-driven growth will increase, because growth in input use is declining as many regions in Asia are reaching high levels of input use. Private investment in agricultural research—which generates significant public benefits—will increase in importance, if policy reforms continue to create and/or improve the incentives for private investments by eliminating price distortions and strengthening property rights. Market failures and social objectives will continue to call for an important role for public investment in agricultural research, however. Agricultural research is often long-term, large-scale, and risky and while the returns to new technologies are often high, the firm responsible for developing the technology may not be able to appropriate the benefits accruing to the innovation—as in the case of improved open-pollinated rice and wheat varieties. The benefits of agricultural research often accrue to consumers (through reduction in commodity prices due to increased supply), rather than to the adopters of the new technology, so social returns may be greater than private returns to research. Therefore, a sustained public role in funding agricultural research will be essential, particularly for crops and regions, such as less favorable environments, that are unlikely to be served by the private sector.

New agricultural technologies in Asia—such as technologies to implement integrated pest management and to improve the nutrient balance and the timing and placement of fertilizer applications—are increasingly complex, knowledge-intensive and location-specific; they demand continued investment to create a better and more decentralized research and extension system. Because new technologies are more demanding for both the farmer and the extension agent, they require more information and skills for successful adoption compared to the initial adoption of modern varieties and fertilizers. Decentralization of the structure of existing extension services that encourage a bottom-up flow from farmers to extension and research could also help farmers cope with the additional complexity of efficiency-enhancing technology. Bottom-up information flows, combined with adaptive, location-specific research, is particularly important in the transfer of complex crop-management technologies. Other modern technologies, such as commercial poultry technology, will be transferred essentially intact from developed countries, without local adaptation, but will similarly require higher levels of education and management skills than traditional livestock operations. Finally, the increasing importance of new, knowledge-intensive technology requires a market-friendly environment for the adoption and adaptation of new technologies and the removal of restrictions on technology imports, which must be encouraged through continued progress in economic liberalization.

Investment in Rural Infrastructure

Infrastructural investments play a crucial role in inducing farmers to move toward a commercial agricultural system and in developing the rural nonfarm economy. Rural towns emerge as focal points in the development of the rural nonfarm economy, with increasing densities of nonfarm activity. To play this role, they need well developed infrastructure, transport, and communications systems, both within their own boundaries and linking them to larger urban areas and to their surrounding hinterland. Village access to rural towns and marketplaces is the key to creating effective demand linkages within rural regions and to spreading the benefits of nonfarm economic growth.

Infrastructure development also affects the supply side of the rural nonfarm economy. Electrification, for example, is especially beneficial to small manufacturing and processing enterprises, shops, and service establishments, giving them a more reliable and cheaper source of power. Rural roads facilitate the movement of raw materials to rural towns and villages and of final products to their main markets, and at lower cost. They enable firms to increase market size by giving them improved access to larger geographic areas; they increase rural labor mobility so that more village-based workers can take advantage of nonfarm employment opportunities in nearby towns. Telecommunications are increasingly important in linking rural firms to their customers and to the larger economy, enabling them to provide better and more timely service.

Priority-setting for infrastructure investment is complicated, because infrastructural investments take time to implement and demand careful attention to sequencing. Given the necessity for targeting, priorities for targeting investment by geographic area and type of infrastructure should be guided by at least three criteria: population density, agricultural development, and potential market integration. Targeting investments is particularly crucial with respect to the development of of resource-poor areas. Less favored environments are often poorly placed to compete in liberalized markets, because of their restricted access to markets and high transport and marketing costs. The public sector has an important role to play in building and maintaining roads in these areas and in promoting the expansion of competitively priced private transport, marketing, input supply, and financial services. Investments in electricity and telecommunications are also needed if the private sector is to grow. Investments in clean water and in rural people's education and health not only increase their productivity in agriculture but also enhance their opportunities to diversify into nonfarm activities, including out-migration to better-paying jobs. Public investments in less favored environments can yield favorable growth as well as poverty-reduction payoffs, so these investments do not have to be a net drain on the national economy.

Property Rights

Secure property rights to land have been a critical stimulus for production efficiency and agricultural growth. Secure rights to land create the incentives farmers need to invest in land improvements that conserve and increase long-term productivity growth. Secure land rights are complementary to policies that aim at liberalizing and integrating financial markets, because secure rights increase the probability that farmers can recoup the benefits of long-term investments and thereby increase their willingness to make them. Because they can act as collateral for loans, secure land rights also increase lender willingness to offer credit, leading to easier financing of purchased inputs and land improvements. Strengthened property rights are important for realizing the growth potential generated by economic liberalization, which places a premium on flexible farmer response in allocation of water, land and other resources in the context of changing prices, comparative advantage, and economic opportunities. If rights to the basic resources such as land and water are poorly secured and enforced, these resources can remain locked into inefficient uses. Moreover, strengthening of property rights, by encouraging on-farm conservation investments, might actually advance soil-conservation efforts in the future.

The lack of clearly defined property rights in the Central Asian transition economies has been a severe constraint on restarting growth. Property rights to land and security of tenure remain uncertain and the transition to autonomous cooperative management or private farming has been slow. Both the boost in agricultural productivity in the East Asian transition economies that came with improvement in land security and the far greater productivity in the small private sector in Central Asia indicate that successful transformation of property rights leading to privatization of state-owned enterprises and establishment or extension of private property would be an important impetus to growth in Central Asia.

The combination of the establishment of the household responsibility system and improved security in land rights was the fundamental reform that spurred rapid productivity growth in agriculture in the PRC after 1978. Despite this tremendous success, property rights in the PRC have not been fully secured; continued weakness in property rights is one reason for the slowdown of growth in the farm sector. The household responsibility system individualized the claim to residual income, but continued to vest land ownership in the collective, thus discouraging farmers from making medium- and long-term investments in land. Granting fully secure individual land rights could boost agricultural productivity in the East Asian transition economies, as well as in Central Asia.

In some cases, informal, indigenous property rights—if locally recognized and enforced—may lend adequate security. Local organizations are needed, particularly where communal property rights make sense as a way for users to spread risk and ensure access to a resource and should therefore be preserved, but also where informal rules guide the way in which privately owned property rights are guaranteed. Such organizations should have clearly defined membership and membership rights and should set rules about monitoring and a range of enforceable penalties of graduated severity matched to the seriousness of the transgression. Sometimes local systems come under stress from changing factors such as greater population pressure; indigenous systems lose their ability to completely guarantee security of access to resources. Even here, however, government policies aimed at bolstering traditional systems or helping them adapt to changed conditions may yield better outcomes than a shift to wholesale reliance on formal, legal systems.

Poor performance in state-owned enterprises in transition economies is also linked to lack of property rights. Corporate governance structure is weak due to the lack of an appropriate legal and regulatory framework, financial discipline, and incentive structures. The lack of financial discipline and poor incentive structures have perpetuated the inefficient managerial and operational practices of centrally planned systems. In order to strengthen market-based incentives, continued progress toward privatization should be accompanied by adoption of effective corporate structures, removal of government subsidies and enforcement of hard budget constraints, introduction of transparent enterprise accounts, and development of bankruptcy-implementation procedures.

Economic Liberalization

Macroeconomic stabilization and trade liberalization have been essential components of the policy-reform process for both agricultural and general economic growth in Asia. Openness to global markets and international trade allows the economy to catch up technologically, according to its comparative advantage, and to adapt the labor force and capital stock to changing factor endowments. Trade liberalization directly boosts trade growth and instills competitive, market-oriented behavior in the transition economies. Asian agriculture has benefited from the general economic liberalization and the reform of trade and macroeconomic-policy regimes that improved its competitive position between the mid-1980s and the early 1990s. The main components of economic liberalization have included reduction in trade restrictions, realignment of macroeconomic policies (reduction of fiscal deficits, elimination of multiple exchange rates, easing of exchange controls), and a liberalization of markets in general, including financial and asset markets. The reform process opened up international trade opportunities and provided improved price signals to guide producer decisions.

This long-term policy evolution was temporarily interrupted in the early 1990s when the competitive position of agriculture and other tradable sectors began to erode due to the dramatic appreciation in real exchange rates in East and Southeast Asia (and a significant but smaller appreciation in South Asia) as a result of macroeconomic policies and the massive influx of short-term foreign capital. However, the sharp depreciation of currencies in several East and Southeast Asian countries during the financial and economic crisis beginning in 1997 eliminated effective taxation of agriculture caused by real exchange-rate overvaluation and provided a significant stimulus to agriculture.

During the 1990s, the highly variable incentive environment for agriculture caused setbacks. Nevertheless, continued reform of trade and macroeconomic and price policies, to create a level playing field across economic sectors and agricultural commodities and to provide a stimulating environment for agricultural exports, will provide further incentives for efficient agricultural growth. International trade agreements, including the Uruguay Round agreements—which covered agriculture for the first time—and the creation of the World Trade Organization, should provide a supportive environment for continued agricultural liberalization in Asia. These agreements provide a framework for reform of agricultural trade and domestic policies, by strengthening the rules governing agricultural trade in order to improve predictability and stability for both importing and exporting countries. International agricultural-product differentiation and international investment and technology transfer will further encourage agricultural trade liberalization and encourage agricultural growth. The dynamic gains to developing countries in Asia from WTO and other liberalization are likely to be substantial, including the improvement of access to international technology and capital, strengthening of investor confidence, and encouragement for unilateral trade liberalization programs.

Apart from some manufacturing activities, the rural nonfarm sector was largely ignored by policymakers until recently; because it depended heavily on agriculture either directly or indirectly for much of its demand, it also suffered as a result of macroeconomic policies that discriminated against the agricultural sector. Recent macroeconomic policy reforms that have benefited the agricultural sector should, therefore, have led to positive growth-multiplier benefits for the rural nonfarm economy. The policy reforms have also favored tradable-goods production in general and this should have been directly beneficial to much rural industry. However, these benefits for the rural nonfarm sector are often limited by a continuing bias toward capital-intensive industry at the expense of trade and services. Apart from India's attempt to protect selected small-scale industries, policies to assist the rural nonfarm economy have generally favored manufacturing rather than service activities and large- rather than small-scale units of production. In many cases, small firms have effectively been placed at a competitive disadvantage against their larger-scaled rivals (for example, they do not receive the same subsidies and tax benefits); this has encouraged more capital-intensive patterns of development than is optimal.

Rural Financial Markets

Along with the increased opportunities for economic liberalization come increased risks for both agriculture and the general economy. Agricultural producers may face increased price volatility, since domestic prices are pegged more closely to international prices. Simultaneous reform to liberalize and integrate domestic financial-capital markets would reduce the costs of increased price variability through risk pooling on an economy-wide basis. Financial integration for risk spreading is critical at the rural household level as well. In order to exploit the income-enhancing potential of the commercialization of agriculture, financial markets must accommodate the increased ability of households to save and build up productive asset bases and improve human resources. Rapid development of rural finance systems at the grass-roots level is thus crucial, particularly since the commercialization of agriculture often leads to large, lumpy payments of cash a few times a year. The process of commercialization itself can provide the critical market size required for efficient, unsubsidized rural banking with low overhead costs. Effective rural financial institutions can in turn assist in spreading the benefits of commercialization more widely across the community and region.

As financial services improve in rural areas, it is possible that larger shares of rural savings will be captured in rural areas (including local towns), and that this will further facilitate the growth of the rural nonfarm economy. The financial needs of agriculture are changing with the transformation of the rural economy. Despite the exodus of labor from agriculture, farms in most of Asia are getting smaller on average, as well as more cash-oriented and more productive per hectare. The vast majority of farm households are also reducing their dependence on agriculture by diversifying into nonfarm sources of income.

This helps raise total household income, leading to higher savings, and gives farms access to more cash income that is less seasonal in nature than agricultural receipts.

Taken together, these changes seem likely to improve seasonal and annual cash flow for most farmers, thereby reducing their need for conventional forms of agricultural credit. Their financial needs are becoming more complex and diverse and include access to deposit and savings accounts and possibly also investment loans for nonfarm business activity as well as for agriculture. There is also increasing demand for financial services by many small-scale and part-time nonfarm businesses, especially in the service sector, as witnessed by the recent explosion in microfinance in rural areas. More flexible and customer-oriented financial services are required to meet these needs.

In the face of these growing needs, government intervention has often had a negative impact on rural financial markets, limiting their ability to serve not just the rural nonfarm economy but even farmers themselves. Government interventions have included lending requirements imposed on banks, refinance schemes, loans at preferential interest rates, credit guarantees, and lending by government-operated development finance institutions. These interventions have in most cases had limited impact on the adoption of new technology or on agricultural production, while seriously impairing the banks, cooperatives, and specialized agricultural development banks that have tried to implement them. Moreover, government interventions in rural financial markets failed to provide savings and other financial services demanded by farmers. Middlemen and banks often captured the subsidies intended for borrowers. Interest rates were low but borrower transaction costs were high, banks earned low returns on their capital, and credit allocation may have worsened income distribution if the credit was skewed in favor of larger firms. Financial discipline was damaged and intermediaries weakened.

The general failure of directed and subsidized credit calls for a new approach that limits the role of financial markets to financial intermediation rather than serving as a tool to stimulate production, compensate for distortions in other markets, and alleviate poverty. The appropriate role for government is to create an environment in which competitive financial institutions can emerge. Among other things, this means macroeconomic stability, reasonably low levels of inflation, procedures to enforce contracts, the protection of property rights, and a regulatory and supervisory system that can ensure prudent financial operations.

Governments also need to avoid the temptation to use financial institutions for social policies such as subsidizing particular economic activities or groups within society. Financial-market interventions are a poor second-best approach for dealing with important social problems that require more direct policies to encourage human capital formation and improve access to productive assets. The new approach is shown by a number of Asian governments that are moving to financial market liberalization, reducing the targeting of loans and setting interest rates high enough to cover costs, but reform of rural financial systems still has a long way to go.

Investment in Education, Health, and Nutrition

Education has been an important source of productivity growth in agriculture. Development of human capital makes investments in physical capital more productive and facilitates the adoption of modern knowledge-intensive technology. Rapid growth in agriculture, the rural economy, and the general economy have in turn placed increasing demands on education and the development of human capital. Absorption of the rural poor into the industrial and service sectors has significant costs in terms of learning new skills; rural people need adequate training if they are to successfully diversify into nonfarm activity. This will be especially true of many service activities, which often depend more on skilled people than on equipment and infrastructure. Investments in general education will be required, as well as targeted training programs to enhance technical, managerial, and service skills.

The rapidly changing nature of agricultural technology also places higher demands upon education. Crop-management technologies to implement integrated pest control and to improve the nutrient balance and the timing and placement of fertilizer applications are highly complex, knowledge-intensive, and location-specific. Because new technologies are more demanding for both the farmer and the extension agent, they require more information and skills for successful adoption than did the initial adoption of modern varieties and fertilizers.

In addition to education, broader social-support services are critically important to increase the benefits of the growth process and and reduce the risk of adverse consequences. Foremost among these support policies are health and nutritional services. Nutritional improvements are determined by both health and food consumption. Negative health effects from poor household and community health and sanitation can overcome potential positive effects of income growth from agricultural growth and commercialization. Increased income and food consumption help to reduce hunger but cannot solve the problem of preschool children's malnutrition, which results from a complex interaction of lack of food and morbidity. Health and sanitation in rural areas must be promoted through improvement of community-level health services to fully exploit the welfare effects of agricultural growth. In addition to causing severe short-term problems, the financial and economic crisis in East and Southeast Asia has revealed policy and institutional shortcomings in social services that will challenge Asian economies well beyond the crisis period. In the short run it is essential to preserve existing economic and social services for the poor, including health, education, and employment, while in the longer term, stronger social safety nets must be built.

COMPLETING THE TRANSFORMATION: REVERSING ENVIRONMENTAL DEGRADATION

Agricultural and economic growth in Asia have had an overwhelmingly positive impact on human well-being in the region. But unintended negative consequences have grown in magnitude and pose serious problems for future policy. Chief among these unintended consequences are environmental degradation and growing regional disparities that threaten to prevent less favorable environments from sharing in growth. Limited amounts of land and water in Asia suitable for agriculture limit the scope for bringing new natural resources on line for food production. In addition, some contraction in land and water resources for agriculture, due to rising pressure to divert resources already in agriculture to nonagricultural uses, may partially offset any expansion. Moreover, environmental degradation of areas already in production can dampen growth in food supplies by eroding the productive capacity of the natural-resource base; any new areas brought under production may be even more susceptible to degradation than are current areas. Lessons for addressing these unintended consequences are reviewed in this and the next section.

Two main types of environmental degradation have occurred in Asia. On the one hand, intensification of agricultural production in irrigated and favorable rainfed environments combined with sometimes-flawed incentives due to inappropriate policies has caused substantial environmental degradation. On the other hand, in resource-poor areas, continuing population growth and a scarcity of good land have forced the expansion of cropped area into forested and woodland areas and onto steeper slopes, increasing soil erosion. A third type of environmental degradation that could expand dramatically is waste-disposal and water-quality problems caused by intensive livestock production.

Agricultural intensification *per se* is not the root cause of lowland resource-base degradation, but rather the policy environment that has encouraged monoculture systems and excessive or unbalanced input use. Trade policies, output price policies, and input subsidies have all contributed to the unsustainable use of Asian lowlands. The dual goals of food self-sufficiency and sustainable resource management often appear mutually incompatible. Policies designed for achieving food self-sufficiency tend to undervalue goods not traded internationally, especially land and labor resources. As a result, food self-sufficiency in countries with an exhausted land frontier has come or could come at a high ecological and environmental cost. Appropriate policy reform, at the macro as well as at the sector level, will go a long way towards arresting and possibly reversing the current degradation trends, but the degree of degradation in many regions will pose severe challenges to policymakers.

In the less favorable areas, mining of soil fertility, soil erosion, deforestation, and loss of biodiversity impose high costs on those who depend on these areas for a living. Soil erosion contributes not only to lower yields on site, but also to siltation problems downstream, reducing the capacity and productivity of reservoir and irrigation schemes and thereby affecting an even broader area. Likewise, deforestation in upper watershed regions can also have broader effects, for example by contributing to flooding problems in lowland areas. These problems are already serious in many "hot spot" areas in Asia such as the foothills of the Himalayas, sloping areas in the southern PRC and Southeast Asia, and the forest margins of Indonesia, Malaysia, Viet Nam, Cambodia, and Lao PDR.

With rapidly increasing demand for meat and livestock products in much of Asia, pressures from livestock production could cause similar or more severe environmental degradation. Modernization of the traditional livestock production systems in many Asian countries will require huge investments to improve feeding potential, ensure a suitable animal environment, and provide other modern production and processing technology. But, as with intensive crop agriculture, the intensification of livestock production poses potentially severe environmental challenges. Production of livestock generates waste by-products that under some conditions can be recycled but, when animal concentrations are high, can become a serious pollution problem. Livestock and feed production use large quantities of water, not only as a direct input but also for waste disposal. The high concentration of industrial livestock production has the potential to produce substantial organic discharges that are in excess of the carrying capacity of the surrounding environment.

Policies that mitigate or even reverse negative environmental effects in the crop sector and help preempt larger problems in the livestock sector include the removal of trade, macroeconomic, and price distortions on input and output markets and the establishment of price incentives or regulations to reduce the production of environmental externalities in both sectors. These environmental problems in higher-potential areas for crop and livestock production will continue to receive attention, since Asia will continue to rely on these areas for its food production and degradation there poses a relatively greater threat to the food supply. For crops, this means particularly those areas already irrigated; for livestock production, increasingly this refers to areas where traditional animal husbandry will be left behind as the sector industrializes.

To mitigate environmental effects that may also be severe in the less favorable areas, perhaps affecting the poor disproportionately, a different type of policy is required. Here, it is the pressure to expand area under production, rather than intensification, that frequently causes degradation. In the short to medium term, intensification for these areas may be the best strategy, but, because these areas are more fragile environmentally, intensification must be undertaken in such a way as to preserve the environment. This will mean greater investment in technologies and policies suited to the diverse conditions that characterize low-potential areas, as well as efforts to link those areas to the broader economy, so that benefits of market reform reach them as well. Environmental degradation has been a by-product both of improper policies in the high-potential areas and of outright neglect in the low-potential areas. In both cases, new strategies to safeguard against or mitigate existing environmental degradation must be brought into line with policies that have been set with other objectives in mind.

Although land degradation is of overriding importance in some geographic regions within Asia, probably the most severe environmental challenges facing Asian developing countries are water scarcity and quality. Water scarcity is increasing and within the next decade or two many Asian countries will approach crisis levels, where there will simply not be enough water to meet all needs for all or part of the year. Growing water scarcity will result largely from rapidly growing demands for agricultural, industrial and household purposes, but the potential for expanding supplies is also diminishing. Water-shortage problems will also be aggravated by worsening environmental conditions related to deteriorating water quality, degradation of irrigated land, insufficient levels of river flow for environmental and navigation purposes, upstream land degradation, and seasonal flooding. Pollution of water from industrial waste, poorly treated sewage, and runoff of agricultural chemicals, combined with poor household and community sanitary conditions, is a major contributor to disease and malnutrition.

These problems are important throughout Asia: water scarcity is more of a seasonal constraint in the monsoon countries of East and Southeast Asia but a year-round problem elsewhere. Water scarcity and quality issues are especially severe in Central Asia (e.g., the Aral Sea) and parts of South Asia. In order to deal with these problems and to avert water scarcities that could depress agricultural production, cause rationing of water to household and industrial sectors, damage the environment, and escalate water-related health problems, new strategies for water development and management are urgently needed.

A large share of the water that is needed to meet new demand must come from water saved from existing uses through comprehensive reform of water policy. Such reform will not be easy, because both long-standing practice and cultural and religious beliefs have treated water as a free good and because entrenched interests benefit from the existing system of subsidies and administered allocations of water. But it should be pointed out that the types of policies needed to improve water management are broadly applicable to other environmental problems as well. In the broadest sense, these are, first, policies to improve the flexibility of resource allocation in agriculture, through removal of subsidies and taxes that distort incentives and encourage misuse of land and water; and, second, the establishment of secure property rights and investments in research, education and training, and public infrastructure.

The most significant reforms in the water sector should include changing the institutional and legal environment in which water is supplied and used to one that empowers water users to make their own decisions regarding use of the resource, while providing correct signals regarding the real scarcity value of water, including environmental externalities. The appropriate combination of new investments and water-management reforms will vary depending on the location, level of institutional and economic development, and degree of water scarcity. But water-policy reforms should include a balancing of improved, integrated water management at the river-basin level, through strengthening of relevant public institutions and improved tools for planning and monitoring purposes, with decentralization and privatization of important sub-basin water management functions to the private sector or community-based water-user groups. Establishment of secure water rights of users and the use of incentives to encourage water conservation, including markets in tradable water rights, pricing reform and reduction in subsidies, and implementation of effluent or pollution charges, would help to reduce water use and the negative environmental consequences of overuse of water. The innovative institutional and policy reforms required for water management require a complex blending of public-sector, market, and civil-society roles in order to address the problems not only of water scarcity and quality, but of the other important environmental challenges.

COMPLETING THE TRANSFORMATION: REACHING LESS FAVORED LANDS

Past agricultural development strategies in Asia have emphasized irrigated agriculture and "high-potential" rainfed lands in an attempt to increase food production and stimulate economic growth. This strategy has been spectacularly successful in many countries and produced the continent's remarkable rural transformation. At the same time, however, large areas of less favored lands have been neglected and lag behind in their economic development. These less favored areas are characterized by lower agricultural potential, often because of poorer soils, shorter growing seasons, and lower and uncertain rainfall, but also because past neglect has left them with limited infrastructure and poor access to markets. Despite some out-migration to more rapidly growing areas, population size continues to grow in many less favored areas and this growth has not been matched by increases in agricultural yields. The result is often worsening poverty and food-insecurity problems, as well as the widespread degradation of natural resources noted in the previous sections.

In order to promote economic growth and redress poverty and environmental problems, Asian policymakers will need to pursue appropriate and sustainable methods of agricultural intensification for both high- and low-potential regions. This dual strategy will be especially challenging if government budgets for investment in agriculture and rural areas continue to remain tight; striking the right investment balance between irrigated and rainfed regions and between high- and low-potential rainfed areas will be particularly important. Investments in irrigated and high-potential rainfed areas cannot be neglected, because these areas still provide much of the food needed to keep prices low and to feed growing urban populations and livestock. On the other hand, the poverty, food-security and environmental problems of many less favored areas are likely to remain serious in the decades ahead as populations continue to grow. While out-migration and economic diversification should become increasingly important in the development of areas with low agricultural potential, agricultural intensification will often offer the only viable way of raising incomes and creating employment on the scale required in the near future.

The successful development of less favored lands will also require new and improved approaches to policy making and institution-building. Successful development often will require stronger partnerships than needed in high-potential areas between agricultural researchers and other agents of change, including local organizations, farmers, community leaders, NGOs, national policymakers, and donors. It will also require time and innovation: new approaches will need to be developed and tried on a small scale before being disseminated more widely and their testing will take time to assess and evaluate. All this will require patience and perseverance on the part of policymakers and donors.

EXTENDING THE TRANSFORMATION: A NEW UNDERSTANDING OF THE ROLE OF GOVERNANCE

Asian societies are changing. With rising incomes and globalization, there is increasing demand for more competitive politics and greater popular participation in government. There is increasing demand for more democratic forms of governance and for greater devolution of the management of public resources to local governments and organizations. Greater participation is an important contributing factor to the quality of life.

The demand for improved governance is also driven by some of the failures of the past, as the East and Southeast Asian financial and economic crisis has exposed serious weaknesses in financial and corporate oversight and corruption in high places. People not only want a greater say in public decisions, but also more accountability in the way funds are spent. These changing expectations about governance have led to an increase in political activity, an increasing visibility for organized civil society, and an increasing importance for NGOs.

At the same time, the nature of many public goods is changing, as are the options for supplying them. As biotechnology becomes more important, for example, more aspects of agricultural research are being privatized; this requires some rethinking about the role of publicly provided agricultural research. The removal of parastatals and the privatization of agricultural marketing and service provision have also redefined the role of the public sector to one of regulation rather than supply. In the case of education and health care, both directly related to the quality of life, household demand for services increases rapidly with income and the private-sector response in provision is already very apparent in Asia's urban areas. There is need to reconfigure the roles of the public and private sectors and of civil society in providing many public goods and services so as to make them more cost-effective and efficient and to better meet the changing needs of rural people. In the case of merit goods such as education, basic health care, and water supply, a public-sector role in provision will need to be maintained for bypassed regions and the rural poor, whose limited consumer wherewithal prevents a satisfactory private-sector response.

Good governance implies the creation of a political and institutional environment in which authority is based on the rule of law, is transparent, is accountable to society, and is based on institutions and not on individuals. Institutional reform to provide good governance is a complex and long-term process that requires both improvement in public administration and public-sector management and movement toward more diversified delivery of services that is responsive to stakeholders. Governance reforms must seek greater transparency and accountability in public-sector activities.

Reform is also necessary in the relationship between the public sector and the recipients of public-sector services. Diversified delivery of services involving government, civil society, and religious institutions would help reduce the risks of relying on only one delivery system. To diversify delivery successfully, it is important also to reform the "demand side" for services. Generation of effective demand for public services and monitoring of public-sector performance is enhanced by a pluralistic society with rights to associate and to organize interest groups that have access to information about government services and programs. Governments would reduce implementation problems and enhance public support for their programs by easing access to information and allowing affected communities the opportunity to voice their concerns.

Decentralization of services to local or community-based institutions can be an important component of improved services but should not be seen as a panacea. Local elites may have weaker technical resources at their disposal than regional or national ones, along with greater opportunities for corruption and lack of transparency.

NGOs and civil society more generally can also play an important role and can be effective in areas more traditionally covered by government, such as poverty relief and health care and nutrition. But mutual distrust between NGOs and government has often become deep-rooted; both parties need to work to develop improved collaboration and provide a better foundation for interaction between government and civil society.

EXTENDING THE TRANSFORMATION: MANAGING A NEW REVOLUTION IN AGRICULTURAL TECHNOLOGY

The unfolding biotechnology revolution in agriculture has the potential to drastically transform agricultural production and processing in the future. Early benefits will be seen in modest yield increases, reduced dependence on agricultural chemicals for pest and weed control, increased drought resistance in crops, and better-quality and more nutritious crops. These could be followed by much more significant breakthroughs in crop and livestock yields, new types of crops, control of major diseases in livestock, nitrogen fixation in cereals, and new types of processed foods. As with the microcomputer revolution of the 1980s, developments may accelerate much faster than the experts may now think.

If successfully tapped, the biotechnology revolution could make an extremely important contribution to future agricultural growth and food security in Asia. In fact, it may offer the only viable way of restoring adequate levels of growth in crop yields in the decades ahead. The green revolution has already run its course in much of Asia; yield growth for major food grains has become sluggish. It is seldom profitable for farmers to aim for more than 50 percent of yield potential as expressed in experimental-station yields, and this level has already been reached in many irrigated areas. Conventional plant breeding is running out of options for providing the needed breakthroughs in yield potentials, but biotechnology is beginning to open up new possibilities. Like many revolutionary developments, however, biotechnology also brings new risks and problems.

Most current agricultural biotechnology research is being undertaken by a handful of multinational companies and caters to the problems of rich farmers and developed-country consumers. Few outputs from this research will be appropriate for most Asian countries. For example, crop varieties with built-in herbicide resistance would require much greater reliance on herbicides than is common in Asia, where most weeding is still done by hand. And crop varieties that incorporate Bt genes for insect pest resistance need to be surrounded by buffer zones of non-Bt varieties if insects are not to become resistant. This may be hard to enforce in many Asian countries.

But the biggest limitation is that hardly any biotechnology research is being undertaken on many of Asia's basic food crops or on the problems of small farmers. Even in Asian countries with the strength to develop biotechnology programs, such as India, research emphasis is often placed on export crops. The private sector is unlikely to change its focus, because it perceives limited potential to reap profits from solving many of these problems. If Asian countries are to tap more fully into the biotechnology revolution, they will need to expand their own national and regional capacity to undertake some of this research.

Greater local capacity will also be essential for forming effective partnerships with relevant multinational companies and biotechnology research centers in developed countries. Several international initiatives are already attempting to improve Asia's access to biotechnology research. For example, the Rockefeller Foundation launched its Rice Biotechnology Network in 1985, with the aim of improving national capacity. After a slow start, some of the International Agricultural Research Centers (e.g., IRRI and the International Crops Research Institute for the Semi-Arid Tropics) are also beginning to become more active players in biotechnology research. They may soon be able to serve an important intermediary role between multinational companies, developed-country research centers, and the needs and capacities of national agricultural research systems in Asia. But these developments are at an early stage and Asian countries need to make a much more concerted effort to tap into the biotechnology revolution. This will require allocation of additional public funds for agricultural research, as well as staffing up for biotechnology research.

The public sector will need to play a particularly important role in ensuring that small and disadvantaged farmers and resource-poor areas are not left further behind by the biotechnology revolution. Private companies have little incentive to work on the problems of these groups, since the latter are the least likely to be able to afford new and improved seeds or to use additional inputs like herbicides. Publicly funded (though not necessarily publicly conducted) research will be crucial for these groups.

Another worry for Asia is that biotechnology is being used in developed countries to genetically engineer substitutes for some of the region's traditional export crops. This could eventually prove costly in terms of lost export earnings. For example, rapeseed plants with more than 35 percent laurate in their oil have now been produced in the US and are expected to provide a cheaper alternative to coconut and palm-kernel oil. Such losses in competitive advantage will take place not only between developed and developing countries, but also between smaller developing countries without biotechnology capacity and those developing countries that have it.

Biotechnology also brings new risks associated with the release of genetically modified material into the environment (e.g., genes "jumping" from genetically modified plants to other plants through cross-pollination, rapid creation of new pest biotypes through adaptation to genetically modified plants) and from the consumption of genetically modified foods (e.g., allergic reactions, toxins). These risks are not well understood and they provoke a great deal of anxiety among some segments of the public. National institutions must have the capacity to evaluate these risks, to adapt breeding and crop-management strategies to minimize these risks, and to implement and rigorously enforce appropriate regulatory systems.

Biotechnology is also associated with a thorny set of intellectual property rights issues. Property rights over genetic resources are needed to reward private companies for their efforts in developing improved varieties. But if these rights are inappropriately defined, they could lead some countries to lose ownership rights over their own indigenous genetic resources. These concerns have been reinforced in recent years by patents issued in the US for frivolous claims to turmeric, neem and basmati rice that essentially gave private companies ownership rights over underlying indigenous genetic material from Asia. Such patents are often overturned when challenged in US courts, but require costly litigation by Asian countries.

Another problem is that as more countries try to assert claims over their indigenous genetic resources (as agreed at the International Convention on Biodiversity), this will impede the free flow of agricultural genetic material between countries. The high-yield varieties of cereal crops associated with the green revolution incorporated genes from a number of countries; these were freely exchanged through public research institutions to the benefit of all countries that could grow the crops. There is a growing danger that it may become increasingly difficult to share genetic material in this way and this could slow or impede future genetic improvements.

The development of an international system of intellectual property rights in agriculture is still in a state of flux, with the US taking an aggressive lead. Asian countries need to take a position that balances the interests of private-sector companies (both foreign and domestic) whose products they would like to use with protection of their own rights of public access to indigenous genetic materials at home and abroad. This probably means they will need to implement patent laws to protect rights to novel and significantly improved genetic material.

EXTENDING THE TRANSFORMATION: MANAGING GLOBALIZATION

Market-oriented policies that have favored economic liberalization, open markets, and integration with the global economy have been enormously successful for both rural and general economic development in Asia. The process of globalization, including increased interlinkages across countries and expanded trade, financial, and information flows, provides new technologies and markets and new sources of finance. But globalization and economic liberalization carry with them risks that have been driven home by the recent Asian financial and economic crisis. Most concretely, the economic crisis raised serious questions about the sequencing of open-economy reforms and about the free convertibility of short-term foreign capital inflows.

Foreign direct investment and other long-term, relatively stable investment have a significant impact on economic growth, but the benefits of short-term international capital are small and uncertain because, unlike foreign direct investment, short-term capital does not bring along technology and management innovations. Moreover, when savings rates are already high and marginal investment is misallocated, short-term capital greatly increases the vulnerability of the economy. Management of international capital flows should therefore focus on the creation of an environment conducive to long-term investments and discouraging to short-term capital inflows. Tax incentives and other distortions that favor short-term inflows over long-term investments should be eliminated.

The importance of sequencing of reforms was shown by the high costs of moving to free convertibility of short-term capital before effective financial intermediation and prudential regulation were in place: the lack of coordination exacerbated all the dire effects of the financial crisis in many Asian countries. Both prudential regulation of currency positions of banks and strengthened enforcement of these regulations and other risk-management procedures are required. In those Asian developing economies that are plagued by weak institutional capacity and financial systems, temporary controls on capital inflows, combined with domestic reforms and greater disclosure, may be necessary to help reduce the frequency and magnitude of shocks. Capital-account restrictions need to be explicit, transparent, and market-oriented, with the Chilean approach described in Chapter IX providing a possible model for market-oriented short-term capital controls.

More generally, both the tremendous successes of the East and Southeast Asian countries with economic liberalization and the recent financial and economic crisis drive home the point that appropriate policies and institutions must be in place if the benefits of globalization and open economic policies are to be reaped. For East and Southeast Asia to correct the problems revealed by the economic crisis and for South Asia and Central Asia to liberalize and open their economies, the continued development of domestic institutions and policies to manage globalization is essential. Full and effective economic liberalization and linkage with the global economy require continued reform of fiscal and financial policies and institutions, property and contract laws that foster modern commerce, flexible and efficient factor and product markets, and continued development of technology and human capital. Future successes in rural and general economic development will be driven by domestic policies and processes, including public and private domestic investment, macroeconomic stability, research-based technological change, education, and human capital development, with globalization serving as an important facilitator and spur to these processes.

ALTERNATIVE FUTURES: NO ROOM FOR COMPLACENCY

The future for Asia, even after the recent economic shocks, looks far brighter than in the desperate days of the 1960s. In fact, if likely trends continue (as captured in Chapter XII's baseline scenario for food and agriculture), both Asia and the world should be able to meet projected food demand with greater production and increased trade, at least in the aggregate. The continuation of likely trends does not call for a passive, sit-back approach, however, but rather for steady progress by governments and the international community in devising and carrying out policies for rural Asia already shown to be cost-effective in terms of agricutural production and its positive multiplier effects in the rural nonfarm economy and beyond. These policies include investment in agricultural research, extension, irrigation and water development, human capital, and rural infrastructure. With these measures, heightened pressure from rising populations and incomes in Asia and around the world will not overwhelm aggregate food supplies, regionally or globally. In fact, Asia will have more food available per person and a stronger overall trade position; real world food prices will be steady or declining slowly for the main food commodities.

As can be seen even from the situation today, however, adequate food supply in the aggregate does not translate into food security for everyone everywhere. Behind even this optimistic picture in the aggregate lies the fact that progress against food insecurity comes neither quickly nor easily. Even with the gains from trade and the overall ability of the world's productive capacity to meet effective demand for food, food security will improve only slowly; massive human suffering will persist in Asia, with over 100 million children still suffering from malnutrition in 2010, approximately 80 percent of these in South Asia.

The numbers of malnourished children could grow substantially, moreover, if complacency wins the day and a more passive policy approach ensues: food security is vulnerable if improvement in any one of the three policy pillars needed for the transformation—agricultural and rural income growth, supporting social services, and declining population growth—lags. Policies that moderately disfavor agriculture and natural resources, reduce social investment, or moderately slow economic reform will lead to much worse net food-security impacts and poorer diets than the baseline.

Conversely, moderate but important reductions in childhood malnutrition can be achieved through relatively small improvements in income growth, investment in agricultural research and irrigation, improved water policy, reductions in environmental degradation, slower population growth, and increases in social investments. Indeed, a gap of 76 million malnourished children separates this high-investment scenario from the scenario where complacency allows investment to drop off. What is more, analysis shows food security responding in almost equal measure to the proposed changes in each of the three pillars noted above.

Even under this scenario, child malnutrition would persist, being particularly intransigent in South Asia, in part due to slower declines in population growth rates there. If income growth and investment in agricultural and social sectors were instead boosted significantly, the effect would be to mount a significant assault on childhood malnutrition in Asia. To bring levels of childhood malnutrition down to near zero would require extremely aggressive policy reform and public investment to raise income growth, agricultural productivity, and social indicators for all of Asia close to the peak levels experienced by Asia's most dynamic economies in the wake of the adoption of green-revolution technologies.

CONCLUSION

On the threshold of the 21st century, Asia stands, if not at a crossroads, then at a point of decision. Unprecedented rates of rural and national economic growth have transformed many parts of rural Asia. But not all Asia shared in the transformation. Moreover, the regional economy has hit a rough spot that some fear may turn the clock back on rapid growth where it occurred and stymie growth elsewhere.

In the face of this challenge, the completion of the rural transformation of Asia will take renewed efforts on the part of governments. Successful economies must not turn away from their market orientation, but rather support the private-sector role where possible and supplement it where not. But meeting the challenge must also involve a renewal of governance itself: transparency, responsiveness, and eradication of corruption are all keys to sustained growth in the next century.

Governments will also have to increase the level of productive investment made in rural infrastructure, agricultural research and extension, education, and health, as well as expand the reach of social safety-net programs. A significant part of these costs could be met in some countries by reducing wasteful public expenditure in rural areas, particularly on subsidies for credit, fertilizers, pesticides, electricity, and irrigation water. These subsidies may have played an important role in launching the green revolution, but today they are rarely needed and can be counterproductive because they create incentives for the overuse of water and farm chemicals, leading to environmental degradation.

There is also considerable scope for "getting more with less" by improving the efficiency of many of the public institutions that implement public investments. This again requires changes in governance structures, with increased transparency and accountability to key stakeholders and greater roles for the private sector, user groups and NGOs where they can better provide the required services. There is also scope for raising more revenues from rural people through user fees and local taxation. These kinds of changes might provide much of the financing needed for rural areas in the future, but they will take time to implement. If poverty and malnutrition are to be seriously reduced within the next generation, then additional allocation of central government funds will almost certainly be required, at least in the near future.

While the specter of famine that hung over Asia in the 1960s has not returned in the 1990s, widespread poverty and malnutrition still coexist with great wealth. Completion of the rural revolution, radical reduction in poverty, and improvement in food security in Asia hang in the balance. They are attainable, if complacency is resisted.

REFERENCES

Abrol, I. P. 1987. Salinity and Food Production in the Indian Sub-Continent. In *Water and Water Policy in World Food Supplies*, edited by W. R. Jordan. College Station, TX: A&M University Press.
ADB (Asian Development Bank). 1997a. *Kyrgyz Republic: Country Economic Review*. Manila: Asian Development Bank.
_____.1997b. *Kazakstan: Country Economic Review*. Manila: Asian Development Bank.
_____.1997c. *Uzbekistan: Country Economic Review*. Manila: Asian Development Bank.
_____.1997d. *Emerging Asia: Changes and Challenges*. Manila: Asian Development Bank.
_____.1998a. *Republic of Uzbekistan: Phase 1 Report—Agricultural Sector Development Project*. Manila: Asian Development Bank.
_____.1998b. *Proposed Loans and Technical Grants to the Kingdom of Thailand for the Social Sector Program*. February 1998. Manila: Asian Development Bank.
_____. 1998c. *Inception Workshop on Social Impact Assessment of the Financial Crisis in Selected DMCs*. Manila: Asian Development Bank.
_____.1998d. *Proposed Loans and Technical Grants to the Republic of Indonesia for the Social Protection Sector Development Program*. June. Manila: Asian Development Bank.
_____. 1998e. *Key Indicators of Developing Asian and Pacific Countries 1998*, Volume XXIX. Hong Kong: Oxford University Press.
_____.1998f. *Anticorruption Policy*. June. Manila: Asian Development Bank.
Adelman, I. 1984. Beyond Export-Led Growth. *World Development* 12 (9): 937–49.

―――. E. Yeldan, A. Harris, and D. Roland-Host. 1989. Optimal Adjustment to Trade Shocks Under Alternative Development Strategies. *Journal of Policy Modeling* 11(4): 451–505.

Ahmed, R., and M. Hossain. 1990. Development Impact of Rural Infrastructure: A Case Study of Bangladesh. *Research Report Number 83.* Washington DC: International Food Policy Research Institute.

Ahuja, V., B. Bidani, F. Ferreira, and M. Walton. 1997. *Everyone's Miracle? Revisiting Poverty and Inequality in East Asia.* Directions in Development Series. Washington, DC: World Bank.

Alexandratos, N. 1996. China's Projected Cereal Deficits in a World Context. *Agricultural Economics* X (15): 1–16.

Ali, K., and A. Hamid. 1996. Technical Change, Technical Efficiency, and their Impact on Input Demand in the Agricultural and Manufacturing Sectors of Pakistan. *The Pakistan Development Review* 35 (3): 215–228.

Ali, M. 1998. *New Paradigms for Vegetable Cultivation in Asia.* Taipei,China: Asian Vegetable Research and Development Center. Mimeo.

―――, and D. Byerlee. 1998. Productivity Growth and Resource Degradation in Pakistan's Punjab: A Decomposition Analysis. World Bank. Mimeo.

Alston, J. M., and P. G. Pardey. 1996. *Making Science Pay: The Economics of Agricultural R&D Policy.* AEI Press: Washington, DC.

―――, B. Craig, and P. Pardey. 1998. Dynamics in the Creation and Depreciation of Knowledge and the Returns to Research. *Environment and Production Technology Division Discussion Paper* No. 35. Washington, DC: International Food Policy Research Institute.

Ammar, Siamwalla, A. 1997. World Agriculture Trade After the Uruguay Round: Disarray Forever? In *The Global Trading System and Developing Asia*, edited by A. Panagariya, M. G. Quibria, and N. Rao. Asian Development Bank, Philippines: Oxford University Press, p. 435–468.

Amsden, Alice. 1994. Why Isn't the Whole World Experimenting With the East Asian Model to Develop? Review of the East Asian Miracle. *World Development* 22 (4):627-633.

Anderson, D., and M. Leiserson. 1980. Rural Non-Farm Employment in Developing Countries. *Economic Development and Cultural Change* 28 (2): 227–248.

Anderson, J. R., and J. A. Roumasset. 1996. Food Insecurity and Stochastic Aspects of Poverty. *Asian Journal of Agricultural Economics* 2 (1): 53–66.

Anderson, K. 1996a. Why the World Needs the GATT/WTO. In *Strengthening the Global Trading System: From GATT to WTO*, edited by K. Anderson. Adelaide, Australia: Education Technology Unit, University of Adelaide. p. 3–12.

_____. (editor). 1996b. *Strengthening the Global Trading System: From GATT to WTO*. Adelaide, Australia: Education Technology Unit, University of Adelaide.

_____, and Y. Hayami. 1986. *The Political Economy of Agricultural Protection: East Asia in International Perspective*. Winchester, MA: Allen & Unwin Inc.

_____, B. Dimaranan, T. Hertel, and W. Martin. 1997. Asia-Pacific Food Markets and Trade in 2005: A Global, Economy-Wide Perspective. *The Australian Journal of Agricultural and Resource Economics* 41 (1): 19–44.

Antle, J. M. and P. L. Pingali. 1994. Pesticides, Productivity, and Farmer Health: A Philippine Case Study. *American Journal of Agricultural Economics* 76 (3): 418–430.

APO (Asian Productivity Organization). 1991. *Public Expenditures on Agriculture in Asia*. APO: Tokyo.

Atinc, T. M., and W. Michael. 1998. East Asia's Social Model after the Crisis. Paper prepared for Asia Development Forum, Manila (March). Washington, DC: World Bank.

Azam, Q. T., E. A. Bloom, and R. E. Evenson. 1991. Agricultural Research Productivity in Pakistan. *Economic Growth Center Discussion Paper* No. 644. New Haven: Yale University Press.

Bach, C. F., W. Martin, and J. Stevens. 1996. China and the WTO: Tariff Offers, Exemptions, and Welfare Implications. *Weltwirtschaftliches Archiv* 132 (3): 409–431.

Balisacan, A. M. 1994. Urban Poverty in the Philippines: Nature, Causes and Policy Measures. *Asian Development Review* 12 (1): 117–152.

_____. 1998. Policy Reforms and Agricultural Development in the Philippines. *ASEAN Economic Bulletin* 15 (1): 77–89.

Ballabh, V., and T. Walker. 1992. Land Fragmentation and Consolidation in Dry Semi-arid Tropics of India. *Artha Vijnana* 34 (4): 363-387.

Bardhan, K. 1993. Women and Rural Poverty: Some Asian Cases. In *Rural Poverty in Asia*, edited by M.G. Quibria. Oxford University Press (for the ADB).

Barker, R., and R. Herdt. 1978. *Interpretive Analysis of Selected Papers from Changes in Rice Farming in Selected Areas of Asia*. Los Baños, Philippines: International Rice Research Institute.

Bautista, R. M. 1990. Development Strategies, Foreign Trade Regimes, and Agricultural Incentives in Asia. *Journal of Asian Economics* 1: 115–134.

──────. 1993. Toward More Rational Trade and Macroeconomic Policies for Agriculture. In *The Bias Against Agriculture: Trade and Macroeconomic Policies in Developing Countries*, edited by R. M. Bautista and A. Valdés. A copublication of the International Center for Economic Growth and the International Food Policy Research Institute. San Francisco: Institute for Contemporary Studies (ICS) Press.

Becker, G., and H. Lewis. 1974. Interaction Between Quantity and Quality of Children. In *Economics of the Family*, edited by T. W. Schultz. Chicago: University of Chicago Press.

Bell, C., P. Hazell, and R. Slade. 1982. *Project Evaluation in Regional Perspective: A Study of an Irrigation Project in Northwest Malaysia*. Baltimore: Johns Hopkins University Press.

Besley, T., and R. Burgess. 1998. Land Reform, Poverty Reduction and Growth: Evidence from India. *The Development Economics Discussion Paper Series* No. 13. London: London School of Economics.

Bhagwati, J. 1997. The World Trading System: The New Challenges. In *The Global Trading System and Developing Asia*, edited by A. Panagariya, M. G. Quibria, and N. Rao. Manila: Asian Development Bank and Oxford University Press. p. 41–77.

Bhalla, G.S., and G. Singh. 1996. *Impact of GATT on Punjab Agriculture*. Monograph Series II. Chandigarh, India: Institute for Development and Communication.

──────, and G. Singh. 1998. Recent Developments in Indian Agriculture: A State Level Analysis. *Economic and Political Weekly* 32 (3): A2–18.

──────, G. K. Chadha, S. P. Kashyap, and R. K. Sharma. 1990. Agricultural Growth and Structural Changes in the Punjab Economy: An Input-Output Analysis. *Research Report* 82. Washington, DC: International Food Policy Research institute.

Bhalla S. 1981. Islands of Growth: A Note on Haryana Experience and Some Possible Implications. *Economic and Political Weekly* 16 (23): 1022–1030.

———. 1991. Report of the Study Group on Employment Generation, National Commission on Rural Labor. New Delhi: Government of India.

———. 1997. The Rise and Fall of Workforce Diversification Process in Rural India. In *Growth, Employment and Poverty Change and Continuity in Rural India*, edited by G. K. Chandha and Alkh N. Sharma. Delhi: Indian Society of Labour Economics.

Bhatia, R., and R. P. S. Malik. 1995. Energy Demand and Supply for Sustainable Agriculture: A Vision for 2020. Washington, DC: International Food Policy Research Institute. Mimeo.

Binswanger, H.P. 1994. Policy Issues: A View from World Bank Research. In *Agricultural Technology: Policy Issues for the International Community*, edited by Jock R. Anderson. Cambridge, UK: CAB International in association with the World Bank.

———, K. Deininger, and G. Feder. 1993. Agricultural Land Relations in the Developing World. *American Journal of Agricultural Economics* 75: 1242–48.

Bird, R. M. 1974. *Taxing Agricultural Land in Developing Countries*. Cambridge, MA: Harvard University Press.

Blarel, B., P. Hazell, F. Place and J. Quiggen. 1992. The Economics of Farm Fragmentation: Evidence from Ghana and Rwanda. *The World Bank Economic Review*, 2 (6): 233–254.

Blomquist, W. 1992. *Dividing the Waters: Governing Groundwater in Southern California*. San Francisco: Institute for Contemporary Studies (ICS) Press.

———, 1995. Institutions for Managing Groundwater Basins in Southern California. In *Water Quantity/Quality Management and Conflict Resolution*, edited by A. Dinar and E.T. Loehman. Westport, CT: Praeger.

Blyn, G. 1983. The Green Revolution Revisited. *Economic Development and Cultural Change* 31 (4): 705–25.

Bongaarts, J., and S. Cotts Watkins. 1996. Social Interactions And Contemporary Fertility Transitions. *Population and Development Review* 22 (4): 639–682.

Bonny, S. 1993. Is Agriculture Using More and More Energy? A French Case Study. *Agricultural Systems* 431: 51–66.

Boone, P., B. Tarvaa, A. Tsend, E. Tsendjav, and N. Unenburen. 1997. Mongolia's Transition to a Democratic Market System. In

Economies in Transition: Comparing Asia and Eastern Europe, edited by W.-T. Woo, S. Parker, and J.D. Sachs. Cambridge and London: MIT Press.

Borensztein, E., and J. D. Ostry. 1996. Accounting for China's Growth Performance. *American Economic Review* 86 (2): 224–8.

Boserup, E. 1965. *The Conditions of Agricultural Growth*. Chicago: Aldine.

Brooks, K., and Z. Lerman. 1994. Farm Reform in the Transition Economies. *Finance & Development* (December).

Boyce, J. 1987. *Agrarian Impasse in Bengal: Institutional Constraints to Technological Change*. Oxford: Oxford University Press.

Brown, L. R., and Hal Kane. 1994. *Full House: Reassessing the Earth's Population Carrying Capacity*. New York, NY: W. W. Norton.

Bumb, B. L. 1995. *Global Fertilizer Perspective, 1980–2000: The Challenges in Structural Transformation*. T-42. Muscle Shoals, AL: International Fertilizer Development Center.

———, and C. A. Baanante. 1996. The Role of Fertilizer in Sustaining Food Security and Protecting the Environment to 2020. *2020 Vision for Food, Agriculture, and the Environment Discussion Paper* No. 17. Washington, DC: International Food Policy Research Institute.

Buringh, P., and R. Dudal. 1987. Agricultural Land Use in Space and Time. In *Land Transformation in Agriculture*, edited by M. G. Wolman and F. G. A. Fournier. New York: John Wiley.

Byerlee, D. 1987. Maintaining the Momentum in Post-Green Revolution Agriculture: A Micro-Level Perspective from Asia. *Michigan State University International Development Paper* No. 10. East Lansing, MI: Michigan State University.

Cai, X. 1999. A Modeling Framework for Sustainable Water Resources Management. Unpublished Ph.D. dissertation, University of Texas at Austin.

Carney, D. (editor). 1998. Livestock in development: the integration of livestock interventions into a sustainable rural livelihoods approach. In *Sustainable rural livelihoods: What contribution can we make?* London: Department for International Development.

Cao, Y. Z., G. Fan, and W. T. Woo. 1997. Chinese Economic Reforms: Past Successes and Future Challenges. In *Economies in Transition: Comparing Asia and Eastern Europe*, edited by Wing Thye Woo, S. Parker, and J. D. Sachs. Cambridge, MA: Massachusetts Institute of Technology Press.

Cassman, K. G., and P. L. Pingali. 1993. Extrapolating Trends from Long-Term Experiments to Farmers Fields: The Case of

Irrigated Rice Systems in Asia. In *Proceedings of the Working Conference on Measuring Sustainability Using Long-Term Experiments*. Rothamsted Experimental Station, funded by the Agricultural Science Division, 28–30 April 1993. New York: The Rockefeller Foundation.

———, and P. L. Pingali. 1995. Extrapolating Trends from Long-Term Experiments to Farmers' Fields: The Case of Irrigated Rice Systems in Asia. In *Agricultural Sustainability in Economic, Environmental, and Statisical Terms*, edited by V. Barnett, R. Payne, and R. Steiner. London: John Wiley and Sons. p. 63–84.

———, S. K. De Datta, D. C. Olk, J. Alcantara, M. Samson, J. Descalsota, and M. Dizon. 1994. Yield Decline and Nitrogen Balance in Long-Term Experiments on Continuous, Irrigated Rice Systems in the Tropics. *Advances in Soil Science*, Special Issue.

Chadha, G. K. 1986. Agricultural Growth and Rural Nonfarm Activities: An Analysis of Indian Experience. In *Rural Industrialization and Nonfarm Activities of Asian Farmers*, edited by Yang-Boo Choe and Fu-Chen Lo. Kuala Lumpur: A.P.D.C.

———. 1993. Nonfarm Sector in India's Rural Economy: Policy, Performance and Growth Prospects. *Visiting Research Fellow Monograph Series* No. 220. Tokyo: Institute of Developing Economies.

Chagnon, J. 1996. *Women in Development. Lao People's Democratic Republic*. ADB Country Briefing Paper. Manila: Asian Development Bank.

Chambers, R. 1988. *Managing Canal Irrigation: Practical Analysis from South Asia*. Cambridge, UK: Press Syndicate of the University of Cambridge.

Chen, E. K.Y. 1997. The Total Factor Productivity Debate: Determinants of Economic Growth in East Asia. *Asian-Pacific Economic Literature* 11 (1): 18–38.

Chen, K., H. Wang, Y. Zheng, G. H. Jefferson, and T. G. Rawski. 1988. Productivity Change in Chinese Industry: 1953–1985. *Journal of Comparative Economics* 12 (12): 570–91.

Choe, C. 1996. Incentive to Work versus Disincentive to Invest: The Case of China's Rural Reform, 1979–84. *Journal of Comparative Economics* 22: 242–66.

Choi, J. S. 1997. Policies promoting rural non-farm activities in rural development programs in Korea after the Uruguay round.

Paper presented at the 23rd Conference of the International Association of Agricultural Economists (IAAE), August 1997, Sacramento, CA, USA.

Clark, J. 1991. *Rural Development: Putting the Last First*. Harlow: Longman.

―――――. 1993. The Relationship Between the State and the Voluntary Sector. *Human Resources Development and Operations Policy (HRO) Working Paper 18*. Washington, DC: The World Bank.

Cohen, J. I. 1994. Biotechnology Priorities, Planning, and Policies: A Framework for Decision Making. *Research Report* No. 6. The Hague, The Netherlands: ISNAR.

Collins, S., and B. Bosworth. 1997. Economic Growth in East Asia: Accumulation versus Assimilation. In *Brookings Papers on Economic Activity*, edited by William Brainard and George Perry. Vol. 2. Washington, DC: Brookings Institution.

Commandeur, P., and G. von Roozendaal. 1993. The impact of biotechnology on developing countries. Opportunities for technology-assessment research and development cooperation. Chapter 3. A study commissioned by the Büro für Technikfolgen-Abschätzung beim Deutschen Bundestag (TAB). Bonn: TAB.

Corsetti, G., P. Pesenti, and N. Roubini. 1998a. Paper Tigers? A model of the Asian Crisis. <http://equity.stern.nyu.edu/~nroubini/referen.htm>. Updated February 1997. Accessed 15 October 1998.

―――――, P. Pesenti, and N. Roubini. 1998b. "What Caused the Asian Currency and Financial Crisis?" <http:equity.stern.nyu.edu/~nroubini/referen.htm>. Updated February 1997. Accessed 15 October 1998.

Craig, B. J., G. P. Pardey, and J. Roseboom. 1991. Internationally Comparable Growth, Development and Research Measures. In *Agricultural Research Policy: International Quantitative Perspectives*, edited by G. P. Pardey, J. Roseboom, and R. J. Anderson. Cambridge: Cambridge University Press.

Crook, F. 1994. Seeds of Change, *China Business Review* (November-December): 20–26.

Crosson, P. 1996. Natural Resources and Environmental Consequences of Rice Production. *In Rice Research and Development Policy: A First Encounter*, edited by R. S. Zeigler. Los Baños, Philippines: International Rice Research Institute.

_____. 1998. The on-farm economic costs of soil erosion. In *Methods for Assessment of Soil Degradation,* edited by R. Lal, W. H. Blum, C. Valentine, and B. A. Stewart. Boca Raton: CRC Press.

_____, and J. R. Anderson. 1992. Resources and Global Prospects: Supply and Demand for Cereals to 2030. *World Bank Technical Paper* No. 184. Washington DC: World Bank.

Darwin, R., M. Tsigas, J. Lewandrowski, and A. Raneses. 1995. World Agriculture and Climate Change: Economic Adaptations. *Agricultural Economic Report* No. 703. Washington, DC: United States Department of Agriculture.

Datt, G. 1998. Poverty in India and Indian States: An Update. *The Indian Journal of Labour Economics* 41: 2.

_____, and M. Ravallion. 1997. Macroeconomic Crises and Poverty Monitoring: A Case Study for India. *Review of Development Economics* 1 (2): 135–152.

David, C. C. 1989. Philippines: Price Policy in Transition. In *Food Price Policy in Asia,* edited by T. Sicular. Ithaca, NY: Cornell University Press.

_____. 1990. Determinants of Rice Price Protection in Asia. *Social Science Division Paper* No. 90-22. Los Baños, Philippines: International Rice Research Institute.

_____, and K. Otsuka (editors). 1994. *Modern Rice Technology and Income Distribution in Asia.* Boulder, CO and London, UK: Lynne Rienner Publishers, Inc. and Manila, Philippines: International Rice Research Institute.

De Broeck, M., and K. Kostial. 1998. Output Decline in Transition: The Case of Kazakhstan. *IMF Working Paper*. Washington, DC: International Monetary Fund.

De Datta, K. S., K. A. Gomez, and J. P. Descalsota. 1988. Changes in Yield Response to Major Nutrients and in Soil Fertility under Intensive Rice Cropping. *Soil Science* 46: 350–358.

De Haan, A. and M. Lipton. 1998. *Poverty in Emerging Asia: Progress and Setback.* UK: University of Sussex.

De Melo, N., C. Denizer, A. Gelb, and S. Tenev. 1997. Circumstances and choice: The role of initial conditions and policies in transition economies. *World Bank Policy Research Working Paper* No. 1866. Washington, DC: World Bank.

_____, C. Denizer, and A. Gelb. 1998. From Plan to Market: Patterns of Transition. Washington, DC: World Bank. Mimeo.

De Menil, G. 1997. Trade Policies in Transition Economies: A Comparison of European and Asian Experiences. In *Economies in Transition:*

Comparing Asia and Eastern Europe, edited by Wing Thye Woo, S. Parker, and J. D. Sachs. Cambridge, MA: Massachusetts Institute of Technology Press.

Deb, N., and M. Hossain. 1984. Demand for Rural Industries Products in Bangladesh. *Bangladesh Development Studies* 12 (1–2): 81–99.

Deininger, K., and L. Squire. 1996. A New Data Set Measuring Income Inequality. *The World Bank Economic Review* 10 (3): 565–91.

Delgado, C., and Ammar Siamwalla. 1997. Rural Economy and Farm Income Diversification in Developing Countries. *Markets and Structural Studies Division Discussion Paper* No. 20. Washington, D.C.: International Food Policy Research Institute.

Delgado, C. L., M. W. Rosegrant, H. Steinfeld, S. Ehui, and C. Courbois. 1999. Livestock to 2020: The Next Food Revolution. *2020 Vision for Food, Agriculture, and the Environment Discussion Paper No. 28*. Washington, DC: International Food Policy Research Institute.

DeRosa, D.A. 1995. Regional Trading Arrangements among Developing Countries: The ASEAN Example. *Research Report* No. 103. Washington, DC: International Food Policy Research Institute.

————. 1998. Regional Integration Arrangements: Static Economic Theory, Quantiative Findings, and Policy Guidelines. *World Bank Policy Research Working Paper* No. 2007. Washington, D.C.: World Bank.

Desai, B. M. 1994. Contributions of Institutional Credit, Self-Finance and Technological Change to Agricutlural Growth in India. *Indian Journal of Agricultural Economics* 49 (3): 457–475.

————, and N. V. Namboodiri. 1997. Determinants of Total Factor Productivity in Indian Agriculture. *Economic and Political Weekly*, 27 December: 165–171.

Desai, G. M. 1986. Fertilizer Use in India: The Next Stage in Policy. *Indian Journal of Agricultural Economics* 41: 248–270.

————. 1988. Development of Fertilizer Markets in South Asia: A Comparative Overview. Washington, DC: International Food Policy Research Institute. Mimeo.

Devendra, C., M. E. Smalley, and H. Li Pun. 1998. Global Agenda for Livestock Research, Proceedings of a Conference on Development of Livestock Research Priorities in Asia, 13–15 May 1997, National Institute of Animal Husbandry, Hanoi, Vietnam. Nairobi, Kenya: International Livestock Research Institute.

Dey, M., and R. E. Evenson. 1991. The Economic Impact of Rice Research in Bangladesh. Economic Growth Center. New Haven: Yale University. Mimeo.

Dimaranan, B., M. W. Rosegrant, and L. Unnevehr. 1996. Supply Response of Philippine Hogs and Poultry: Modeling Industries in Transition. *Asian Journal of Agricultural Economics* 2 (1): 91–98.

Dogra, B. 1986. The Indian Experience with Large Dams. In *The Social and Environmental Effects of Large Dams*, Vol. 2, edited by E. Goldsmith and N. Hildyard. London: Wadebridge Ecological Center.

Dollar, D. 1994. Macroeconomic Management and the Transition to the Market in Vietnam. *Journal of Comparative Economics* 18 (3): 357–375.

Dorosh, P., and A. Valdés. 1990. Effects of Exchange Rate and Trade Policies on Agriculture in Pakistan. *Research Report* No. 84. Washington, DC: International Food Policy Research Institute.

Downing, T. E. 1993. The Effects of Climate Change on Agriculture and Food Security. *Renewable Energy* 3 (4/5): 491–97.

Dregne, H. E., and N. T. Chou. 1992. Global Desertification Dimensions and Costs. In *Degradation and Restoration of Arid Lands*, edited by H. E. Dregne. Lubbock, TX: Texas University.

Drysdale, P., and Y. Huang. 1995. Technological Catch-up and Productivity Growth in East Asia. Canberra: The Australian National University. Mimeo.

Esty, D. C. 1997. Environmental Protection During the Transition to a Market Economy. In *Economies in Transition: Comparing Asia and Eastern Europe*, edited by Wing Thye Woo, S. Parker, and J. D. Sachs. Cambridge, MA: Massachusetts Institute of Technology Press.

Evans, H. E. 1990. Rural-urban Linkages and Structural Transformation. *Infrastructure and Urban Development Department Report* INU 71. Washington, DC: World Bank.

Evans, P. 1998. Transferable Lessons? Re-examining the Institutional Prerequisites of East Asian Economic Policies. *Journal of Development Studies* 34 (6): 66–86.

Evenson, R. E. 1987. The International Agricultural Research Centers: Their Impact on Spending for National Agricultural Research and Extension. Yale University. Mimeo.

―――, and D. Gollin. 1994. Genetic Resources, International Organizations, and Rice Varietal Improvement. *Economic*

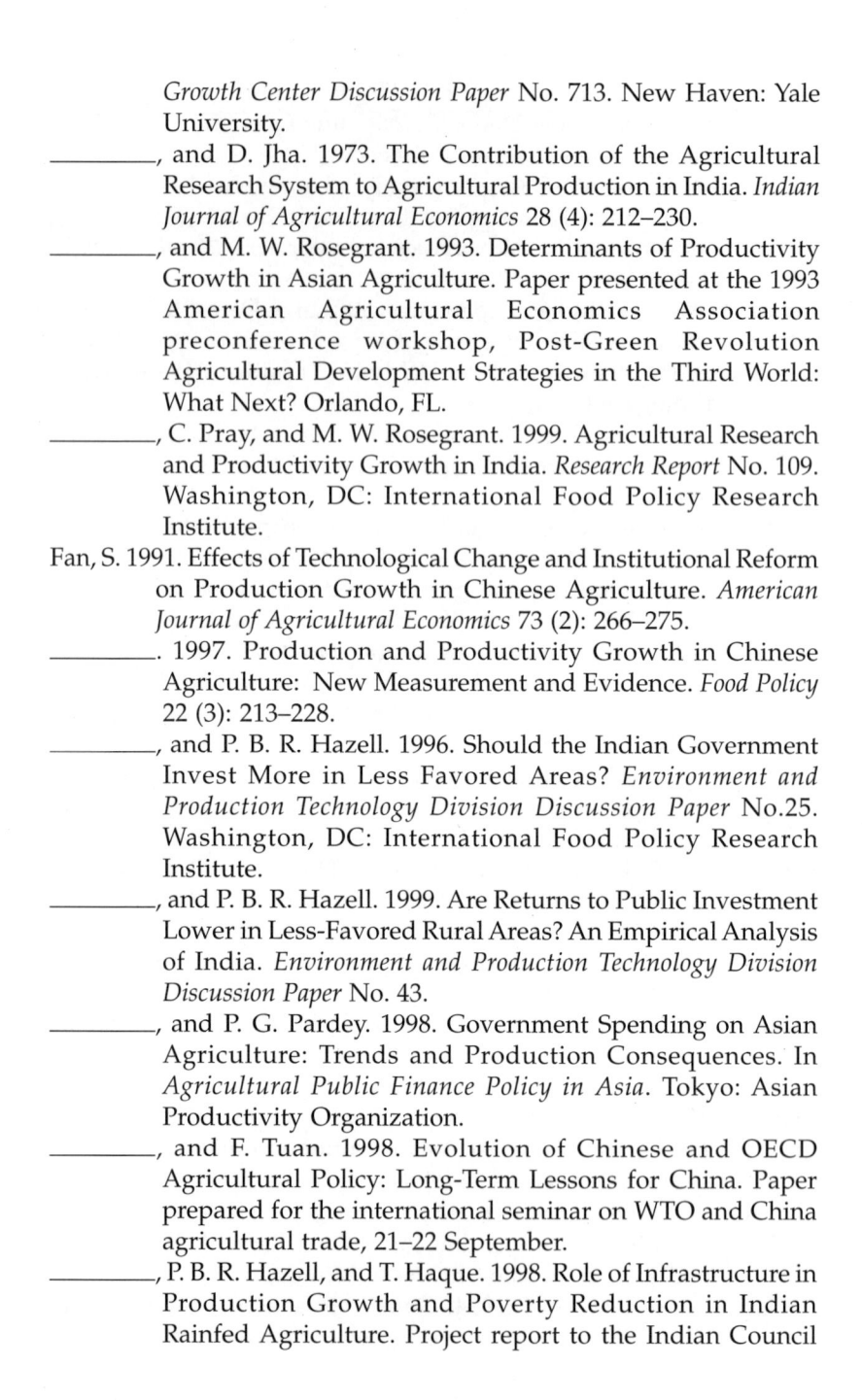

Growth Center Discussion Paper No. 713. New Haven: Yale University.

──────, and D. Jha. 1973. The Contribution of the Agricultural Research System to Agricultural Production in India. *Indian Journal of Agricultural Economics* 28 (4): 212–230.

──────, and M. W. Rosegrant. 1993. Determinants of Productivity Growth in Asian Agriculture. Paper presented at the 1993 American Agricultural Economics Association preconference workshop, Post-Green Revolution Agricultural Development Strategies in the Third World: What Next? Orlando, FL.

──────, C. Pray, and M. W. Rosegrant. 1999. Agricultural Research and Productivity Growth in India. *Research Report* No. 109. Washington, DC: International Food Policy Research Institute.

Fan, S. 1991. Effects of Technological Change and Institutional Reform on Production Growth in Chinese Agriculture. *American Journal of Agricultural Economics* 73 (2): 266–275.

──────. 1997. Production and Productivity Growth in Chinese Agriculture: New Measurement and Evidence. *Food Policy* 22 (3): 213–228.

──────, and P. B. R. Hazell. 1996. Should the Indian Government Invest More in Less Favored Areas? *Environment and Production Technology Division Discussion Paper* No.25. Washington, DC: International Food Policy Research Institute.

──────, and P. B. R. Hazell. 1999. Are Returns to Public Investment Lower in Less-Favored Rural Areas? An Empirical Analysis of India. *Environment and Production Technology Division Discussion Paper* No. 43.

──────, and P. G. Pardey. 1998. Government Spending on Asian Agriculture: Trends and Production Consequences. In *Agricultural Public Finance Policy in Asia*. Tokyo: Asian Productivity Organization.

──────, and F. Tuan. 1998. Evolution of Chinese and OECD Agricultural Policy: Long-Term Lessons for China. Paper prepared for the international seminar on WTO and China agricultural trade, 21–22 September.

──────, P. B. R. Hazell, and T. Haque. 1998. Role of Infrastructure in Production Growth and Poverty Reduction in Indian Rainfed Agriculture. Project report to the Indian Council for Agricultural Research and the World Bank. Washington, DC: International Food Policy Research Institute.

———, P. B. R. Hazell, and S. Thorat. 1998. Government Spending, Growth and Poverty: An Analysis of Interlinkages in Rural India. *Environment and Production Technology Division Discussion Paper* No. 33. Washington, DC: International Food Policy Research Institute.

Fane, G. 1998. The Role of Prudential Regulation. In *East Asia in Crisis: From Being a Miracle to Needing One?*, edited by R. H. McLeod and R. Garnaut. London and New York: Routledge.

FAO (Food and Agriculture Organization of the United Nations) 1994. *Land Degradation in South Asia: Its Severity, Causes and Effects upon the People.* Rome and Nairobi: FAO and UNEP.

———1996. Report of a Meeting of Experts on Agricultural Price Instability. Rome: FAO.

———1997. FAOSTAT Database. <http://apps.fao.org/default.htm>. Accessed Fall 1997.

———FAOSTAT 1998 <http://faostat.fao.org/default.htm>. Accessed during June–December, 1998.

Farmer, B. H. (editor). 1977. *Green Revolution?* London: Macmillan Press.

———(editor). 1986. *Green Revolution? Technology and Change in Rice-growing Areas of Tamil Nadu and Sri Lanka.* Cambridge: Cambridge University Press.

Feder, G., T. Onchan, Y. Chalamwong, and C. Hongladarom. 1988. *Land Policies and Farm Productivity in Thailand.* Baltimore: Johns Hopkins University Press.

———, L. J. Lau, J. Y. Lin, and X. Luo. 1992. The Determinants of Farm Investment and Residential Construction in Post-Reform China. *Economic Development and Cultural Change* 41: 1–26.

Fforde, A., and S. de Vylder, 1996. Viet Nam. In *From Centrally Planned to Market Economies: The Asia Approach*, edited by Pradumna B. Rana and Naved Hamid. Hong Kong: Oxford University Press. p. 335–446.

Finance and Development. 1998. Mitigating the social costs of the Asian crisis. Finance and Development 35 (3). http://www.imf.org/external/pubs/ft/fandd/1998/09/imfstaf2.htm.

Fischer, S., R. Sahay and C. Vegh. 1998. From Transition to Market; Evidence and Growth Prospects. *International Monetary Fund Discussion Paper.* Washington, DC: International Monetary Fund.

Fishlow, A., C. Gwin, S. Haggard, D. Rodrik, and R. Wade. 1994. *Miracle or Design? Lessons from the East Asian Experience*. Washington, DC: Overseas Development Council.

Francois, J. F., B. McDonald, and H. Nordstrom. 1995. Assessing the Uruguay Round. In *The Uruguay Round and the Developing Countries, Discussion Paper* No. 307, edited by W. Martin and L. A. Winters. Washington, DC: The World Bank.

Frankel, F. R. 1976. *India's Green Revolution: Economic Gains and Political Costs*. Princeton: Princeton University Press.

Frankel, J. A, and S.-J. Wei. 1997. The new regionalism and Asia: Impact and options. In *The global trading system and developing Asia*, edited by A. Panagariya, M. G. Quibria, and N. Rao. An Asian Development Bank Publication. Oxford University Press.

Frederick, K. D. 1991. The Disappearing Aral Sea. *Resources for the Future* 102 (Winter): 11–13.

Gaiha, R. 1994. *Design of poverty alleviation strategy in rural areas*. Rome: FAO.

Garnaut, R. 1998a. The East Asian Crisis. In *East Asia in Crisis: From Being a Miracle to Needing One?*, edited by R. H. McLeod and R. Garnaut. London and New York: Routledge.

Garnaut, R. 1998b. Economic Lessons. In *East Asia in Crisis: From Being a Miracle to Needing One?*, edited by R. H. McLeod and R. Garnaut. London and New York: Routledge.

GATT (General Agreement on Tariffs and Trade). 1994. News of the Uruguay Round of Multilateral Trade Negotiations: The Final Act. Press Summary. *Uruguay Round Newsletter* No. 084. Geneva: Information and Media Relations Division, GATT.

Gibb, A. 1974. *Agricultural Modernization, Non-Farm Employment and Low Level Urbanization: A Case Study of a Central Luzon Sub-Region*. Ph.D. thesis, University of Michigan.

Goletti, F., N. Minot, and P. Berry 1997. Marketing Constraints on Rice Exports from Vietnam. *Markets and Structural Studies Division Discussion Paper* No. 15. Washington, DC: International Food Policy Research Institute.

Golubev, G. N. 1993. State and Perspectives of Aral Sea Problem. In *Water for Sustainable Development in the Twenty-first Century*, edited by A. K. Biswas, M. Jellali, and G. Stout. Water Resources Management Series, Oxford University Press. pp. 244–253.

Gonzales, L. A., F. Kasryno, N. Perez, and M. Rosegrant. 1993. Economic Incentives and Comparative Advantage in Indonesian Food

Crop Production. *Research Report* No. 93. Washington, DC: International Food Policy Research Institute.

Graham C. 1997. Addressing the Social Costs of Market Transition. In *Economies In Transition: Comparing Asia and Eastern Europe*, edited by W.-T. Woo, S. Parker, and J. Sachs Cambridge, MA: Massachusetts Institute of Technology Press.

Green, D. J., and R. W. A. Vokes. 1997. Agriculture and the Transition to the Market in Asia. *Journal of Comparative Economics* 25: 256–280.

Griffin, K. 1972. *The Green Revolution: An Economic Analysis*. Geneva: United Nations Research Institute for Social Development.

————. 1979. *The Political Economy of Agrarian Change*. London: MacMillan.

Grootaert, C., and J. Braithwaite. 1998. Poverty correlates and indicator-based targeting in Eastern Europe and the Former Soviet Union. *World Bank Policy Research Working Paper* No. 1942. Washington, DC: The World Bank.

Grossman, G. M., and E. Helpman. 1994. Endogenous Innovation in the Theory of Growth. *Journal of Economic Perspectives* 8 (1): 23–44.

————, and E. Helpman. 1995. Trade Wars and Trade Talks. *Journal of Political Economy* 103 (4): 675–708.

Haggblade, S., and D. C. Mead. 1998. An Overview of Policies and Programs for Promoting Growth of the Rural Nonfarm Economy. Paper presented at an IFPRI/World Bank-sponsored Workshop on Strategies for Stimulating Growth of the Rural Nonfarm Economy in Developing Countries, Airlie House Conference Center, Warrenton, VA, USA, May, 1998.

————, J. Hammer, and P. Hazell. 1991. Modeling Agricultural Growth Multipliers. *American Journal of Agricultural Economics* 73 (2): 361–74.

————, P. Hazell, and J. Brown. 1989. Farm-Nonfarm Linkages in Rural Sub-Saharan Africa. *World Development* 17 (8): 1173–1201.

Hamid, N. 1995. Agricultural Reform. In *From Centrally Planned to Market Economies: The Asian Approach*, Vol. 1, *An Overview*, edited by P. B. Rana and N. Hamid. Oxford: Oxford University Press and Manila: Asian Development Bank.

Han, K. 1992. Future Development of Feed Industry in Asia and Pacific Region. In *Proceedings of the Sixth AAAP Animal Science Congress*, edited by P. Bunyavajchewin, S. Sangdid, and K. Hansanet. Bangkok: Kasetsart University.

Harrington, L. W., M. Morris, P. R. Hobbs, V. Pal Singh, H. C. Sharma, R. P. Singh, M. K. Chaudhary, and S. D. Dhiman (editors). 1992. *Wheat and Rice in Karnal and Kurukshetra Districts, Haryana, India—Practices, Problems and an Agenda for Action. Report from an Exploratory Survey, 22–29 March 1992.* Haryana Agricultural University, Indian Council for Agricultural Research, International Maize and Wheat Improvement Center, International Rice Research Institute.

Hayami, Y., (editor) 1998. *Toward the Rural-Based Development of Commerce and Industry: Selected Experiences from East Asia.* EDI Learning Resource Series. Washington DC: Economic Development Institute, World Bank.

Hazell, P. 1992. The Appropriate Role of Agricultural Insurance in Developing Countries. *Journal of International Development* 4 (6): 567–581.

_____, and J. L. Garrett. 1996. Reducing Poverty and Protecting the Environment: The Overlooked Potential of Less-favored Lands. *2020 Brief* 39. Washington, DC: International Food Policy Research Institute.

_____, and S. Haggblade. 1991. Rural-Urban Growth Linkages in India. *Indian Journal of Agricultural Economics* 46 (4): 515–529.

_____, and S. Haggblade. 1993. Farm-Nonfarm Growth Linkages and the Welfare of the Poor. In *Including the Poor*, edited by. M. Lipton and J. van der Gaag. Washington, DC: The World Bank. p. 190–204.

_____, and C. Ramasamy. 1991. *The Green Revolution Reconsidered: The Impact of High-yielding Rice Varieties in South India.* Baltimore: Johns Hopkins University Press.

_____, and T. Reardon. 1998. Interactions Among the Rural Nonfarm Economy, Poverty, and the Environment in Resource-poor Areas. Paper presented at an IFPRI/World Bank-sponsored Workshop on Strategies for Stimulating Growth of the Rural Nonfarm Economy in Developing Countries, Airlie House Conference Center, Warrenton, VA, USA, May, 1998.

_____, and A. Roell. 1983. Rural Growth Linkages: Household Expenditure Patterns in Malaysia and Nigeria. *Research Report* No. 41. Washington, DC: International Food Policy Research Institute.

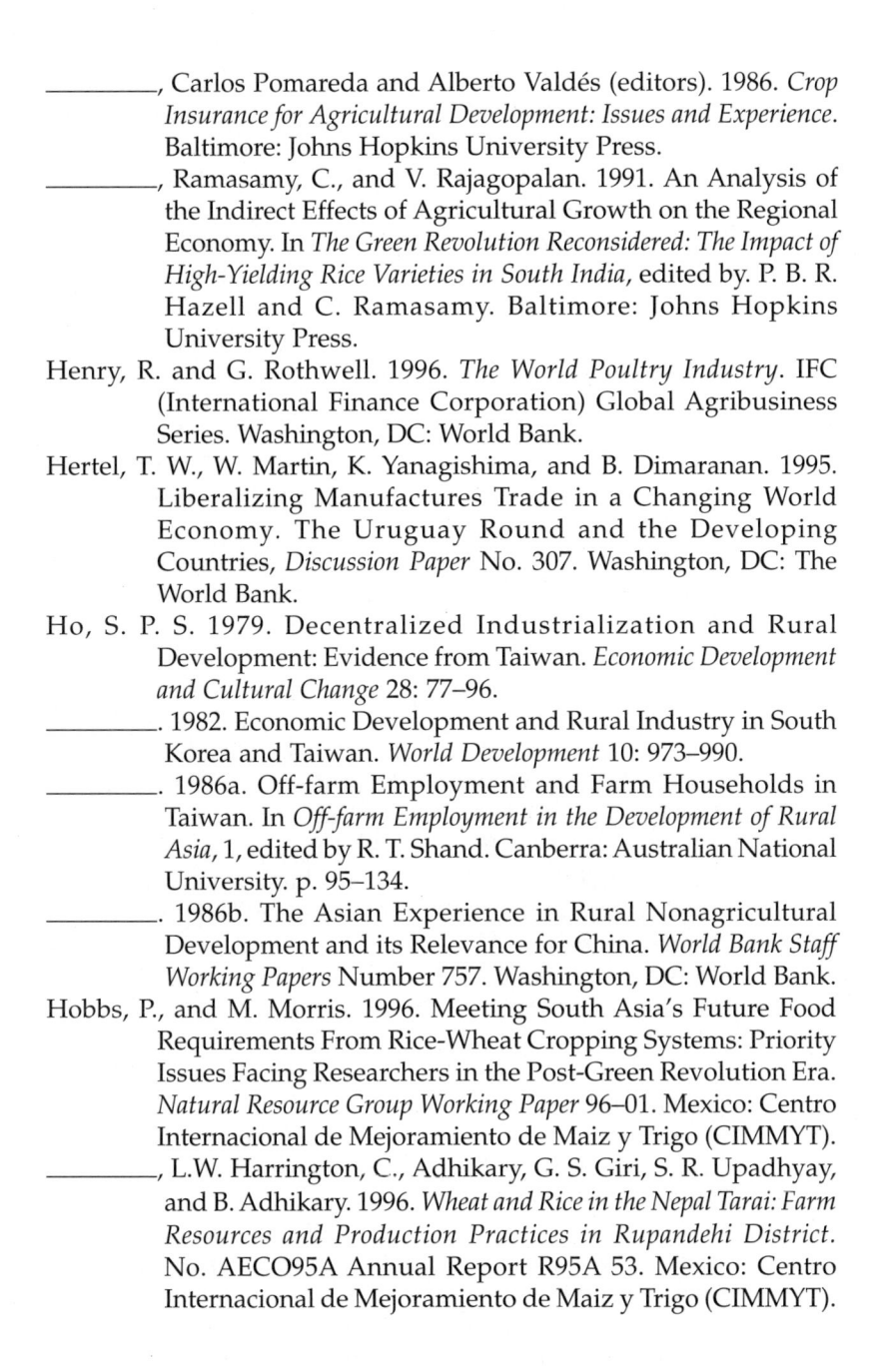

_____, Carlos Pomareda and Alberto Valdés (editors). 1986. *Crop Insurance for Agricultural Development: Issues and Experience*. Baltimore: Johns Hopkins University Press.

_____, Ramasamy, C., and V. Rajagopalan. 1991. An Analysis of the Indirect Effects of Agricultural Growth on the Regional Economy. In *The Green Revolution Reconsidered: The Impact of High-Yielding Rice Varieties in South India*, edited by. P. B. R. Hazell and C. Ramasamy. Baltimore: Johns Hopkins University Press.

Henry, R. and G. Rothwell. 1996. *The World Poultry Industry*. IFC (International Finance Corporation) Global Agribusiness Series. Washington, DC: World Bank.

Hertel, T. W., W. Martin, K. Yanagishima, and B. Dimaranan. 1995. Liberalizing Manufactures Trade in a Changing World Economy. The Uruguay Round and the Developing Countries, *Discussion Paper* No. 307. Washington, DC: The World Bank.

Ho, S. P. S. 1979. Decentralized Industrialization and Rural Development: Evidence from Taiwan. *Economic Development and Cultural Change* 28: 77–96.

_____. 1982. Economic Development and Rural Industry in South Korea and Taiwan. *World Development* 10: 973–990.

_____. 1986a. Off-farm Employment and Farm Households in Taiwan. In *Off-farm Employment in the Development of Rural Asia*, 1, edited by R. T. Shand. Canberra: Australian National University. p. 95–134.

_____. 1986b. The Asian Experience in Rural Nonagricultural Development and its Relevance for China. *World Bank Staff Working Papers* Number 757. Washington, DC: World Bank.

Hobbs, P., and M. Morris. 1996. Meeting South Asia's Future Food Requirements From Rice-Wheat Cropping Systems: Priority Issues Facing Researchers in the Post-Green Revolution Era. *Natural Resource Group Working Paper* 96–01. Mexico: Centro Internacional de Mejoramiento de Maiz y Trigo (CIMMYT).

_____, L.W. Harrington, C., Adhikary, G. S. Giri, S. R. Upadhyay, and B. Adhikary. 1996. *Wheat and Rice in the Nepal Tarai: Farm Resources and Production Practices in Rupandehi District*. No. AECO95A Annual Report R95A 53. Mexico: Centro Internacional de Mejoramiento de Maiz y Trigo (CIMMYT).

Hossain, M. 1988. Nature and Impact of the Green Revolution in Bangladesh. *Research Report* No. 67. Washington, DC: International Food Policy Research Institute.

_____,1995. Sustaining Food Security For Fragile Environments in Asia: Achievements, Challenges, and Implications for Rice Research. *Fragile Lives in Fragile Ecosystems; Proceedings of the International Rice Research Conference.* Los Baños, Philippines: International Rice Research Institute.

Hossain, S. I. 1997. Making Education in China Equitable and Efficient. *Policy Research Working Paper* No. 1814. Washington, DC: World Bank.

Huang, J. and H. Bouis. 1996. Structural Changes in the Demand for Food in Asia: Food, Agriculture, and the Environment, *Discussion Paper* No. 11. Washington, DC: International Food Policy Research Institute.

_____, and S. Rozelle. 1995. Environmental Stress and Grain Yields in China. *American Journal of Agricultural Economics* 77 (11): 853–864.

_____, M. W. Rosegrant and S. Rozelle. 1996. Public Investment, Technological Change and Reform: A Comprehensive Accounting of Chinese Agricultural Growth. Washington, D.C.: International Food Policy Research Institute. Mimeo.

_____, S. Rozelle, and M. W. Rosegrant. 1995. Public Investment, Technological Change and Agricultural Growth in China. Paper presented at the Final Workshop of the International Cooperative Research Project on Projections and Policy Implications of Medium and Long-Term Rice Supply and Demand, organized by International Food Policy Research Institute, International Rice Research Institute, and CCER, Beijing, China, April 23–26, 1995.

Hyman, E. L. 1998. NGO Roles in Developing and Commercializing Technologies for the Rural Nonfarm Economy. Paper presented at an IFPRI/World Bank-sponsored Workshop on Strategies for Stimulating Growth of the Rural Nonfarm Economy in Developing Countries, Airlie House Conference Center, Warrenton, VA, USA, May, 1998.

IFAD (International Fund for Agricultural Development). 1995. *The State of World Rural Poverty: A Profile of Asia.* Rome: International Fund for Agricultural Development.

IFPRI (International Food Policy Research Institute). 1999. Sociopolitical Effects of New Biotechnologies in Developing Countries. *2020*

Vision Brief 35, July 1996. http://www.cgiar.org/ifpri/2020/briefs/number35.htm/. Accessed 1 February 1999.

ILO (International Labor Organization). 1977. *Poverty and Landlessness in Rural Asia.* Geneva: ILO.

_____.1998. The Social Impact of the Asian Financial Crisis. Technical report for discussion at the high-level tripartite meeting on social responses to the financial crisis in East and South-East Asian countries, Bangkok, Thailand, 22–24 April.

IMF (International Monetary Fund). 1998a. *World Economic Outlook*, October Washington, DC: IMF

_____.1998b. *International Financial Statistics, December 1998*, Washington, DC: IMF.

Intal, J. P., M. Milo, C. Reyes, and L. Basilio. 1998. The Philippines. In *East Asia in Crisis: From Being a Miracle to Needing One?*, edited by R. H. McLeod and R. Garnaut. London and New York: Routledge.

IPPC (Intergovernmental Panel on Climate Change). 1990. *Climate Change: The IPPC Response Strategies.* Geneva: WMO and UNEP.

Islam, N., and S. Thomas. 1996. Foodgain Price Stabilization in Developing Countries. *Food Policy Review* 3. Washington, DC: International Food Policy Research Institute.

Jamison, D. and L. Lau. 1982. *Farmer Education and Farm Efficiency.* Baltimore: Johns Hopkins University Press.

Jayaraman, R. and P. Lanjouw. 1998. The Evolution of Poverty and Inequality in Indian Villages. *Policy Research Working Paper* 1870. Washington, DC: Development Research Group, World Bank,.

Jin, Z., D. Ge, H. Chen, and J. Fang. 1995. Effects of Climate Change on Rice Production and Strategies for Adaptation in Southern China. In *Climate Change and Agriculture: Analysis of Potential International Impacts*, edited by C. Rosenzweig, L. H. Allen, Jr., L. A. Harper, S. E. Hollinger, and J. W. Jones. *ASA Special Publication* 59: 307–323. Madison, WI: American Society of Agronomy.

Johnston, B. F. and P. Kilby. 1975. *Agriculture and Structural Transformation: Economic Strategies in Late-developing Countries.* London: Oxford University Press.

_____, and J. W. Mellor. 1961. The Role of Agriculture in Economic Development. *American Economic Review* 51 (4): 566–593.

Jomo, K. S. 1996. Lessons from Growth and Structural Change in the Second-tier South East Asian Newly Industrializing Countries. *East Asian Development: Lessons for a New Global Environment Study* No. 4. Geneva: United Nations Conference on Trade and Development, United Nations.

Joshi, V. 1998. Fiscal Stabilization and Economic Reform in India. In *India's Economic Reforms and Development: Essays for Manmohan Singh*, edited by I. J. Ahluwalia and I. M. D. Little. New York: Oxford University Press.

Josling, T. E., S. Tangerman, and T. K. Warley. 1996. *Agriculture in the GATT*. New York: St. Martin's Press, Inc.

Kada, R. 1986. Off-farm Employment and the Rural-urban Interface in Japanese Economic Development. In *Off-farm Employment in the Development of Rural Asia*, 1, edited by R.T. Shand. Canberra: Australian National University. p. 75–94

Kalirajan, K. P., and R. T. Shand. 1997. Sources of Output Growth in Indian Agriculture. *Indian Journal of Agricultural Economics* 52 (4): 693–706.

Karim, Z., M. Ahmed, S. G. Hussain, and Kh. B. Rashid. 1994. Impact of Climate Change on the Production of Modern Rice in Bangladesh. *Implications of Climate Change for International Agriculture: Crop Modeling Study*. EPA 230-B-94-003. Washington, DC: U.S. Environmental Protecion Agency.

Kasryno, F., P. Simatupang, I. W. Rusastra, A. Djatihati, and B. Irawan. 1989. *Government Incentives and Comparative Advantage in the Livestock and Feedstuff Subsectors in Indonesia*. Bogor, Indonesia: Center for Agro-Economic Research. Mimeo.

Kathen, A. de. 1996. *Gentechnik in Entwicklungsländern: Ein Überblick: Landwirtschaft*. Berlin, Germany: Umweltbundesamt.

Ke, B. 1997. Recent development in the livestock sector of China and changes in livestock/feed relationship. Unpublished Food and Agriculture Organization of the United Nations report, Animal and Health Division (LPT2). Rome. Mimeo.

Kendall, H. W., and D. Pimentel. 1994. Constraints on the Expansion of the Global Food Supply. *Ambio* 23 (3): 198–205.

Kerr, J. 1996. Sustainable Development of Rainfed Agriculture in India. *Environment and Production Technology Division Discussion Paper* No. 20. Washington, DC: International Food Policy Research Institute.

Khan, S.M. 1997. South Asia: Free trade area and trade liberalization. *Journal of Asian Economics* 8 (1): 165-177.

Kim, J. Il, and L. J. Lau. 1994. The Sources of Economic Growth of the East Asian Newly Industrialized Countries. *Journal of the Japanese and International Economies* 8 (3): 235–71.

King, R. P., and D. Byerlee. 1978. Factor Intensities and Locational Linkages of Rural Consumption Patterns in Rural Sierra Leone. *American Journal of Agricultural Economics* 60 (2): 197–206.

Knack, S., and P. Keefer. 1995. Institutions and Economic Performance: Cross-country Tests Using Alternative Institutional Measures. *Economics & Politics* 7 (3).

Knight, J., and L. Song. 1992. Income inequality in rural China: Communities, households and resource mobility. *Oxford Applied Economics Discussion Paper Series* No. 150. Oxford: University of Oxford Institute of Economics and Statistics.

Kochlar, K., P. Loungani, and M.R. Stone. 1998. The East Asian Crisis: Macroeconomic Developments and Policy Lessons. *IMF Working Paper* WP/98/128. Washington, DC: International Monetary Fund.

Korea Statistical Yearbook. 1996. Seoul: Republic of Korea National Bureau of Statistics.

Krueger, A. 1995. NAFTA: Strengthening or Weakening the International Trading System. *The Dangerous Drift to Preferential Trade Arrangements.* Washington, DC: AEI Press. p. 19–33.

Krugman, P. 1994. The Myth of Asia's Miracle. *Foreign Affairs* 73 (6): 62–78.

Kumar, P., and M. W. Rosegrant. 1994. Productivity and Sources of Growth for Rice in India. *Economic and Political Weekly*, 31 December: 183–188.

Kumar, P., and M. W. Rosegrant. 1997. Dynamic Supply Response of Cereals and Supply Projections: A 2020 Vision. *Agricultural Economics Research Review* 10 (1): 3–33.

Kumar, P., and Mruthyunjaya. 1992. Measurement and Analysis of Total Factor Productivity Growth in Wheat. *Indian Journal of Agricultural Economics* 477: 451–458.

Kumar, Vinod. 1998. Study Meeting on Changing Food Demand and Agricultural Diversification. *Country Paper* STM-06-98 (November). New Delhi: Government of India.

Lee, J. W. 1992. Government Interventions and Productivity Growth in Korean Manufacturing Industries. Unpublished Paper. Washington, D. C.: International Monetary Fund.

Lee, Teng-hui. 1971. *Intersectoral Capital Flows in the Economic Development of Taiwan, 1895–1960*. Ithaca, NY: Cornell University Press.

Leisinger, K. M. 1995. Sociopolitical Effects of New Biotechnologies in Developing Countries. *2020 Vision for Food, Agriculture, and the Environment Discussion Paper* No. 2. Washington, DC: International Food Policy Research Institute.

Leuck, D., S. Haley, P. Liapis, and B. McDonald. 1995. The EU Nitrate Directive and CAP Reform: Effects on Agricultural Production, Trade, and Residual Soil Nitrogen. *Report* 225. Washington, DC: US Department of Agriculture, Economic Research Service.

Lewis, J. D., and S. Robinson. 1996. Partners or Predators? The Impact of Regional Trade Liberalization on Indonesia. *Policy Research Working Paper* No. 1626. Country Operations Division. Washington, DC: The World Bank.

Liedholm, C. 1988. The Role of the Nonfarm Activities in the Rural Economies of the Asia-Pacific region. Paper prepared for Conference on Directions and Strategies of Agricultural Development in the Asia-Pacific Region. Taipei, Taiwan, January 1988.

Lin, J. Y. 1992. Rural Reforms and Agricultural Growth in China. *American Economic Review* 82 (3): 34–51.

———.1997. Institutional Reforms and Dynamics of Agricultural Growth in China. *Food Policy* 22 (3): 201–212.

Lindert, P. H. 1991. Historical Patterns of Agricultural Policy. In *Agriculture and the State: Growth, Employment, and Poverty in Developing Countries*, edited by C. P. Timmer. Ithaca, NY: Cornell University Press. p. 29–83.

———.1996. The Bad Earth? China's Agricultural Soils since the 1930s. *Working Paper Series* No. 83. Davis, CA: Agricultural History Center, University of California, Davis.

Linneman, H., J. De Hoogh, M. A. Keyser, and H. D. J. Van Heemst. 1979. Potential World Food Production. *MOIRA (Model of International Relations in Agriculture) Report of the Project Group on Food for a Doubling World Population*. Amsterdam: North Holland Publishing Co. p. 34–74.

Lipton, M. 1994. Agricultural Research Investment Themes and Issues: A View from Social-Science Research. In *Agricultural Technology: Policy Issues for the International Community*, edited by Jock R. Anderson. Cambridge, UK: CAB International in association with the World Bank. p. 601–616.

―――――. 1998. *Successes in anti-poverty*. Geneva: International Labor Organization publication.

―――――, and S. Sinha. 1998. Issues Paper for Discussion. Prepared for Brainstorming Workshop on IFAD's Strategic Focus on Poverty, Rome, October 20–21. Rome: International Fund for Agricultural Development.

Liu, S. B. 1995. Animal Production Development in China. In *Supply of Livestock Products to Rapidly Expanding Urban Populations*. Proceedings of the Joint FAO/WAAP/KSAS Symposium, Hoam Faculty Club, Seoul National University, Seoul, Korea, 16–20 May, 1995. Rome, Italy: FAO. p. 121–127.

Livernash, R. 1996. Agricultural Biotechnology in Developing Countries. Washington, DC: International Food Policy Research Institute. Mimeo.

Lucas, R. E. 1993. Making a Miracle. *Econometrica* 61 (2): 251-272.

Lucas, R. E., Jr. 1988. On the Mechanics of Economic Development. *Journal of Monetary Economics* 22: 3–42.

Magrath, W., and P. Arens. 1989. The Costs of Soil Erosion on Java: A Natural Resource Accounting Approach. *Environmental Department Working Paper* No. 18. Washington DC: World Bank.

Mao, W., and W. W. Koo. 1997. Productivity Growth, Technological Progress, and Efficiency Change in Chinese Agriculture After Rural Economic Reforms: A DEA Approach. *China Economic Review* 8 (2): 157–174.

Marcoux, A. 1996. Population change, natural resources, environmental linkages in East and Southeast Asia. Department for Economic and Social Information and Policy Analysis. Population Information Network of the United Nations Population Division. FAO Population Programme Service. Rome: Food and Agriculture Organization.

Martin, W., and P. G. Warr. 1991. Agriculture's Decline in Indonesia: Supply or Demand Determined? *Policy Research Working Paper International Trade* No. WPS 798. Washington, DC: World Bank.

―――――, and P. G. Warr. 1992. The Declining Economic Importance of Agriculture: A Supply Side Analysis for Thailand. *Working Paper in Trade and Development* No. 92/1. Research School of Pacific Studies, Department of Economics and National Centre for Development Studies. Canberra: The Australian National University.

Matthews, R.B., M. J. Kropff, D. Bachelet, and H. H. van Laar. 1995. *Modelling the Impact of Climate Change on Rice Production in Asia.* Oxford: CAB International in association with the International Rice Research Institute.

McDonald, B. J., S. W. Martinez, M. Otradovsky, and J. V. Stout. 1991. *A Global Analysis Of Energy Prices and Agriculture.* Washington, DC: United States Department of Agriculture Economic Research Service, Agricultural and Trade Analysis Division.

McIntire, J., D. Bourzat, and P. L. Pingali. 1992. *Crop Livestock Interaction in Sub-Saharan Africa.* Washington, DC: World Bank.

McKibbin, W., and W. Martin. 1998. The East Asian Crisis: Investigating Causes and Policy Consequences. *Working Papers in Trade and Development*, No. 98/6. Research School of Pacific and Asian Studies. Department of Economics Canberra: Australian National University.

McLeod, R. H. 1998. Indonesia. In *East Asia in Crisis: From Being a Miracle to Needing One?*, edited by R. H. McLeod and R. Garnaut. London and New York: Routledge.

_____, and R. Garnaut (editors). 1998. *East Asia in Crisis: From Being a Miracle to Needing One?* London and New York: Routledge.

McMillan, J., J. Whalley, and L. Zhu, 1989. Impact of China's economic reforms on agricultural productivity growth. *Journal of Political Economy* 97 (4): 781–807.

McNeely, J. A., K. R. Miller, W. V. Reid, R. A. Mittermeir, and T. B. Werner. 1990. *Conserving the World's Biological Diversity.* Gland, Switzerland, and Washington, DC: International Union for the Conservation of Nature and Natural Resources, World Resources Institute, Conservation International, World Wildlife Fund, World Bank.

Mearns, R. 1999. Access to Land in Rural India; Policy Issues and Options. *Policy Research Working Paper* No. 2123. Washington DC : World Bank.

_____, and S. Sinha. 1999. Social Exclusion and Land Administration in Orissa, India. *Policy Research Working Paper.* Washington DC: World Bank.

Mellor, J. W. 1973. Accelerated Growth in Agricultural Production and the Intersectoral Transfer of Resources. *Economic Development and Cultural Change* 22 (1).

_____. 1976. *The New Economics of Growth: A Strategy for India and the Developing World.* Ithaca, NY: Cornell University Press _____(editor). 1995. *Agriculture on the Road to Industrialization.* Baltimore: Johns Hopkins University Press..

_____, and U. Lele. 1972. Growth linkages of the new food grain technologies. *Indian Journal of Agricultural Economics* 18 (1): 35–55.

Meyer, Richard L., and Geetha Nagarajan. 1999. *Rural Finance in Asia: Policies, Paradigms, and Performance.* Hong Kong, China: Oxford University Press (China).

Mingsarn Santikarn Kaosa-ard, and Benjavan Rerkasem 1999. *The Growth and Sustainability of Agriculture in Asia.* Hong Kong, China: Oxford University Press (China).

Mishra, P. K. 1996. *Agricultural Risk, Insurance and Income: A Study of the Impact and Design of India's Comprehensive Crop Insurance Scheme.* Aldershot, England: Avebury.

Monke, E., and S. Pearson. 1991. Introduction. In Postplanting weed control in direct-seeded rice, edited by S. Pearson, W. Falcon, P. Heytens, E. Monke, and K. Moody. In *Proceedings of the National Workshop on Direct Seeding Practices and Productivity,* edited by A. H. Ali. Penang: Malaysian Agricultural Research and Development Institute.

Moormann, F.R., and N. van Breemen. 1978. *Rice: Soil, Water, Land.* Los Baños, Philippines: International Rice Research Institute.

Müller, K. 1995. *The Political Economy of Agricultural Trade of the ASEAN Countries.* Bonn, Germany: Verlag M. Wehle.

Mustafa, U. 1991. Economic Impact of Land Degradation, Salt Affected and Waterlogged Soils on Rice Production in Pakistan's Punjab. Ph.D. dissertation, College of Economics and Management, University of the Philippines at Los Baños, Los Baños, Philippines.

Nagy, J. G. 1985. Overall rate of return to agricultural research and extensions investments in Pakistan. *Pakistan Journal of Applied Economics* 4 (1): 17-28.

Neiss, H. 1998. In Defense of the IMF's Emergency Role in East Asia. *International Herald Tribune,* 8 October.

Nellor, D. C. L. 1998. The Role of the International Monetary Fund. In *East Asia in Crisis: From Being a Miracle to Needing One?* Part V, 15, edited by Ross H. McLeod and Ross Garnaut. London and New York: Routledge. p. 248–265.

Noland, M., L.-G. Liu, S. Robinson, and Z. Wang. 1998. *Global Economic Effects of the Asian Currency Devaluations.* Washington, DC: Institute for International Economics.

Oldeman, L. R. 1992. Global Assessment of Soil Degradation. *State of the Environment Report* 1992. Wageningen, The Netherlands: International Soil Reference and Information Centre.
_____. 1998. Soil Degradation: A Threat to Food Security. Report 98/01. Wageningen, The Netherlands: International Soil Reference and Information Centre.
_____, T. A. Hakkeling, and W. G. Sombroek. 1990. *World Map of the Status of Human-Induced Soil Degradation: An Explanatory Note*. Wageningen, The Netherlands: International Soil Reference and Information Center, and Nairobi: United Nations Environment Programme.
Oram, P. 1990. Agricultural Productivity Growth and the Structure and Organization of Agricultural Research. Policy brief presented at the International Food Policy Research Institute policy seminar on technology for sustainable agricultural growth, The Hague, 2–3 July 1990.
Osaka, H. 1997. Productivity growth analysis for selected Asian countries. *Asia Pacific Development Journal* 4 (2): 93-110.
Oshima, H. T. 1986. Off-farm Employment and Incomes in Postwar East Asian Growth. In *Off-farm Employment in the Development of Rural Asia* 1, edited by R.T. Shand. Canberra: Australian National University. p. 25–74.
Ostrom, E. 1994. Neither Market nor State: Governance of Common-pool Resources in the Twenty-first Century. Lecture Series No. 2. Washington, DC: International Food Policy Research Institute.
Otsuka, K. 1998. Rural Industrialization in East Asia. In *The Institutional Foundation of East Asian Economic Development*, edited by Y. Hayami and M. Aoki. London: Macmillan.
_____, and T. Reardon. 1998. Lessons from Rural Industrialization in East Asia: Are They Applicable to Africa? Paper presented at an IFPRI/World Bank-sponsored Workshop on Strategies for Stimulating Growth of the Rural Nonfarm Economy in Developing Countries, Airlie House Conference Center, Warrenton, VA, USA, May, 1998.
_____, S. Suyanto, and T. Tomich. 1997. Does Land Tenure Insecurity Discourage Tree Planting? Evolution of Customary Land Tenure and Agroforestry Management in Sumatra. *Environment and Production Technology Division Discussion Paper* No. 31. Washington, DC: International Food Policy Research Institute.

Paarlberg, R. 1997. Feeding China: A Confident View. *Food Policy* 22 (3): 269–279.
Panagiriya, A., M.G. Quibria, and N. Rao (editors). *The Global Trading System and Developing Asia*. Manila: Asian Development Bank and Oxford University Press.
Pardey, P. G., J. Roseboom, and S. Fan. 1998. Trends in Financing Asian and Australian Agricultural Research. In *Financing Agricultural Research, a Sourcebook*, edited by S. Tabor, W. Janssen, and H. Bruneau. The Hague: International Service for National Agricultural Research.
Park, A., S. Wang, and G. Wu. 1998. Assessing China's War on Poverty. *Department of Economics Working Paper*. Ann Arbor, MI: University of Michigan.
Park, F. K. 1986. Off-farm Employment in Korea: Current Status and Future Prospects. In *Off-farm Employment in the Development of Rural Asia* 1, edited by R.T. Shand. Canberra: Australian National University. p. 59–106.
Parker, S., G. Tritt, and W. T. Woo. 1997. Some Lessons Learned from the Comparison of Transitions in Asia and Eastern Europe. In *Economies in Transition: Comparing Asia and Eastern Europe*, edited by Wing Thye Woo, S. Parker, and J. D. Sachs. Cambridge, MA: Massachusetts Institute of Technology Press.
Parry, M. L., M. Blantran de Rozari, A. L. Chong, and S. Panich (editors). 1992. *The Potential Socio-Economic Effects of Climate Change in South-East Asia*. Nairobi: United Nations Environment Programme.
Pearse, A. 1980. *Seeds of Plenty, Seeds of Want*. London: Oxford University Press.
Pender, J., and J. Kerr. 1996. Determinants of Farmers' Indigenous Soil and Water Conservation Investments in India's Semi-Arid Tropics. *Environment and Production Technology Division Discussion Paper* No.17. Washington, DC: International Food Policy Research Institute.
_____, F. Place, and S. Ehui. 1998. Strategies for Sustainable Agricultural Development in the East African Highlands. *Environment and Production Technology Division Discussion Paper* No. 41. Washington, DC: International Food Policy Research Institute.
Penning de Vries, F. W. T., F. Agus, and J. Kerr. 1998. *Soil Erosion at Multiple Scales: Principles and Methods for Assessing Causes and Impacts*. Wallingford, England: CABI Publishing.

———, H. Van Keulen, R. Rabbinge, and J. C. Luyten. 1995. Biophysical Limits to Global Food Production, *2020 Vision Brief* No. 18. Washington, DC: International Food Policy Research Institute.

Pernia, E. M., and J. C. Knowles. 1998. Assessing the Social Impact of the Financial Crisis in Asia. *EDRC Briefing Notes* No. 6. Manila: Asian Development Bank.

Persley, G. J. 1994. Biotechnology's Promise. In *Agricultural Technology: Policy Issues for the International Community*, edited by. J. R. Anderson. Wallingford, UK: CAB International.

Pingali, P. L., and M. Rosegrant. 1994. Confronting the Environmental Consequences of the Green Revolution in Asia. *Environment and Production Technology Division Discussion Paper* No. 2. Washington, DC: International Food Policy Research Institute.

———, and M. W. Rosegrant. 1995. Agricultural Commercialization and Diversification: Processes and Policies. *Food Policy* 20: 171–185.

———, and M. W. Rosegrant. 1998. Intensive Food Systems in Asia: Can the Degradation Problems be Reversed? Paper presented at the Pre-Conference Workshop, Agricultural Intensification, Economic Development and the Environment, of the Annual Meeting of the American Agricultural Economics Association, Salt Lake City, Utah, July 31–August 1, 1998.

———, and V. T. Xuan. 1992. Vietnam: Decollectivitization and Rice Productivity Growth. *Economic Development and Cultural Change* 40 (4): 697–719.

———, and P. A. Roger (editors). 1995. *Impact of Pesticides on Farmer Health and the Rice Environment*. Boston: Kluwer Academic Publishers, and Los Baños, Philippines: The International Rice Research Institute.

———, Hossain, M., and R.V. Gerpacio. 1997. *Asian Rice Bowls: The Returning Crisis?* New York: CAB International, in association with Manila, Philippines: International Rice Research Institute.

Pinstrup-Andersen, P., and P. B. R. Hazell. 1985. The Impact of the Green Revolution and Prospects for the Future. *Food Reviews International* 1 (1).

Plucknett, D. L. 1995. Prospects of Meeting Future Food needs through New Technology. In *Population and Food in the Early Twenty-First Century: Meeting Future Food Demand of an Increasing Population*, edited by N. Islam. Washington, DC: International Food Policy Research Institute.

Poppele, J., S. Sumarto, and L. Pritchett. 1998. Social Impacts of the Indonesian Crisis: New Data and Policy Implications. Mimeo.

Postel, S. 1989. Water for Agriculture: Facing the Limits. *Worldwatch Paper* 93.

Postel, S. 1993. Water and Agriculture. In *Water in Crisis: A Guide to the World's Fresh Water Resources*, edited by P. H. Gleick. New York: Oxford University Press.

Prosterman, R. L., T. Hanstad, and L. Ping. 1996. Can China Feed Itself? *Scientific American* (November): 90–96.

Quibria, M. G., and T. N. Srinivasan. 1993. Introduction. In *Rural Poverty in Asia: Priority Issues and Policy Options*, edited by M. G. Quibria. Hong Kong: Oxford University Press.

Radelet, S., and J. Sachs. 1998. The Onset of the East Asian Financial Crisis. Cambridge, MA: Harvard Institute for International Development. Mimeo.

Rae, A. N. 1992. Interaction between Livestock and Feeds Policies: Evidence from Southeast Asia. *Agricultural Economics* 6 (7): 25–37.

Rahman, M. 1997. Recent policy of trade liberalization in Bangladesh and issues of regional cooperation in South Asia. *Journal of Asian Economics* 8 (1): 117–141.

Rahman, S. H. 1994. The Impact of Trade and Exchange Rate Policies on Economic Incentives in Bangladesh Agriculture. *Working Papers on Food Policy in Bangladesh* No. 8. Washington, DC: International Food Policy Research Institute.

Rajapakse, P., and N. Arunatilake. 1997. Would a reduction in trade barriers promote intra-SAARC trade? A Sri Lankan perspective. *Journal of Asian Economics* 8 (1): 95-115.

Rana, P. B. 1995a. Introduction: The Asian Approach to Reforming Transitional Economies. In *From Centrally Planned to Market Economies: The Asian Approach*, Vol. 1, An Overview, edited by P. B. Rana and N. Hamid. New York: Oxford University Press.

_____, 1995b. Industrial Enterprise Reform in the Transitional Economies of Asia. In *From Centrally Planned to Market Economies: The Asian Approach*, Vol. 1, An Overview, edited by P. B. Rana and N. Hamid. New York: Oxford University Press.

Ranade, A., and S. M. Dev. 1997. Agriculture and Rural Development: Stock, Subsides and Food Security. *India Development Report 1997*. Indira Gandhi Institute of Development Research, India: Oxford University Press

Rangarajan, C. 1982. Agricultural Growth and Industrial Performance in India. *Research Report* 33. Washington DC: International Food Policy Research Institute.

Ranis, G., and F. Stewart. 1998. The Asian Crisis and Human Development. Paper prepared for the IDS Seminar on the Asian Crisis, 13–14 July (Originally written for the United Nations Development Programme).

_____, F. Stewart, and E. Angeles-Reyes. 1990. *Linkages in Developing Economies: A Philippine Study.* Sector Studies No. 1. International Center for Economic Growth. San Francisco: ICS Press.

Rasmussen, L.N. and R. Meinzen-Dick. 1995. Local Organizations for Natural Resource Management: Lessons From Theoretical and Empirical Literature. *Environment and Production Technology Division Discussion Paper* No. 11. Washington, DC: International Food Policy Research Institute.

Ravallion, M., and S. Chen. 1997. What Can New Survey Data Tell Us about Recent Changes in Distribution and Poverty? *The World Bank Economic Review* 11 (2): 357–82.

_____, and G. Datt. 1996. How Important to India's Poor is the Sectoral Composition of Economic Growth? *The World Bank Economic Review* 10 (1): 1–25.

_____, and Q. Wodon. 1997. What are a Poor Farmer's Prospects in the Rural Non-farm Sector? Unpublished manuscript, Policy Research Department. Washington, DC: World Bank.

Reardon, T., K. Stamoulis, M. Elena Cruz, A. Balisacan, J. Berdugue, and K. Savadogo. 1998. Diversification of household incomes into nonfarm sources: Patterns, determinants and effects. Paper presented at the IFPRI/World Bank conference on Strategies for Stimulating Growth of the Rural Nonfarm Economy in Developing Countries, Airlie House, Virginia, May 1998.

Reilly, J. 1995. Climate Change and Agriculture—Research Findings and Policy Considerations. In *Population and Food in the Early Twenty-First Century: Meeting Future Food Demand of an Increasing Population*, edited by N. Islam. Washington, DC: International Food Policy Research Institute.

Robinson, S., T. L. Roe, and A. Erinç Yeldan. 1998. Macroeconomic Policies and the Rural Non-Farm Economy: A General Equilibrium Investigation of Alternative Development Strategies in an Archetype Model for Africa and South Asia.

Paper presented at an IFPRI/World Bank-sponsored Workshop on Strategies for Stimulating Growth of the Rural Nonfarm Economy in Developing Countries, Airlie House Conference Center, Warrenton, VA, USA, May, 1998.

Rodrik, D. 1994. King Kong Meets Godzilla: The World Bank and the East Asian Miracle. In *Miracle or Design? Lessons from the East Asian Experience*, edited by A. Fishlow, C. Gwin, S. Haggard, D. Rodrik, and R. Wade. Washington, DC: Overseas Development Council.

_____. 1995. Trade Strategy, Investment and Exports: Another Look at East Asia. *Working Paper* 5339. Cambridge, MA: National Bureau of Economic Research.

_____. 1997. TFPG Controversies, Institutions, and Economic Performance in East Asia. *Working Paper* 5914. Cambridge, MA: National Bureau of Economic Research Working Paper Series.

Rohwer, J. 1995. *Asia Rising: Why America will Prosper as Asia's Economies Boom*. New York: Touchstone.

Rola, A. C., and P. L. Pingali. 1993. *Pesticides, Rice Productivity, and Farmers Health—An Economic Assessment*. Los Baños, Philippines: IRRI-World Resources Institute.

Rosegrant, M. W. 1995. Water Resources in the 21st Century: Increasing Scarcity, Declining Quality, and Implications for Action. Paper presented at the conference on the sustainable future of the global system organized by the United Nations University and the National Institute for Environmental Studies, Japan, 15–18 October.

_____. 1997. Water Resources in the 21st Century: Challenges and Implications for Action. *2020 Vision for Food, Agriculture, and the Environment Discussion Paper* No. 20. Washington, DC: International Food Policy Research Institute.

_____, and R. E. Evenson. 1993. Agricultural Productivity Growth in Pakistan and India: A Comparative Analysis. *The Pakistan Development Review* 32 (4, part 1): 433–451.

_____, and R. Meinzen-Dick. 1996. Water Resources in the Asia-Pacific Region: Managing Scarcity. *Asia-Pacific Economic Literature* 10 (2): 32–53.

_____, and P. L. Pingali. 1994. Policy and Technology for Rice Productivity Growth in Asia. *Journal of International Development* 6 (6): 665–688.

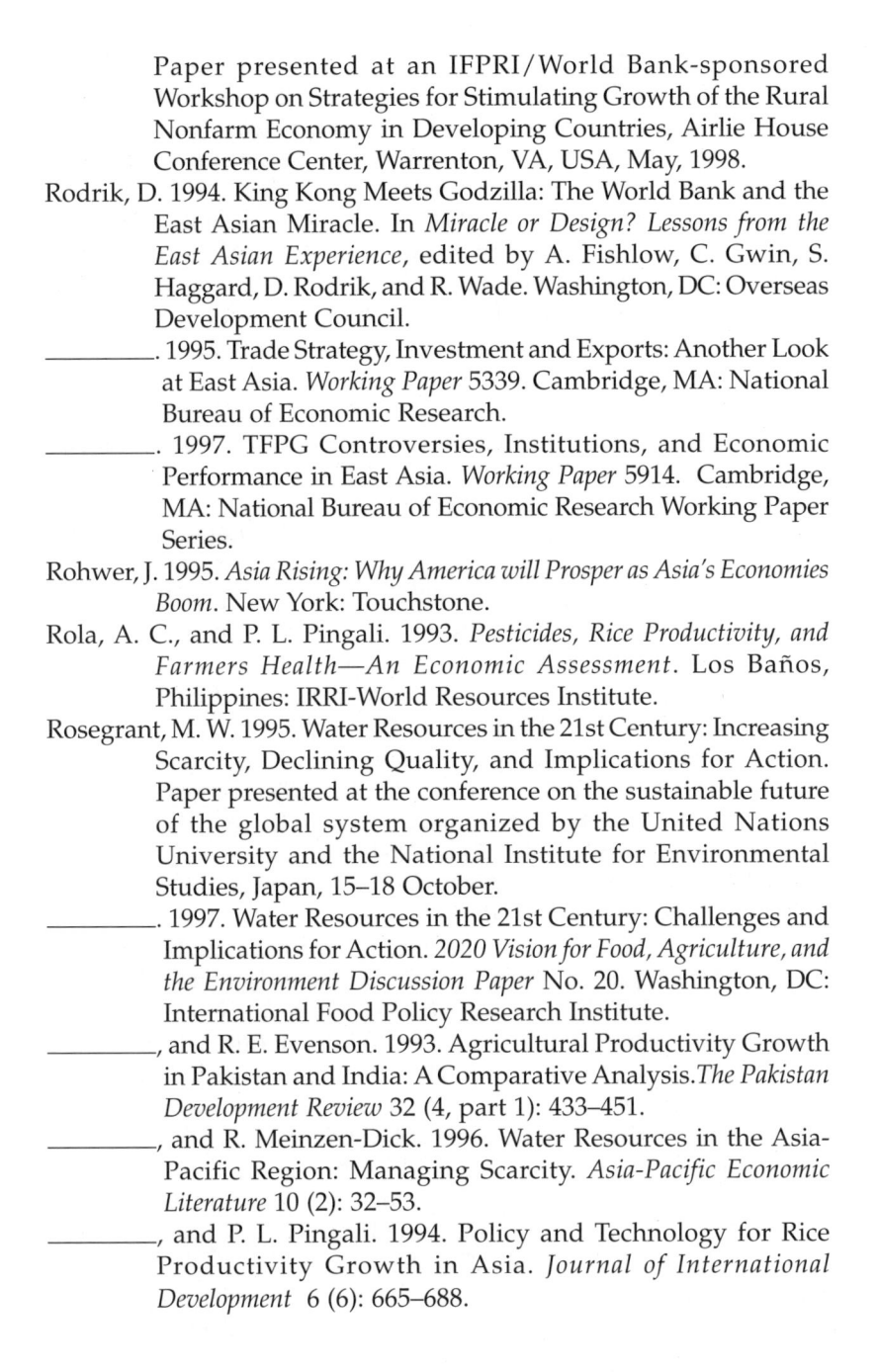

———, and C. Ringler. 1998. Asian Economic Crisis and the Long-Term Global Food Situation. Paper prepared for the international agricultural trade research consortium symposium on policy reform, market stability, and food security, Alexandria, Virginia, 26–27 June.

———, and S. Shetty. 1994. Production and Income Benefits from Improved Irrigation Efficiency: What is the Potential? *Irrigation and Drainage Systems* 8 (4): 251–270.

———, and M. Svendsen. 1993. Asian Food Production in the 1990s: Irrigation Investment and Management Policy. *Food Policy* 18 (2): 13–32.

———, R. Gazmuri Schleyer, and S. Yadav. 1995. Water Policy for Efficient Agricultural Diversification: Market-Based Approaches, *Food Policy* 203: 203–223.

———, F. Kasryno, and N. D. Perez. 1998. Output Response to Prices and Public Investment in Agriculture: Indonesian Food Crops. *Journal of Development Economics* 55 (2): 333–352.

Rosenzweig, C., and D. Hillel. 1998. *Climate Change and the Global Harvest: Potential Impacts of the Greenhouse Effect on Agriculture.* New York: Oxford University Press.

———, M. L. Parry, G. Fischer, and K. Frohberg. 1993. Climate Change and World Food Supply. *Research Report* No. 3. University of Oxford, Great Britain: Environmental Change Unit.

Rozelle, S., A. Park, V. Benziger, and C. Ren. 1998. Targeted Poverty Investments and Economic Growth in China. *World Development* 26 (12): 2137–2151.

Sachs, J. D., and A. Warner. 1995. Economic Reform and the Process of Global Integration. *BPEA* 1: 1–95.

Sahota, G. S., M. Huq, N. Hossain, and K. K. Sanyal. 1991. South Asian Development Model and Productivity in Bangladesh. *The Bangladesh Development Studies* 19 (1 & 2): 51–87.

Saith, A. 1987. Contrasting Experiences in Rural Industrialization: Are the East Asian Experiences Transferable? In *Rural Industrialization and Employment in Asia*, edited by R. Islam. New Delhi: International Labor Office. p. 241–301.

Samad, M., D. Merrey, D. Vermillion, M. Fuchs-Carsch, K. Mohtadullah, and R. Lenton. 1992. Irrigation Management Strategies for Improving the Performance of Irrigated Agriculture. *Outlook on Agriculture* 21 (4): 279–286.

Samarappuli, I. N., A. Ekanayake, L. Samarappuli and N. Yogaratnam. 1997. Modelling the Effect of Land Degradation on Yield of

Rubber. *MPI-LTU Sri Lanka Land Degradation Research Project Technical Survey Report Series.*

San, N.N., and M.W. Rosegrant. 1998. Indonesian agriculture in transition; Projections of alternative futures. *Journal of Asian Economies* 9 (3): 445-65.

Sarris, A. H. 1998. World Cereal Price Instability and a Market Based Scheme for Managing the Risk of Developing Country Cereal Imports. Paper presented at the international agriculture trade research consortium meeting held in Alexandria, Virginia on 26–27 June, 1998. Draft.

Scherr, S. J. 1999. Soil Degradation: A Threat to Developing-Country Food Security in 2020? *Food, Agriculture, and the Environment Discussion Paper.* Washington, D.C.: International Food Policy Research Institute.

_____, and P.B.R. Hazell. 1994. Sustainable Agricultural Development Strategies in Fragile Lands. *Environment and Production Technology Division Discussion Paper* No. 1. Washington, DC: International Food Policy Research Institute.

_____, and S. Yadav. 1995. Land Degradation in the Developing World: Implications for Food, Agriculture and the Environment to 2020. *2020 Vision for Food, Agriculture, and the Environment Discussion Paper* No. 14. Washington, DC:. International Food Policy Research Institute.

Schimmelpfennig, D., J. Lewandrowski, J. Reilly, M. Tsigas, and I. Parry. 1996. Agricultural Adaptation to Climate Change: Issues of Longrun Sustainability. *Agricultural Economic Report* Number 740. Washington, DC: U.S. Department of Agriculture.

Schultz, T. P. 1981. *Economics of Population.* Reading, MA: Addison-Wesley Publishing Company.

Schwartz, S. J., and D. H. Brooks. 1990. *Thailand's Feed and Livestock Industry to the Year 2000.* Washington, DC: United States Department of Agriculture.

Seckler, D. 1990. Private Sector Irrigation in Africa. In *Irrigation in Sub-Saharan Africa: The Development of Public and Private Systems,* edited by S. M. Barghouti and G. Le Moigne. World Bank Technical Paper Number 123. Washington, D.C.: World Bank.

_____. 1996. The New Era of Water Resource Management: From Dry to Wet Water Savings. *Research Report* No. 1. Colombo, Sri Lanka: IIMI.

Sen, Abhijit. 1996. Economic Reforms, Employment and Poverty: Trends and Options. *Economic and Political Weekly* Special Number, Vol. XXXI (35, 36, and 37): 2459-2477.

Sengupta, N., and A. Banik. 1997. Regional trade and investment: case of SAARC. *Economic and Political Weekly* 32 (November 15-21): 2930-2931.

Setboonsarng, S., and R. E. Evenson. 1991. Technology, Infrastructure, Output Supply, and Factor Demand in Thai Agriculture. In *Research and Productivity in Asian Agriculture*, edited by Robert E. Evenson and Carl E. Pray. Ithaca, NY, and London: Cornell University Press. p. 206–216.

Severino, J.-M. 1999. *Is Asia Rising? An Update.* Notes Prepared for a Presentation to the World Bank Board of Executive Directors. July 13. http://www.worldbank.org/html/extdr/offrep/eap/.

Sharma, M., M. Garcia, A. Qureshi, and L. Brown. 1996. Overcoming Malnutrition: Is There an Ecoregional Dimension? *Food, Agriculture, and the Environment Discussion Paper* No. 10. Washington DC: International Food Policy Research Institute.

Singh, I. 1990. *The Great Ascent: The Rural Poor in South Asia.* Baltimore, MD: Johns Hopkins University Press.

Singh, R., and P. Hazell. 1993. Rural poverty in the semi-arid tropics of India. *Economic and Political Weekly* 28. Agriculture Supplement.

Smale, M. 1996. Understanding global trends in use of wheat diversity and international flows of wheat genetic resources. *Economics Working Paper* 96–02. Mexico, DF: Centro Internacional de Mejoramiento de Maiz y Trigo (CIMMYT).

―――, and T. McBride. 1996. Understanding Global Trends in the Use of Wheat Diversity and International Flows of Wheat Genetic Resources. Part 1 of *CIMMYT 1995/96 World Wheat Facts and Trends: Understanding Global Trends in the Use of Wheat Diversity and International Flows of Wheat Genetic Resources.* Mexico, DF: Centro Internacional de Mejoramiento de Maiz y Trigo (CIMMYT).

Soesastro, H. 1998. Long-Term Implications for Developing Countries. In *East Asia in Crisis: From Being a Miracle to Needing One?*, edited by R. H. McLeod and R. Garnaut. London and New York: Routledge.

Srinivasan, T. N. 1994. Indian Agriculture: Policy and Performance. In *Agriculture and Trade in China and India: Policies and*

Performance Since 1950. San Francisco, CA: Institute for Contemporary Studies Press.

———. 1998. *Developing Countries and the Multilateral Trading System. From the GATT to the Uruguay Round and the Future.* Boulder, CO: Westview Press.

———, and G. Canonero. 1995. Preferential Trading Arrangements in South Asia: Theory, Empirics and Policy. New Haven: Yale University. Unpublished manuscript.

Steinfeld, H., C. de Haan, and H. Blackgrun. 1997. *Livestock-environment Interactions: Issues and Options.* Report of a study coordinated by the Food and Agriculture Organization of the United Nations, the United States Agency for International Development, and the World Bank. Brussels: European Commission Directorate-General for Development.

Stiglitz, J. E. 1998. Boats, Planes and Capital Flows. *Financial Times,* March 25.

Stoeckel, A., S. Fisher, W. McKibbin, and B. Borrell. 1998. *Asia's Meltdown and Agriculture.* Canberra, Australia: Centre for International Economics.

Stone, B. 1988. Relative Prices in the People's Republic of China: Rural Taxation through Public Monopsony. In *Agricultural Price Policy for Developing Countries,* edited by J. W. Mellor and R. Ahmed. Baltimore: Johns Hopkins University Press.

Subbarao, K., J. Braithwaite, and J. Jalan. 1995. Protecting the poor during adjustment and transitions. *Human Capital Development and Operations Policy (HCO) Working Paper* No. 58. Washington, DC: World Bank.

———, A. Bonnerjee, J. Braithwaite, S. Carbalho, K. Ezemenari, C. Graham, and A. Thompson. 1997. *Safety Net Programs and Poverty Reduction: Lessons from Cross-Country Experience.* Washington D.C.: World Bank.

Tabatabai, H. 1996. Statistics on Poverty and Income Distribution: An ILO Compendium of Data. Geneva, Switzerland: International Labour Organization.

TAC (Technical Advisory Committee). 1996. Report of the Study on CGIAR Research Priorities for Marginal Lands. Paper prepared for the Seventy-First Meeting of the Technical Advisory Committee. Consultative Group for International Agricultural Research, 26–28 November. Addis Ababa, Ethiopia.

Taube, G., and J. Zettelmeyer. 1998. Output Decline and Recovery in Uzbekistan: Past Performance and Future Prospects. *IMF Working Paper* WP/98/132. Washington, DC: International Monetary Fund.

Thapa, G. B., and M. W. Rosegrant. 1995. Projections and Policy Implications of Food Supply and Demand in Nepal to the Year 2020. *Research Report Series* No. 30. Kathmandu, Nepal: Winrock International.

Thiesenhusen, W. and J. Melmed-Sanjak. 1990. Brazil's agrarian structure: Changes from 1970 through 1980. *World Development* 18 (3).

Thimm, H. V. 1990. Agricultural Extension and the Diffusion of New Technologies. Policy brief presented at the International Food Policy Research Institute policy seminar on technology for sustainable agricultural growth, The Hague, 2–3 July.

Thomas, V., and Y. Wang. 1997. East Asian Lessons from Economic Reforms. In *Economies in Transition: Comparing Asia and Eastern Europe*, edited by Wing Thye Woo, S. Parker, and J. D. Sachs. Cambridge, MA: Massachusetts Institute of Technology Press.

Timmer, C. P. 1988. The Agricultural Transformation. *Handbook of Development Economics*. Vol. 1., edited by H. B. Chenery and T. N. Srinivasan. Amsterdam: North-Holland.

―――. 1997. *Food Security Strategies: The Asian Experience*. FAO Agricultural Policy and Economic Development Series. Rome, Italy: Food and Agriculture Organization.

Tomich, T. P., P. Kilby, and B. F. Johnston. 1995. *Transforming Agrarian Economies: Opportunities Seized, Opportunities Missed*. Ithaca, NY: Cornell University Press.

Tyagi, D. S. 1981. Growth of Agricultural Output and Labor Absorption in India. *Journal of Development Studies* 18 (1): 104–114.

Tyers, R., and K. Anderson. 1992. *Disarray in World Food Markets: A Quantitative Assessment*. Cambridge and New York: Cambridge University Press.

UN (United Nations). 1996. *World Population Prospects: The 1996 Revision*. New York: Population Division, Department for Economic and Social Information and Policy Analysis.

UNDP (United Nations Development Programme). 1998. *Poverty in Transition?* New York: Regional Bureau for Europe and the CIS.

UNICEF (United Nations Children's Fund). 1998. *The Impact of the Crisis on Food Security in Indonesia*. Jakarta, Indonesia: UNICEF. Mimeo.

Uphoff, N. 1986. *Local Institutional Development: An Analytical Source Book With Cases.* West Hartford, CT: Kumarian Press.

USAID (United States Agency for International Development). 1988. Urbanization in the Developing Countries. Interim report to Congress. Washington, DC.

Van de Valle, D. 1998. Protecting the Poor in Vietnam's emerging market economy. *Policy Research Working Paper* No. 1969. Washington DC: World Bank.

Van Lynden, G. W. J., and L. R. Oldeman. 1997. *Soil Degradation in South and Southeast Asia.* Wageningen, The Netherlands: International Soil Reference and Information Centre for UNEP.

Van Zyl, J., J. Kirsten, and H. Binswanger. 1996. *Agricultural Land Reform in South Africa: Policies, Markets and Mechanisms.* Cape Town: Oxford University Press.

Vokes, R., and Fabella, A. 1996. Lao PDR. In *From Centrally Planned to Market Economies: The Asian Approach*, edited by Pladumna B. Rana and Naved Hamid. Vol. 3. Hong Kong: Oxford University Press. p. 3–148.

Von Braun, J. 1995. Agricultural Commercialization: Impacts on Income and Nutrition and Implications for Policy. *Food Policy* 20: 187–202.

_____, and E. Kennedy. 1994. *Agricultural Commercialization, Economic Development, and Nutrition.* Baltimore, MD: Johns Hopkins University Press.

Vyas, V. S., and G. Mathai. 1978. Farm and Non-farm Employment in Rural Areas: A Perspective for Planning. *Economic and Political Weekly* 13 (6 & 7): 333–347.

Wade, R. 1994. Selective Industrial Policies in East Asia: Is the East Asian Miracle Right? in *Miracle or Design? Lessons from the East Asian Experience*, by A. Fishlow, C. Gwin, S. Haggard, D. Rodrik, and R. Wade. Washington, D.C.: Overseas Development Council.

_____. 1998. The Asian Debt-and-Development Crisis of 1997: Causes and Consequences. *World Development* 26: 1535–53.

Walker, T., and J. Ryan. 1990. *Village and Household Economies in India's Semi-arid Tropics.* Baltimore: Johns Hopkins University Press.

Wanmali, S. 1983. *Service Centres in Rural India.* Delhi: B. R. Publishing Corporation.

Warr, P. G. 1998. Thailand. In *East Asia in Crisis: From Being a Miracle to Needing One?*, edited by R. H. McLeod and R. Garnaut. London and New York: Routledge.

WDI (World Development Indicators) CD-ROM. 1997. Washington, D. C.: The World Bank.
―――――.CD-ROM. 1998. Washington, D. C.: The World Bank.
―――――.1996. *World Resources 1996–97*. New York: Oxford University Press.
―――――.1998. *World Resources 1998–99*. New York: Oxford University Press.
Wen, G. J. 1995. The Land Tenure System and Its Saving and Investment Mechanism: The Case of Modern China. *Asian Economic Journal* 9 (3): 233–59.
Wiemer. 1994. State Policy and Rural Resource Allocation in China as Seen through a Hebei Province Township, 1970–85. *World Development* 22 (6): 935–947.
Wilkes, G. 1992. Strategies for sustaining crop germplasm preservation, enhancement, and use. *Issues in Agriculture* 5, 62. Washington, DC: Consultative Group on International Agricultural Research (CGIAR) Secretariat.
Wodon, Q.T. 1999. Microdeterminants of consumption, poverty, growth, and inequality in Bangladesh. *Policy Research Working Paper* No. 2076. Washington, D.C.: World Bank.
Wolfe, D. 1996. Potential Impact of Climate Change on Agriculture and Food Supply. In *Sustainable Development and Global Climate Change: Conflicts and Connections*, edited by James White, William R. Wagner, and Wendy H. Petry. Proceedings of a conference sponsored by the Center for Environmental Information, Inc., Washington, DC, 4–5 December 1995, and published with the assistance of the U.S. Global Change Research Program (USGCRP).
Wong, C. P. W. 1996. People's Republic of China. In *From Centrally planned to market economies: the Asian approach. Vol 2: People's Republic of China and Mongolia*, edited by P. B. Rana and N. Hamid. Oxford, New York: Oxford University Press and Asian Development Bank.
Wood, S., K. Sebastian, F. Nachtergaele, D. Nielson, and A. Dai 1998. Spatial Aspects of the Design and Targeting of Development Strategies for Fragile Lands. Paper presented at an IFPRI/DSE/NARO/EC sponsored International Conference on Strategies for Poverty Alleviation and Sustainable Resource Management in the Fragile Lands of Sub-Saharan Africa, Entebbe, Uganda, May, 1998.

World Bank. 1990. *Poverty: World Development Indicators*. World Development Report 1990. New York: Oxford University Press.
──────.1993a. *The East Asian Miracle: Economic Growth and Public Policy*. A World Bank Policy Research Report. New York: Oxford University Press.
──────.1993b. *Water Resources Management*. A World Bank Policy Study. Washington, DC: World Bank.
──────.1996a. *From Plan to Market: World Development Report 1996*. Washington, DC: World Bank.
──────.1996b. *Commodity Markets and the Developing Countries: A World Bank Quarterly*. International Trade Division (November). Washington, DC: World Bank.
──────.1996c. *Poverty Reduction and the World Bank: Progress and Challenges in the 1990s*. Washington, DC: The International Bank for Reconstruction and Development/The World Bank.
──────.1998a. Kyrgyz Republic. <http://www.worldbank.org/html/extdr/offrep/eca/ kyr2.htm> Accessed 30 November 1998.
──────.1998b. Kazakhstan. <http://www.worldbank.org/html/extdr/offrep/eca /kz2.htm> Accessed 3 December 1998.
──────.1998c. Mongolia-Fiscal Technical Assistance Project. Project Information Document. <http://www.worldbank.org/pics/pid/mn51855.txt>
──────.1998d. Uzebekistan. <http://www.worldbank.org/html/extdr/offrep/eca /uz2.htm> Accessed 3 December 1998.
──────.1998e. *East Asia: The Road to Recovery*. Washington, DC: World Bank.
──────.1998f Global Economic Prospects: 1998/99. <http://www.worldbank. org/prospects/gep98-99/>. Accessed 9 December 1998.
──────/FAO/UNIDO/Industry Fertilizer Working Group. 1994. World and Regional Supply and Demand Balances for Nitrogen, Phosphate, and Potash, 1992/93–1998/99. Washington, DC: World Bank.
WRI (World Resource Institute), 1998. *World Resources 1998–99*. New York: Oxford University Press.
Wright, B. D. 1996. Crop Genetic Resource Policy: Towards a Research Agenda, *EPTD Environment and Production Technology Division Discussion Paper* No. 19. Washington, DC: International Food Policy Research Institute.

WTO (World Trade Organization), 1998. Accession to the WTO. State of Play. Note by the Secretariat WT/GC/W/100. http://www/wto/about/GCW100.htm.

Yao, Y. 1995. *Institutional Arrangements, Tenure Insecurity, and Agricultural Productivity in Post-Reform Rural China.* Madison, WI: University of Wisconsin, Department of Agricultural Economics.

Yeldan, E. 1989. Structural Adjustment and Trade in Turkey: Investigating the Alternatives Beyond Export-Led Growth. *Journal of Policy Modeling* 11 (2): 273-296.

Young, A. 1994. Lessons from the East Asian NICs: A Contrarian View. *European Economic Review* 38 (3–4): 964–73.

Young, A. 1995. The Tyranny of Numbers: Confronting the Statistical Realities of the East Asian Growth Experience. *Quarterly Journal of Economics* 110 (3): 641–80.

Yudelman, M.. 1996. The Impact of Biotechnology on Crop Production in Developing Countries. Washington, DC: International Food Policy Research Institute. Mimeo.

Zeigler, R., M. Hossain, and P. S. Teng. 1994. Sustainable Agricultural Development of Asian Tropics: Emerging Trends and Potential. Paper presented at an IFPRI conference on ecoregions of the developing world, a lens for assessing food, agriculture and the environment to the year 2000, Airlie House Conference Center, Warrenton, VA, USA. November 1994.

Zettelmeyer, J. 1998. The Uzbek Growth Puzzle. *IMF Working Paper* WP/98/133. Washington, DC: International Monetary Fund.

APPENDIX

Impact Baseline Scenario for Asia: Trends to 2010 in Food Supply, Demand, Trade, and Child Malnutrition

With rapid growth in income, rapid dietary changes due to increasing incomes and urbanization, and the increasing openness of Asian economies, Asia will continue to grow in importance to the global economy and particularly to world food markets. In this appendix, prospects are examined for food supply, demand, and trade to 2010 for Asian countries, in the context of global food supply and demand projections and the international price trends that they imply. Implications for one important barometer of well-being—malnutrition among children—are drawn and discussed. Despite the recent financial and economic crisis in East and Southeast Asia, developing countries there should continue their rapid economic development, with enhanced food security stemming from lower food prices and either greater food surpluses or cheaper food imports. Food insecurity in countries with slower economic growth and persistently high population growth, particularly in South Asia, will be far less amenable to improvement.

ASIA IN THE GLOBAL AGRICULTURAL ECONOMY

Global food policy has been driven by the need to feed an increasing population and to support diversified consumption patterns as incomes rise. Agricultural production growth has been able to meet these goals: in the past three decades, effective demand has been met, while real food prices have declined dramatically. In 1961, the world produced 867 million metric tons (mt) of cereals for 3.1 billion people; by 1995 production was 1.9 billion mt for 5.7 billion people—2.2 times the grain for 1.8 times the population (FAO 1997). At the same time, real prices of major cereals declined: between 1982 and 1995, real world wheat prices dropped by 28 percent, rice prices by 42 percent, and maize prices by 43 percent.

The International Food Policy Research Institute's (IFPRI), International Model for Policy Analysis of Agricultural Commodities and Trade (IMPACT) a tool that takes account of current best estimates for such critical trends as income and population growth, as well as agricultural productivity improvements or declines, allows a country- and/or region-specific look at likely future food supply, demand, and trade. What trends are likely to continue, at what pace, and with what implications within specific regions or countries of the developing world, particularly Asia? Following a brief description of the model, IMPACT projections for supply and demand of cereals and livestock products are examined, and implications for international prices and trade explored. Overall, global trends are presented, but results are highlighted for Asia and specific subregions or countries within Asia. In order to place Asia within the context of global developments in supply and demand, key results for other regions of the world are also presented.

IFPRI'S IMPACT GLOBAL FOOD MODEL: MODEL CHARACTERISTICS

IMPACT covers 37 countries and regions (which account for virtually all of world food production and consumption), and 18 commodities, including all cereals, soybeans, roots and tubers, meats, milk, eggs, oils, and oilcakes and meals. The model is specified as a set of country-level supply and demand equations. Each country model is linked to the rest of the world through trade. World food prices are determined annually at levels that clear international commodity markets. Demand is a function of prices, income, and population growth. Growth in crop production in each country is determined by crop prices and by the rate of productivity growth. Future productivity growth is estimated by its component sources, including crop-management research, conventional plant breeding, wide-crossing and hybridization breeding, and biotechnology and transgenic breeding. Other sources of growth considered include private-sector agricultural research and development, agricultural extension and education, markets, infrastructure, and irrigation.

A key assumption of the baseline productivity projections is that future trends in public investment in Asia and other developing countries in agricultural research, irrigation, and the social sector (education, total social expenditures, and water and sanitation) will continue at the already reduced rates of growth prevailing in the early 1990s. Water availability also has a significant impact on area and yield assumptions. The increase in municipal and industrial water demand between 1995 and 2010 in Asia is projected to be equivalent to 19 percent of 1995 agricultural water demand (Rosegrant, Gazmuri Schleyer, and Yadav 1997). Under the baseline, it is assumed that just under one half of this is a net loss to agricultural water use, representing a 9 percent reduction in water available for agriculture between 1995 and 2010. The methodology of IMPACT is described in Rosegrant, Gazmuri Schleyer, and Yadav (1995).

The results presented here are generated from a revised and updated version of IMPACT. The updated model incorporates additional features in its structure and input data that improve its capability on both the supply and demand sides. Modifications of the model structure are reflected primarily in the supply and demand equations. In the supply equation, the marginal contribution of further expansion of irrigated area is incorporated in two places. It appears in the area function through potential increases in cropping intensities and in the yield function through the addition of a yield differential between irrigated and nonirrigated crops that represents the improvement that will be realized with the conversion of farm areas into irrigated ecosystems. The demand side of IMPACT incorporates the dynamic adjustment of income elasticities with respect to growth in income.

In addition to the modifications in the model structure, the baseline data on which the projections are made are updated from 1990 to 1993 (with 1993 representing the three-year average value for 1992–94). The revised IMPACT also includes the November 1996 revised population projections from the United Nations (UN 1996) and updated information on investment in agricultural research. The baseline projection income-growth rates for Asia have been adjusted to reflect the estimated impact on income growth of the Asian financial and economic crisis. Projections for income and population growth for the IMPACT countries and regions during the 1993–2010 period are reported in Table A.1.

PROJECTIONS FOR CEREALS

Projected Cereal Demand and its Composition: The Relative Rise of Feed vs. Food Demand

Changing patterns of demand are apparent in the projected per capita and total food demand for cereals shown in Tables A.2 and A.3, the projected growth rates in feed demand and feed-demand levels (Tables A.4 and A.5), and total cereal demand (Table A.6). Per capita food consumption of all cereals will be virtually constant on a global basis, with slightly declining consumption of cereals in higher-income countries balancing the slightly increasing demands of lower-income countries. Global food consumption of cereals, driven mainly by population growth, will increase by 243 million mt (27 percent), from 918 million mt in 1993 to 1,161 million mt

Table A.1: Income and Population Growth, 1993-2010
(percnt per year)

Countries/Regions	Population Growth	Income Growth
USA	0.78	2.2
EC12	0.09	2.2
Other Western Europe	0.35	2.3
Japan	0.12	2.2
Australia	1.05	2.2
Other developed countries	1.59	2.2
Eastern Europe	-0.01	1.6
CIS	0.08	1.6
Argentina	1.20	3.2
Brazil	1.22	3.2
Colombia	1.55	3.2
Mexico	1.48	3.2
Other Latin America	1.72	3.0
Madagascar	3.11	3.8
Nigeria	2.80	3.2
Northern Sub-Saharan Africa	2.89	3.3
Central & Western SSA	2.78	3.8
Southern Sub-Saharan Africa	2.53	3.2
Eastern Sub-Saharan Africa	2.68	4.5
Egypt	1.75	3.2
Turkey	1.40	4.5
Other West Asia & North Africa	2.52	3.2
India	1.48	5.1
Pakistan	2.63	4.6
Bangladesh	1.66	4.5
Other South Asia	2.53	4.6
Indonesia	1.32	5.1
Malaysia	1.85	5.4
Myanmar	1.64	4.0
Philippines	1.86	5.0
Thailand	0.72	5.6
Viet Nam	1.56	5.0
Other Southeast Asian countries	2.12	4.0
People's Republic of China	0.79	5.6
Republic of Korea	0.75	3.9
Other East Asian countries	1.21	2.4
Rest of the World	1.37	4.9

Source: IFPRI IMPACT.

Table A.2: Per Capita Food Demand for Cereals, 1993 and Projected 2010 (kilogram per capita)

	Wheat 1993	Wheat 2010	Maize 1993	Maize 2010	Rice 1993	Rice 2010	Other Grains 1993	Other Grains 2010	All Cereals 1993	All Cereals 2010
East Asia	79.5	86.2	25.6	22.5	94.9	94.9	9.5	8.9	209.5	212.5
PRC	82.6	89.8	25.8	22.6	95.8	96.2	9.5	8.9	213.7	217.5
Republic of Korea	48.3	51.8	17.6	15.6	105.6	102.6	13.2	12.5	184.7	182.5
Other East Asia	36.4	37.7	27.3	26.3	64.2	61.7	6.2	6.1	134.2	131.8
South Asia	59.1	66.4	7.9	7.7	77.4	79.7	17.9	17.0	162.2	170.9
India	54.6	61.1	8.4	8.1	77.5	81.6	22.7	22.1	163.2	172.9
Pakistan	128.8	135.5	5.5	5.2	15.7	16.1	3.2	3.1	153.1	160.0
Bangladesh	19.9	22.4	0.0	0.0	143.2	146.1	0.6	0.7	163.8	169.2
Other South Asia	39.2	43.0	27.6	28.2	91.4	94.1	7.6	7.8	165.9	173.0
Southeast Asia	13.8	16.0	12.9	12.5	141.2	141.6	1.1	1.1	169.1	171.1
Indonesia	14.1	16.4	20.9	20.4	151.7	151.3	0.2	0.2	186.9	188.4
Thailand	9.3	11.0	0.5	0.4	102.1	100.3	1.6	1.4	113.4	113.1
Malaysia	36.5	41.5	3.6	3.2	96.5	95.8	1.7	1.5	138.3	142.0
Philippines	30.6	34.0	18.1	16.6	89.3	90.6	3.2	3.1	141.2	144.2
Viet Nam	4.7	5.5	7.2	7.0	154.3	156.6	0.4	0.4	166.5	169.5
Myanmar	3.0	3.3	2.2	2.1	211.0	210.0	2.8	2.8	218.9	218.3
Other Southeast Asia	0.9	1.0	5.5	5.5	172.5	172.3	0.1	0.1	179.1	178.9
Latin America	49.4	49.7	44.8	45.2	26.9	28.0	6.5	6.3	127.6	129.2
WANA	157.4	158.9	14.8	13.3	20.4	20.6	21.5	21.8	214.1	214.5
Sub-Saharan Africa	12.7	13.7	39.4	39.4	16.7	17.5	43.1	45.3	111.8	116.0
Developing	62.1	65.6	22.1	21.2	71.8	69.9	15.7	16.7	171.6	173.4
Developed	98.2	98.1	11.8	11.6	12.0	11.9	22.2	21.1	144.1	142.7
Japan	43.9	48.6	23.2	20.8	70.6	68.6	8.9	8.3	146.6	146.3
World	70.4	72.0	19.7	19.3	58.1	58.6	17.2	17.5	165.3	167.4

Source: IFPRI IMPACT.

Appendix 459

Table A.3: Food Demand for Cereals, 1993 and Projected 2010 (million metric tons)

	Wheat 1993	Wheat 2010	Maize 1993	Maize 2010	Rice 1993	Rice 2010	Other Grains 1993	Other Grains 2010	All Cereals 1993	All Cereals 2010
East Asia	101.2	125.7	32.6	32.9	120.6	138.3	12.0	13.0	266.4	309.9
PRC	97.1	120.7	30.4	30.4	112.6	129.2	11.1	12.0	251.2	292.3
Republic of Korea	2.1	2.6	0.8	0.8	4.7	5.1	0.6	0.6	8.1	9.1
Other East Asia	1.9	2.4	1.4	1.7	3.4	4.0	0.3	0.4	7.0	8.5
South Asia	70.2	104.8	9.4	12.1	92.0	125.7	21.2	26.8	192.8	269.4
India	49.2	70.8	7.6	9.3	69.9	94.5	20.4	25.6	147.1	200.3
Pakistan	17.1	28.0	0.7	1.1	2.1	3.3	0.4	0.6	20.4	33.1
Bangladesh	2.3	3.4	0.0	0.0	16.5	22.3	0.1	0.1	18.9	25.8
Other South Asia	1.5	2.6	1.1	1.7	3.6	5.6	0.3	0.5	6.5	10.3
Southeast Asia	6.4	9.5	6.0	7.4	65.5	83.8	0.5	0.7	78.4	101.3
Indonesia	2.7	3.9	4.0	4.9	29.1	36.3	0.0	0.1	35.8	45.2
Thailand	0.5	0.7	0.0	0.0	5.9	6.5	0.1	0.1	6.5	7.4
Malaysia	0.7	1.1	0.1	0.1	1.9	2.5	0.0	0.0	2.7	3.7
Philippines	2.0	3.0	1.2	1.5	5.8	8.0	0.2	0.3	9.1	12.8
Viet Nam	0.3	0.5	0.5	0.7	11.0	14.5	0.0	0.0	11.9	15.7
Myanmar	0.1	0.2	0.1	0.1	9.4	12.4	0.1	0.2	9.8	12.8
Other Southeast Asia	0.0	0.0	0.1	0.1	2.5	3.5	0.0	0.0	2.6	3.7
Latin America	22.8	29.4	20.7	26.7	12.4	16.5	3.0	3.7	58.9	76.3
WANA	58.1	85.3	5.5	7.1	7.5	11.0	7.9	11.7	79.0	115.1
Sub-Saharan Africa	6.5	11.2	20.2	32.3	8.6	14.3	22.1	37.1	57.4	94.9
Developing	265.4	366.1	94.3	118.4	306.9	390.2	66.9	93.0	733.5	967.7
Developed	125.4	133.0	15.1	15.7	15.3	16.1	28.3	28.6	184.2	193.4
Japan	5.5	6.2	2.9	2.6	8.8	8.7	1.1	1.1	18.3	18.6
World	390.8	499.1	109.4	134.1	322.3	406.3	95.2	121.6	917.7	1,161.1

Source: IFPRI IMPACT.

Table A.4: Projected Annual Growth Rates of Feed Demand for Cereals, 1993 - 2010 (percent per year)

	Wheat	Maize	Rice	Other Grains	All Cereals
East Asia	3.52	3.75	0.79	3.20	3.65
PRC	3.67	3.90	0.79	3.24	3.78
Republic of Korea	2.92	2.75	—	3.38	2.81
Other East Asia	2.60	2.60	—	2.73	2.60
South Asia	3.60	6.61	1.94	3.56	4.46
India	3.82	7.61	1.79	3.62	4.96
Pakistan	3.27	3.29	—	3.41	3.29
Bangladesh	—	3.26	—	3.02	3.16
Other South Asia	2.75	3.05	2.71	2.89	2.86
Southeast Asia	3.12	3.34	1.41	3.22	3.15
Indonesia	—	3.42	1.31	—	3.10
Thailand	—	3.09	0.52	3.13	2.93
Malaysia	3.12	3.00	1.74	3.16	2.98
Philippines	—	3.71	1.95	3.61	3.52
Viet Nam	—	3.22	1.66	3.43	2.67
Myanmar	—	3.05	—	3.64	3.12
Other Southeast Asia	—	2.62	—	3.07	2.64
Latin America	2.29	2.17	—	2.28	2.20
WANA	2.64	2.57	—	2.68	2.65
Sub-Saharan Africa	3.74	3.63	—	3.68	3.65
Developing	3.11	3.25	1.15	2.65	3.09
Developed	0.47	0.78	0.09	0.61	0.67
Japan	0.34	0.20	—	0.40	0.27
World	1.11	1.86	1.15	1.05	1.51

Source: IFPRI IMPACT.

Table A.5: Feed Demand for Cereals, 1993 and Projected 2010 (million metric tons)

	Wheat 1993	Wheat 2010	Maize 1993	Maize 2010	Rice 1993	Rice 2010	Other Grains 1993	Other Grains 2010	All Cereals 1993	All Cereals 2010
East Asia	8.0	14.3	70.1	131.0	2.1	2.4	3.7	6.3	83.8	154.0
PRC	6.4	11.8	60.9	116.8	2.1	2.4	3.2	5.5	72.6	136.5
Republic of Korea	1.2	2.0	3.5	5.6	—	—	0.1	0.2	4.8	7.8
Other East Asia	0.3	0.5	5.6	8.7	—	—	0.3	0.5	6.3	9.8
South Asia	2.9	5.2	1.5	4.3	0.4	0.5	0.7	1.3	5.4	11.3
India	1.8	3.4	1.0	3.6	0.3	0.4	0.6	1.0	3.7	8.5
Pakistan	0.9	1.6	0.3	0.5	0.0	0.0	0.1	0.2	1.3	2.3
Bangladesh	0.0	0.0	0.0	0.0	0.0	0.0	0.0	0.0	0.0	0.0
Other South Asia	0.2	0.2	0.1	0.2	0.1	0.1	0.0	0.0	0.4	0.6
Southeast Asia	0.1	0.2	12.7	22.2	1.6	2.1	0.2	0.4	14.7	24.8
Indonesia	0.0	0.0	3.2	5.6	0.6	0.8	0.0	0.0	3.8	6.4
Thailand	0.0	0.0	3.6	6.1	0.3	0.3	0.2	0.3	4.1	6.7
Malaysia	0.1	0.2	1.9	3.1	0.0	0.1	0.0	0.0	2.1	3.4
Philippines	0.0	0.0	3.6	6.8	0.5	0.7	0.0	0.0	4.1	7.5
Viet Nam	0.0	0.0	0.2	0.4	0.1	0.2	0.0	0.0	0.4	0.6
Myanmar	0.0	0.0	0.1	0.2	0.0	0.0	0.0	0.0	0.1	0.2
Other Southeast Asia	0.0	0.0	0.0	0.1	0.0	0.0	0.0	0.0	0.0	0.1
Latin America	2.1	3.0	39.1	56.3	—	—	12.4	18.2	53.5	77.5
WANA	7.1	11.1	8.9	13.7	—	—	18.3	28.7	34.3	53.5
Sub-Saharan Africa	0.0	0.0	1.3	2.3	—	—	0.9	1.6	2.2	4.0
Developing	20.2	34.0	133.4	229.8	4.1	5.0	36.2	56.4	193.9	325.2
Developed	77.8	84.3	209.2	238.8	0.0	0.0	155.3	172.3	442.3	495.4
Japan	0.5	0.6	12.1	12.5	—	—	5.5	5.9	18.1	19.0
World	98.0	118.3	342.6	468.6	4.1	5.0	191.5	228.7	636.1	820.6

Source: IFPRI IMPACT.

Table A.6: Total Cereal Demand, 1993 and Projected 2010 (million metric tons)

	Wheat 1993	Wheat 2010	Maize 1993	Maize 2010	Rice 1993	Rice 2010	Other Grains 1993	Other Grains 2010	All Cereals 1993	All Cereals 2010
East Asia	118.1	151.6	106.4	169.4	133.6	153.1	18.2	22.4	376.3	496.6
PRC	111.2	142.3	91.5	147.4	125.0	143.4	16.6	20.3	344.3	453.4
Republic of Korea	4.6	6.3	6.3	9.3	5.1	5.6	0.7	0.9	16.6	22.0
Other East Asia	2.3	3.0	8.7	12.8	3.6	4.2	0.9	1.2	15.4	21.2
South Asia	79.9	120.1	12.2	18.5	101.7	139.0	24.3	31.2	218.1	308.8
India	56.9	82.8	9.7	14.5	77.4	104.7	23.3	29.6	167.3	231.5
Pakistan	18.8	30.1	1.2	2.0	2.3	3.7	0.6	0.9	22.9	37.4
Bangladesh	2.5	3.6	0.0	0.0	17.9	24.2	0.1	0.1	20.5	28.0
Other South Asia	1.7	2.9	1.3	2.1	4.0	6.3	0.3	0.5	7.4	11.8
Southeast Asia	6.8	10.1	19.4	30.7	73.6	93.9	0.9	1.2	100.7	136.0
Indonesia	2.8	4.1	7.6	11.1	32.8	40.9	0.0	0.1	43.2	56.2
Thailand	0.6	0.7	3.7	6.2	7.6	8.5	0.3	0.5	12.2	15.9
Malaysia	1.0	1.5	2.0	3.3	1.9	2.6	0.1	0.1	5.0	7.6
Philippines	2.0	3.0	5.0	8.5	6.5	9.1	0.2	0.3	13.7	20.9
Viet Nam	0.3	0.5	0.8	1.1	12.8	16.9	0.0	0.1	14.0	18.6
Myanmar	0.1	0.2	0.2	0.3	9.4	12.4	0.1	0.2	9.9	13.1
Other Southeast Asia	0.0	0.0	0.1	0.2	2.5	3.5	0.0	0.0	2.6	3.8
Latin America	27.4	35.7	69.3	96.1	14.4	19.2	17.7	25.1	128.8	176.1
WANA	72.9	107.2	15.8	22.9	8.2	11.9	29.8	45.7	126.5	187.9
Sub-Saharan Africa	6.9	11.8	26.3	42.3	9.6	16.2	27.6	46.4	70.4	116.7
Developing	312.2	437.0	249.4	380.0	341.4	433.8	118.5	172.0	1,021.5	1,422.9
Developed	240.1	256.9	276.8	314.5	17.8	18.8	216.8	237.0	751.5	827.2
Japan	6.5	7.3	16.6	16.8	9.6	9.5	6.9	7.2	39.6	40.8
World	552.2	693.9	526.2	694.5	359.2	452.6	335.3	409.1	1,773.0	2,250.1

Source: IFPRI IMPACT.

in 2010. Increased food demand for wheat will account for 44 percent of the total increase in food demand, followed by rice with 35 percent (Table A.3).

Developed countries as a group have higher per capita wheat consumption levels (led by the former Soviet Union and Eastern Europe, which averaged close to 150 and 120 kilograms (kg) per capita, respectively, in 1993) than developing countries. In this latter group, countries in West Asia and North Africa (WANA) have the highest levels (Table A.2). Still, while in the group of developed countries per capita wheat consumption is projected to stagnate at 98 kg through 2010, levels in the developing world are projected to rise from 62 kg in 1993 to 66 kg in 2010 (Table A.2). At the global level, this translates into slightly slowing growth in per capita food demand for wheat compared to recent trends, but at rates still significantly faster than for maize or rice. Within the developing-country group, per capita food consumption of wheat will remain flat in Latin America and WANA, but is expected to move up in the group of Asian developing countries (see the Tables in Chapter XII for these figures) from 61 kg to 66 kg in the same period. In South Asia, growth in wheat consumption from 59 kg to 66 kg per capita consumption will be driven by solid income growth. In the PRC, per capita wheat demand is expected to increase from 83 kg to 90 kg. Modest growth is projected for the Southeast Asian country group from 14 to 16 kg per capita over the period; this masks a substantial jump in per capita wheat consumption of the region's largest wheat consumer, Malaysia (moving from 37 kg per capita in 1993 to 42 kg in 2010).

Per capita food demand for rice is projected to decline only slightly from 1993 to 2010, as rapid income growth in Asia drives a diversification of diets in developing countries. As a group, developing countries will reduce per capita consumption of rice as food from 72 kg in 1993 to 70 kg in 2010; per capita consumption will virtually stagnate in the group of Asian developing countries, increasing only slightly, from 95 to 96 kg. At the same time, per capita demand for rice will fall for the PRC and most of the Southeast Asian countries (the only projected increases, and these small, coming in the Philippines and Viet Nam). In some developing countries, however, rice consumption will rise due to a dietary shift to rice from maize and other coarse grains. India will increase its per capita food demand for rice from 78 kg in 1993 to 82 kg in 2010, and slight increases in per capita consumption are projected for the regions of Latin America, Sub-Saharan Africa, and WANA. In developed countries, global per capita consumption levels will be virtually constant. Population growth worldwide, however, will spur an increase in total food demand for rice of 84 million mt (26 percent) over the 1993 level of 322 million mt.

Per capita food demand for maize and other coarse grains will continue to be virtually stagnant or declining in both developing and developed countries, due to the dietary shifts described above. Developing Asia will experience substantial declines in per capita demand for maize as food, whereas demand will increase slightly in Latin America, where the major maize consumer, Mexico, is located. The increase in per capita food demand for other grains in developing countries is mainly due to Sub-Saharan Africa, where these grains are a primary cereal staple. For the world as a whole, food consumption of other coarse grains is stagnant at around 17 kg per capita (Table A.2). Global food demand for maize and other coarse grains will increase by 25 million mt and by 26 million mt, respectively; most of this increase (49 percent and 57 percent, respectively) will occur in Sub-Saharan Africa (Table A.3).

Tables A.4 and A.5 spotlight the fact that the rapid growth in demand for cereals—in particular maize—as animal feed in developing countries, already seen as important for Asia, (see Chapter V) will proceed apace. Demand for maize for feed is projected to grow during 1993–2010 at a strong 7.6 percent per year in India (though from a very low base), and at about 4 percent per year in the PRC and Sub-Saharan Africa. The average growth for all developing countries is projected at 3.3 percent per year, and for the group of Asian developing economies at 3.8 percent annually. This high growth is due to the rapid expansion of the livestock industry, especially in the more rapidly growing developing economies, where consumption of livestock products (as described throughout this appendix, exclusive of dairy products such as milk) will expand dramatically. Demand for wheat and other coarse grains as feed will also experience high rates of growth in developing countries, at 3.1 and 2.7 percent per year, respectively. Again, demand growth for these commodities will be even higher in the group of Asian developing countries, at 3.5 and 3.3 percent per year, respectively.

Looking at all cereals, demand for feed will increase during 1993–2010 by 3.6 percent per year, on average, in Asian developing countries and by 3.1 percent annually in the group of developing countries. Growth in cereal feed demand will be much lower in developed countries, at 0.7 percent per year in 1993–2010, but slightly higher than the growth rate prevailing since 1982. Global demand for cereals as animal feed will increase by 185 million mt (29 percent), more than 70 percent of which will be accounted for by developing countries (Table A.5). Cereal feed demand will more than double in India (from very low levels) and nearly double in the PRC, to the point that the PRC will overtake the United States as the largest consumer of cereal feeds by 2010.

How will these demand components translate into total demand requirements? Table A.6 shows the projected levels of total demand for cereals in 1993 and 2010. Global cereal demand will increase by 477 million mt (27 percent), compared to the 1993 level of 1,773 million mt. The group of developing countries will account for 84 percent of the global increase in cereal demand, and the Asian developing economies will account for 52 percent. The PRC alone will account for 23 percent and India for another 13 percent. Cereal demand will increase by 66 percent in Sub-Saharan Africa; by 49 percent in WANA; by 37 percent in Latin America; and by 35 percent in Southeast Asia.

The largest increase in demand will be for maize (35 percent), driven by the rapid growth in demand for feed, followed by wheat (30 percent). Eighty-eight percent of the increase in wheat demand will come from the developing world, where both population and income growth are higher than in the developed economies. Even more striking, 99 percent of the increase in rice demand will be accounted for by the group of (primarily Asian) developing countries. However, rice demand will grow more slowly than wheat and maize because of changing consumption patterns and the limited role of rice in animal feed. Overall rice demand is expected to increase by 93 million mt, with the most significant increases in India (27 million mt), Southeast Asia (20 million mt), and the PRC (18 million mt).

Projected Production Growth in Cereals: Burden on Yield vs. Area to Meet Demand?

How will this growth in demand be met? As shown in Table A.7, world cereal production is projected to grow at an average rate of 1.4 percent per annum, virtually the same annual growth as was achieved during 1982–1993. Annual growth in specific cereals will not maintain its 1982–93 pace: wheat production is projected at 1.4 percent during 1993–2010, compared to 1.5 percent during 1982–93; maize production at 1.7 percent compared to 2.2 percent, and rice production at 1.4 percent compared to 2.0 percent in the former period. Production of other coarse grains will recover from negative growth between 1982 and 1993, to a growth rate of 1.2 percent per year in 1993–2010.

Annual growth in crop production will be higher in the group of developing countries, at 1.8 percent, than in the developed countries, at 1.0 percent. Growth in cereal crop production is projected at 1.5 percent annually, on average, in the group of Asian developing countries. Sub-Saharan Africa will experience the highest production growth rate for each of the cereals; this is mainly due to its past low production levels. In the group of developed countries, Australia is expected to experience the highest cereal-production growth rate, at 2.0 percent, while Japan will face the lowest growth in production at a negative 0.7 percent per year during 1993–2010.

Production trends can better be understood by looking at their component parts, area and yield. Table A.8 provides an overview on the levels of cereal-crop area in 1993 and 2010, Table

Table A.7: Projected Annual Production Growth Rates for Cereals, 1993–2010 (percent per year)

	Wheat	Maize	Rice	Other Grains	All Cereals
East Asia	0.95	1.97	0.71	0.40	1.15
PRC	0.95	1.97	0.74	0.39	1.17
Republic of Korea	—	1.64	0.20	0.10	0.22
Other East Asia	1.44	2.21	0.10	1.12	1.13
South Asia	1.91	2.11	1.79	1.08	1.78
India	1.93	2.09	1.75	1.10	1.75
Pakistan	1.91	2.38	2.09	0.62	1.94
Bangladesh	1.04	1.80	1.77	0.77	1.73
Other South Asia	1.72	1.95	2.42	0.60	2.15
Southeast Asia	0.75	2.27	1.63	0.63	1.74
Indonesia	—	1.85	1.15	—	1.28
Thailand	1.12	2.15	1.17	0.49	1.39
Malaysia	—	1.67	1.56	—	1.56
Philippines	—	3.04	2.35	—	2.66
Viet Nam	—	1.87	1.87	1.39	1.87
Myanmar	0.75	1.93	2.52	0.85	2.46
Other Southeast Asia	—	1.81	2.80	—	2.75
Latin America	2.61	2.17	2.39	1.77	2.23
WANA	2.10	2.07	2.42	2.76	2.30
Sub-Saharan Africa	2.85	3.09	3.20	3.08	3.09
Developing	1.64	2.19	1.41	2.03	1.75
Developed	1.10	1.19	0.46	0.74	1.02
Japan	0.60	0.82	-0.77	-0.21	-0.65
World	1.35	1.65	1.37	1.18	1.41

Source: IFPRI IMPACT.

A.9 presents the projected annual cereal yield growth rates for the same period, and Table A.10 shows the projected cereal yield levels. Only 43 million hectares (ha) (6 percent) will be added to the global cereal crop area of 700 million ha between 1993 and 2010. The largest increases will be in maize (13 million ha) and other coarse grains (15 million ha). Rice area, on the other hand, will be virtually stagnant, with an increase of only 5 million ha by 2010. More than half of this increase will be in the group of Asian developing countries. The major crop-area expansion will occur in the group of developing countries at 35 million ha. Sub-Saharan Africa will experience the lion's share of area expansion with 19 million ha, 44 percent of the global increase, with the additional area mainly planted to other coarse grains (62 percent) and maize (27 percent). Crop area will increase by 6 million ha each in developing Asia and Latin America, and by 3 million ha in WANA. In the group of developed countries, cereal-crop area will stabilize after falling significantly after 1982 due to declining cereal prices, policies to remove land from production, and the economic collapse in much of Eastern Europe and the former Soviet Union. The annual area growth rate will be 0.2 percent for wheat and 0.3 percent for maize. These projected expansions in cereal-crop area are small compared to the projected increases in production.

The projected slow growth in the expansion of crop area places the burden of meeting future cereal demands on crop yield growth. Although yield growth will vary considerably by commodity and country, in general, rates of growth in crop yields are projected to continue to decline compared to the already reduced rates of the 1982–93 period. The global cereal yield-growth rate is projected to decline from 1.7 percent per year in 1982–93 to 1.1 percent in 1993–2010.

In the group of developed countries, yield growth will slow down or stagnate for each of the cereals. Wheat yields are projected to increase by 0.89 percent per year in 1993–2010, on average, compared with an annual growth of 1.74 percent during 1982–93. Maize yield growth will slow from 1.35 percent per year in 1982–93 to 0.87 percent per year in 1993–2010, and yields for other coarse grains will decline from 1.29 percent to 0.67

Table A.8: Cereal Crop Area, 1993 and Projected 2010 (million hectares)

	Wheat 1993	Wheat 2010	Maize 1993	Maize 2010	Rice 1993	Rice 2010	Other Grains 1993	Other Grains 2010	All Cereals 1993	All Cereals 2010
East Asia	30.5	30.4	21.7	23.2	33.0	32.4	7.2	7.1	92.4	93.1
PRC	29.9	29.8	21.0	22.4	30.9	30.4	6.9	6.8	88.6	89.5
Republic of Korea	—	—	0.0	0.0	1.1	1.0	0.1	0.1	1.3	1.2
Other East Asia	0.6	0.6	0.7	0.7	1.0	1.0	0.2	0.2	2.5	2.5
South Asia	33.5	35.0	7.7	7.9	56.2	58.0	28.7	28.7	126.1	129.6
India	24.3	25.3	6.0	6.1	41.8	43.1	27.3	27.4	99.4	101.8
Pakistan	8.1	8.5	0.9	0.9	2.2	2.3	1.0	1.0	12.1	12.7
Bangladesh	0.6	0.6	0.0	0.0	10.0	10.2	0.1	0.1	10.7	11.0
Other South Asia	0.6	0.6	0.8	0.8	2.2	2.4	0.3	0.3	3.9	4.1
Southeast Asia	0.1	0.2	8.5	8.8	38.2	39.6	0.4	0.4	47.2	48.9
Indonesia	0.0	0.0	3.2	3.3	11.0	11.1	0.0	0.0	14.2	14.4
Thailand	0.0	0.0	1.3	1.3	8.9	9.1	0.2	0.2	10.3	10.6
Malaysia	0.0	0.0	0.0	0.0	0.7	0.7	0.0	0.0	0.7	0.7
Philippines	0.0	0.0	3.2	3.3	3.4	3.6	0.0	0.0	6.6	6.9
Viet Nam	0.0	0.0	0.5	0.5	6.5	6.8	0.0	0.0	7.1	7.4
Myanmar	0.1	0.2	0.1	0.2	5.4	5.9	0.2	0.3	5.9	6.5
Other Southeast Asia	0.0	0.0	0.1	0.1	2.3	2.4	0.0	0.0	2.3	2.5
Latin America	8.3	9.7	27.8	31.5	6.8	7.5	4.9	5.6	47.9	54.3
WANA	28.6	30.0	2.4	2.6	1.5	1.6	23.1	24.9	55.6	59.1
Sub-Saharan Africa	1.1	1.4	20.1	25.2	6.1	7.8	35.2	46.7	62.4	81.1
Developing	102.2	106.7	88.2	99.0	141.7	147.0	99.5	113.5	431.6	466.2
Developed	118.2	122.5	47.0	49.5	4.4	4.2	98.9	100.1	268.5	276.3
Japan	0.2	0.2	0.0	0.0	2.2	1.8	0.1	0.1	2.4	2.1
World	220.4	229.2	135.1	148.5	146.1	151.2	198.4	213.6	700.0	742.5

Source: IFPRI IMPACT.

Table A.9: Projected Annual Cereal Yield Growth Rates, 1993–2010 (percent per year)

	Wheat	Maize	Rice	Other Grains	All Cereals
East Asia	0.96	1.56	0.82	0.45	1.10
PRC	0.96	1.56	0.83	0.45	1.11
Republic of Korea	—	1.55	0.67	0.99	0.72
Other East Asia	1.23	1.81	0.58	0.53	1.11
South Asia	1.65	2.01	1.60	1.07	1.61
India	1.69	2.05	1.57	1.08	1.60
Pakistan	1.59	2.16	1.69	0.79	1.65
Bangladesh	0.95	1.45	1.61	0.79	1.57
Other South Asia	1.51	1.60	2.05	0.71	1.85
Southeast Asia	0.04	2.05	1.41	0.44	1.52
Indonesia	—	1.63	1.10	—	1.19
Thailand	0.99	2.03	1.02	0.91	1.25
Malaysia	—	1.34	1.33	—	1.33
Philippines	—	2.81	2.10	—	2.41
Viet Nam	—	1.46	1.62	0.67	1.60
Myanmar	0.03	1.45	2.01	0.25	1.94
Other Southeast Asia	—	1.22	2.44	—	2.38
Latin America	1.68	1.44	1.78	0.98	1.47
WANA	1.83	1.60	1.85	2.32	1.94
Sub-Saharan Africa	1.43	1.74	1.68	1.37	1.52
Developing	1.38	1.49	1.20	1.24	1.28
Developed	0.89	0.87	0.72	0.67	0.85
Japan	0.67	1.35	0.35	0.21	0.36
World	1.12	1.08	1.17	0.74	1.06

Source: IFPRI IMPACT.

Table A.10: Cereal Crop Yields, 1993 and Projected 2010 (metric tons per hectare)

	Wheat 1993	Wheat 2010	Maize 1993	Maize 2010	Rice 1993	Rice 2010	Other Grains 1993	Other Grains 2010	All Cereals 1993	All Cereals 2010
East Asia	3.4	4.0	4.7	6.1	4.1	4.7	2.3	2.5	3.8	4.6
PRC	3.4	4.0	4.7	6.2	4.1	4.7	2.3	2.5	3.9	4.7
Republic of Korea	—	—	4.2	5.4	4.4	4.9	3.4	4.0	4.3	4.9
Other East Asia	1.0	1.2	3.5	4.8	3.1	3.4	1.9	2.0	2.6	3.2
South Asia	2.2	3.0	1.6	2.2	1.8	2.4	0.9	1.0	1.7	2.2
India	2.4	3.1	1.6	2.3	1.9	2.4	0.9	1.0	1.7	2.2
Pakistan	1.9	2.5	1.4	2.0	1.7	2.2	0.6	0.7	1.7	2.3
Bangladesh	1.8	2.2	3.7	4.7	1.8	2.3	0.7	0.8	1.8	2.3
Other South Asia	1.3	1.7	1.6	2.1	1.7	2.4	1.1	1.2	1.6	2.1
Southeast Asia	1.0	1.0	2.0	2.8	2.1	2.7	1.0	1.1	2.1	2.7
Indonesia	0.0	0.0	2.2	2.9	3.0	3.6	0.0	0.0	2.8	3.4
Thailand	1.0	1.2	2.9	4.1	1.5	1.8	1.5	1.8	1.7	2.0
Malaysia	0.0	0.0	1.8	2.3	2.0	2.5	0.0	0.0	2.0	2.5
Philippines	0.0	0.0	1.5	2.4	1.9	2.6	0.0	0.0	1.7	2.5
Viet Nam	0.0	0.0	1.7	2.2	2.3	3.0	1.5	1.7	2.2	2.9
Myanmar	1.0	1.0	1.6	2.0	1.8	2.5	0.6	0.7	1.7	2.4
Other Southeast Asia	0.0	0.0	1.9	2.3	1.0	1.6	0.0	0.0	1.1	1.6
Latin America	2.3	3.0	2.4	3.1	1.9	2.5	2.7	3.2	2.4	3.0
WANA	1.8	2.4	3.8	5.0	3.3	4.5	1.0	1.5	1.6	2.2
Sub-Saharan Africa	1.5	2.0	1.2	1.6	1.0	1.4	0.8	1.0	0.9	1.2
Developing	2.4	3.1	2.6	3.4	2.4	2.9	1.1	1.3	2.2	2.7
Developed	2.6	3.0	6.3	7.3	4.0	4.6	2.3	2.6	3.1	3.6
Japan	3.6	4.0	1.0	1.3	4.3	4.6	5.5	5.7	4.3	4.6
World	2.5	3.0	3.9	4.7	2.5	3.0	1.7	1.9	2.5	3.0

Source: IFPRI IMPACT.

percent annually in the same period. In 2010, however, a yield gap in cereals between developed and developing countries will persist (Table A.10): yields for maize and other coarse grains in developed countries will still be about twice the levels of developing-country yields; rice yields are expected to be 60 percent higher than in developing countries. Wheat yield levels in 2010 are projected to vary widely in both developing and developed countries, with only a slight convergence in yield levels over the projection period.

Compared to developed-country yield-growth rates, growth rates will be much stronger in developing countries, albeit starting from lower levels. However, they will experience a drop-off similar to that seen in developed countries from 1982–93 levels: wheat yields are projected to grow at 1.38 percent per year, compared to 2.64 percent per year in 1982–93; maize yields will increase at 1.49 percent annually, compared to 2.21 percent per year in the earlier period; and rice yields will grow at 1.20 percent per year, compared to 1.86 percent per year in 1982–93. Cereal-crop yields will increase, on average, from 2.2 mt per ha in 1993 to 2.7 mt per ha in 2010; the biggest increases will be in maize yields, from 2.6 mt per ha to 3.4 mt per ha, and wheat yields, from 2.4 mt per ha to 3.1 mt per ha. In the group of developing Asian countries, cereal yields are projected to increase from 2.5 mt per ha in 1993 to 3.1 mt per ha in 2010 and rice yields from 2.5 mt per ha to 3.0 mt per ha during the same period.

PROJECTIONS FOR LIVESTOCK PRODUCTS

Projected Demand for Livestock: Upward Trends Continue to Close the Gap Between Developed and Developing Countries

As shown in Table A.11, per capita demand for meat will grow rapidly in developing countries, particularly in the PRC and East and Southeast Asia. The PRC's per capita meat demand is projected to increase to 51 kg in 2010, from 33 kg in 1993. This level of meat consumption is substantially higher than the level projected for Japan (46 kg per capita), and is closing the gap between it and and the consumption levels of the developed countries, where per capita meat consumption is projected to increase only slightly, from 78 kg to 82 kg. For developing countries as a whole, per capita meat demand is projected to increase from 21 kg to 27 kg.

The strongest projected increase will be in pork and poultry consumption.The per capita meat-consumption levels in the developed countries, however, will still be three times those of the levels of the developing countries by 2010. The United States remains the highest per capita consumer of livestock products at 122 kg in 2010, with an average annual per capita consumption of 42 kg of beef, 30 kg of pork, 48 kg of poultry and 1 kg of sheep and goat meat. Per capita meat consumption is also important in Australia, the only country with significant sheep and goat meat consumption, at 24 kg per capita by 2010.

With continued population growth and strong per capita demand growth, the total demand for livestock products will grow very rapidly in developing countries. The PRC, South Asia, Southeast Asia, Sub-Saharan Africa, and WANA will all experience growth rates in demand at or above 3 percent per year during 1993–2010, with an average growth for all developing countries of 3.16 percent per year, compared to the 0.66 percent per year for the developed countries. These rates of growth are lower than the 5.49 percent per year during 1982–93, but are still substantial. In 1993, developing countries accounted for 47 percent of world meat consumption (Table A.12); in 2010, this figure is projected to jump to 58 percent; developing Asia's share in world meat demand alone will be 40 percent in 2010. In developed countries, on the other hand, demand for livestock products will only increase by 12 million mt. Growth rates in demand for beef and sheep and goat meat will be slightly higher in the 1993–2010 period than in the recent past, whereas the growth rate for poultry and pork will be slower than in 1982–93.

Table A.11: Per Capita Demand For All Meat Products, 1993 and Projected 2010 (kilograms per capita)

	Beef 1993	Beef 2010	Pork 1993	Pork 2010	Sheep & Goat 1993	Sheep & Goat 2010	Poultry 1993	Poultry 2010	All Meats 1993	All Meats 2010
East Asia	2.5	4.0	24.3	36.6	1.2	1.5	5.7	9.4	33.7	51.4
PRC	2.1	3.6	24.5	37.4	1.2	1.4	5.0	8.6	32.8	51.1
Republic of Korea	9.3	12.7	17.6	25.7	0.3	0.4	8.7	13.8	35.9	52.5
Other East Asia	5.2	5.8	25.4	28.8	2.5	2.7	17.7	20.8	50.8	58.2
South Asia	2.9	3.9	0.4	0.5	1.1	1.4	0.7	1.0	5.0	6.7
India	2.6	3.6	0.5	0.6	0.7	0.8	0.5	0.8	4.3	5.8
Pakistan	5.8	7.0	0.0	0.0	4.5	4.8	1.6	2.1	11.8	13.8
Bangladesh	1.3	1.6	0.0	0.0	0.8	0.9	0.8	1.3	2.9	3.8
Other South Asia	3.8	4.5	0.3	0.3	0.9	1.0	0.6	0.8	5.7	6.6
Southeast Asia	2.6	3.7	6.7	8.9	0.4	0.4	5.5	7.5	15.1	20.5
Indonesia	1.8	2.6	3.2	4.3	0.5	0.6	3.5	4.8	9.0	12.3
Thailand	4.7	7.8	5.8	8.3	0.0	0.0	10.6	15.8	21.2	32.0
Malaysia	4.2	6.3	11.3	14.4	0.5	0.5	30.8	37.0	46.8	58.2
Philippines	2.6	3.6	13.7	17.4	0.8	0.9	5.4	7.7	22.5	29.6
Viet Nam	2.4	3.7	12.1	16.0	0.0	0.1	2.5	3.8	17.0	23.5
Myanmar	2.5	3.3	2.0	2.5	0.2	0.2	2.1	2.9	6.8	8.8
Other Southeast Asia	3.2	3.9	6.5	7.3	0.0	0.0	2.2	2.8	11.9	14.0
Latin America	22.2	25.7	7.5	8.7	1.0	1.2	15.2	19.2	45.9	54.7
WANA	6.2	7.0	0.1	0.1	5.4	5.9	8.0	9.4	19.7	22.5
Sub-Saharan Africa	4.1	5.0	1.2	1.4	1.7	1.8	1.9	2.2	8.8	10.4
Developing	5.3	6.7	9.0	11.8	1.5	1.8	5.0	6.8	20.8	27.0
Developed	25.2	25.9	29.4	29.7	2.8	3.0	20.3	23.4	77.7	81.9
Japan	9.8	11.1	15.6	16.6	0.5	0.5	14.4	17.7	40.3	45.9
World	9.8	10.4	13.7	15.3	1.8	2.0	8.5	10.0	33.9	37.8

Source: IFPRI IMPACT.

Appendix 475

Table A.12: Demand for Livestock Products, 1993 and Projected 2010 (million metric tons)

	Beef 1993	Beef 2010	Pork 1993	Pork 2010	Sheep & Goat 1993	Sheep & Goat 2010	Poultry 1993	Poultry 2010	All Meats 1993	All Meats 2010
East Asia	3.1	5.9	30.9	53.4	1.6	2.1	7.2	13.6	42.8	75.0
PRC	2.4	4.9	28.8	50.2	1.4	1.9	5.9	11.6	38.6	68.6
Republic of Korea	0.4	0.6	0.8	1.3	0.0	0.0	0.4	0.7	1.6	2.6
Other East Asia	0.3	0.4	1.3	1.9	0.1	0.2	0.9	1.3	2.7	3.7
South Asia	3.4	6.2	0.4	0.7	1.3	2.1	0.8	1.6	6.0	10.6
India	2.4	4.2	0.4	0.7	0.6	0.9	0.5	0.9	3.8	6.8
Pakistan	0.8	1.4	0.0	0.0	0.6	1.0	0.2	0.4	1.6	2.8
Bangladesh	0.1	0.2	0.0	0.0	0.1	0.1	0.1	0.2	0.3	0.6
Other South Asia	0.1	0.3	0.0	0.0	0.0	0.1	0.0	0.0	0.2	0.4
Southeast Asia	1.2	2.2	3.1	5.3	0.2	0.2	2.5	4.4	7.0	12.2
Indonesia	0.3	0.6	0.6	1.0	0.1	0.1	0.7	1.2	1.7	2.9
Thailand	0.3	0.5	0.3	0.5	0.0	0.0	0.6	1.0	1.2	2.1
Malaysia	0.1	0.2	0.2	0.4	0.0	0.0	0.6	1.0	0.9	1.5
Philippines	0.2	0.3	0.9	1.5	0.0	0.1	0.4	0.7	1.5	2.6
Viet Nam	0.2	0.3	0.9	1.5	0.0	0.0	0.2	0.4	1.2	2.2
Myanmar	0.1	0.2	0.1	0.1	0.0	0.0	0.1	0.2	0.3	0.5
Other Southeast Asia	0.0	0.1	0.1	0.1	0.0	0.0	0.0	0.1	0.2	0.3
Latin America	10.3	15.2	3.5	5.1	0.5	0.7	7.0	11.4	21.2	32.3
WANA	2.3	3.7	0.0	0.1	2.0	3.2	2.9	5.0	7.3	12.0
Sub-Saharan Africa	2.1	4.1	0.6	1.1	0.8	1.4	1.0	1.8	4.5	8.5
Developing	22.4	37.3	38.6	65.7	6.4	9.9	21.5	37.9	88.9	150.9
Developed	32.2	35.1	37.5	40.3	3.6	4.1	25.9	31.6	99.3	111.0
Japan	1.2	1.4	1.9	2.1	0.1	0.1	1.8	2.3	5.0	5.8
World	54.6	72.4	76.1	106.0	10.1	14.0	47.4	69.6	188.2	261.9

Source: IFPRI IMPACT.

Projected Production Growth in Livestock Products: Importance of Numbers over Yield

Driven by the rapid increases in demand for livestock products, meat production is expected to increase much more rapidly than cereal production in the coming decades. However, as with the case of cereals, growth in global meat production will slow down compared to past trends, growing at 1.96 percent per year in 1993–2010 (Table A.13), compared to 2.82 percent annually in 1982–93. Global meat production is projected to increase by 74 million mt, from 188 million mt in 1993 to 262 million mt in 2010. In developing countries, the annual rate of growth of meat production will decline from the rapid 5.51 percent in 1982–93 to a still strong 3.08 percent in 1993–2010. This will be accounted for mainly by pork (46 percent) and poultry (27 percent). The group of Southeast Asian countries will show the greatest production growth, at 3.58 percent, with some countries, notably Indonesia and the Philippines, reaching rates close to 4 percent. Growth will be rapid in all developing Asian economies, at 3.36 percent per year, on average. In the developed countries, the production growth rate is expected to decline from 1.05 percent per year in 1982–93 to 0.78 percent per year in 1993–2010, with a livestock production increase of 14 million mt.

More than 80 percent of the increase in global pork production, more than 70 percent of the growth in global beef and poultry production, and more than 60 percent of the increase in global production of sheep and goat meat during 1993–2010 will be accounted for by growth in the numbers of animals slaughtered (Table A.14). Numbers growth is driven mainly by the rapid expansion of developing-country livestock production. Several Asian countries will experience dramatic annual growth in the numbers of livestock slaughtered: in Southeast Asia, numbers growth for beef is projected at 2.77 percent per year; for pork, at 2.72 percent annually; and for poultry, at 2.95 percent per year. The PRC will experience growth in the numbers of cattle and chicken slaughtered in excess of 3 percent annually. In the Republic of Korea, numbers growth is expected to be largest for pork, at 2.77 percent per year during 1993–2010.

Table A.13: Projected Annual Growth Rates of Livestock Production, 1993–2010 (percent per year)

	Beef	Pork	Sheep & Goat	Poultry	All Meats
East Asia	3.80	3.21	1.63	3.92	3.32
PRC	3.94	3.23	1.68	4.06	3.36
Republic of Korea	2.81	3.17	0.14	3.58	3.23
Other East Asia	2.12	2.75	0.99	2.54	2.57
South Asia	3.50	2.46	2.61	3.69	3.27
India	3.61	2.50	2.06	4.04	3.33
Pakistan	3.88	—	3.24	3.16	3.55
Bangladesh	1.37	—	2.48	3.22	2.25
Other South Asia	1.06	1.33	1.31	2.97	1.35
Southeast Asia	3.50	3.44	2.34	3.85	3.58
Indonesia	4.13	3.48	3.08	4.13	3.81
Thailand	3.34	3.54	1.09	3.75	3.62
Malaysia	2.97	3.15	—	3.42	3.32
Philippines	3.49	3.82	1.24	4.60	3.92
Viet Nam	3.12	3.18	0.14	3.88	3.27
Myanmar	3.06	3.14	0.28	3.23	3.08
Other Southeast Asia	2.94	2.55	—	2.44	2.65
Latin America	2.11	2.35	2.24	2.68	2.34
WANA	2.58	1.85	2.64	2.84	2.71
Sub-Saharan Africa	3.66	3.58	2.70	3.72	3.48
Developing	2.84	3.15	2.36	3.37	3.08
Developed	0.74	0.40	1.27	1.26	0.78
Japan	0.43	0.11	—	0.69	0.41
World	1.67	1.97	1.95	2.28	1.96

Source: IFPRI IMPACT.

Table A.14: Projected Annual Numbers and Yield Growth Rates for Livestock Products, 1993–2010 (percent per year)

	Beef Num.	Beef Yield	Pork Num.	Pork Yield	Sheep & Goat Num.	Sheep & Goat Yield	Poultry Num.	Poultry Yield
East Asia	3.05	0.73	2.75	0.45	1.24	0.38	2.98	0.91
PRC	3.26	0.66	2.77	0.44	1.28	0.39	3.10	0.93
Republic of Korea	1.89	0.91	2.77	0.40	-0.13	0.27	2.34	1.21
Other East Asia	1.33	0.78	2.24	0.50	0.47	0.52	1.76	0.77
South Asia	2.54	0.94	1.56	0.89	1.30	1.29	2.71	0.96
India	2.84	0.75	1.60	0.88	1.34	0.71	3.10	0.92
Pakistan	2.99	0.86	—	—	1.31	1.91	2.02	1.12
Bangladesh	0.60	0.77	—	—	1.26	1.21	2.48	0.72
Other South Asia	0.43	0.63	0.34	0.98	1.17	0.13	2.08	0.88
Southeast Asia	2.77	0.71	2.72	0.70	1.53	0.80	2.95	0.87
Indonesia	3.34	0.76	2.64	0.82	1.92	1.14	2.97	1.13
Thailand	2.68	0.65	2.40	1.11	0.38	0.72	2.86	0.86
Malaysia	2.11	0.84	3.06	0.10	0.39	—	2.70	0.70
Philippines	2.48	0.99	3.40	0.41	0.39	0.85	3.77	0.80
Viet Nam	2.42	0.68	2.54	0.63	0.05	0.09	3.29	0.57
Myanmar	2.46	0.59	2.33	0.79	0.06	0.23	2.65	0.57
Other Southeast Asia	2.22	0.70	1.71	0.83	0.05	—	1.92	0.51
Latin America	1.60	0.50	1.40	0.94	1.10	1.13	1.87	0.80
WANA	1.66	0.91	1.47	0.37	1.83	0.79	1.59	1.23
Sub-Saharan Africa	2.53	1.10	2.68	0.88	1.99	0.69	2.71	0.98
Developing	2.15	0.67	2.60	0.54	1.53	0.83	2.45	0.90
Developed	0.27	0.47	0.17	0.23	0.54	0.73	0.76	0.50
Japan	0.25	0.18	-0.17	0.28	-0.00	—	0.40	0.29
World	1.20	0.47	1.63	0.33	1.22	0.72	1.65	0.62

Source: IFPRI IMPACT.

In developed countries, on the other hand, weight growth (carcass weight per animal) will drive production growth, at 0.47 percent annually for beef, and 0.23 percent annually for pork during 1993–2010. Numbers growth outweighs yield growth only for poultry, at 0.76 percent per year.

PROJECTIONS FOR WORLD FOOD PRICES AND INTERNATIONAL TRADE

International Prices for Cereals and Livestock Products: Light to Moderate Downturns

The world food-price implications of these projected outcomes are summarized in Table A.15. The baseline projection results of IMPACT indicate that food production in the world will grow fast enough for real world prices of food to be falling (but very slowly), or to increase only slightly, compared to the past two decades. Projected real cereal prices will be particularly strong through the year 2010. Over the 17-year period, world wheat prices are projected to decline by 1.4 percent, compared to the much bigger drop of 28 percent between 1982 and 1995. Rice prices will increase by 5.9 percent compared to the earlier decline of 42 percent, but when the projections are extended to 2020, the rice price begins to fall again and will decline by 7.3 percent to 2020 compared to 1993. Maize prices will also increase to 2010 by 0.8 percent (and fall by 3.1 percent to 2020), compared to the 43 percent drop between 1982 and 1995. Finally, prices for other coarse grains will experience the biggest decline, 4.9 percent from 1993 values. The weighted average price for all cereals will increase by 1.2 percent by 2010 (but fall by 7.9 percent over the 1993–2020 period). It is only after 2010 that the combination of continued declines in the rate of population growth and declining income elasticities for cereals will reduce demand growth enough to cause real cereal prices to decline more sharply.

Table A.15: Real World Prices for Selected Commodities, 1993 and 2010
(US dollars per metric ton)

	World prices 1993	World prices 2010	Percent Change 1993–2010
Wheat	148	146	-1.35
Maize	126	127	0.79
Rice	286	303	5.94
Other Grains	122	116	-4.92
All Cereals	164	166	1.22
Beef	2,023	1,835	-9.29
Pork	1,366	1,260	-7.76
Sheep & Goat	2,032	1,915	-5.76
Poultry	1,300	1,175	-9.62
All Meat	1,576	1,431	-9.20

Source: IFPRI IMPACT.

Prices for livestock commodities will drop more sharply throughout the 1993–2010 period, with price declines in the range of 7.5 to 10 percent. The average meat price will decline by 9.2 percent between 1993 and 2010. Beef and pork prices both declined by 22 percent between 1982 and 1991, but will decline by only 9.3 and 7.8 percent, respectively, during 1993–2010. The price for sheep and goat meat is expected to decrease by 5.8 percent during 1993–2010, compared to a drop in the proxy price of lamb of 33 percent between 1982 and 1991. The price of poultry is expected to decrease by 9.6 percent, compared to a decline of 18 percent between 1982 and 1991.

International Trade in Cereals and Livestock Products: A Higher Volume of Trade With More Cereal Imports and Some Livestock Exports for Asia

The slow decline in cereal and livestock prices will be accompanied by rapidly increasing world trade in food, with the primary impetus for expanded trade being generated by the group of developing countries increasing its food imports from developed countries. World trade in cereals is projected to increase by 73 million mt, from 185 million mt in 1993 to 259 million mt in 2010. Among the cereals, wheat imports will show the biggest absolute increase, 40 million mt, from 91 million mt in 1993 to 131 million mt in 2010. Net cereal imports by developing countries will increase by nearly 90 percent, from 94 million mt in 1993 to 177 million mt in 2010, with Asia accounting for the largest increase (74 percent), followed by WANA (23 percent) (Table A.16).

The increase in exports of maize from developed to developing countries is the highest in percentage terms, with a more than 150 percent increase from 18 million mt in 1993 to 46 million mt in 2010. It should be noted that trade among developed countries causes the aggregate exports of maize (and other commodities) from developed countries to be lower than the exports of some individual developed countries. The developed countries will also substantially increase exports of wheat and other coarse grains to 2010 and increase their imports of rice from developing countries by 5 million mt.

The major beneficiaries of increased cereal import demand from developing countries will be the main cereal exporters, particularly the United States, whose cereal exports are expected to increase by 38 percent, from 85 million mt in 1993 to 117 million mt in 2010. In addition, Eastern Europe and the former Soviet Union are expected to take advantage of the increased import demands of developing countries, shifting from a large net importing position of 27 million mt in 1993 to become net exporters (of 9 million mt) by 2010. Removal of food subsidies and other price distortions, combined with sharply lower incomes, have already resulted in falling per capita cereal consumption in these regions. Feeding-efficiency improvements in the livestock industry and a projected gradual recovery in incomes will cause production growth to outstrip demand growth.

World net trade in livestock products will expand proportionally even more rapidly than the cereal trade, although from much lower levels. Global net trade is projected to increase from 8 million mt in 1993 to 11.5 million mt in 2010, an increase of 42 percent over 1993 levels. Developing countries will dramatically increase their imports of livestock products.

Table A.16: Net Trade for All Cereal Crops, 1993 and Projected 2010 (million metric tons)

	1993					2010				
	Wheat	Maize	Rice	O Grains	Total	Wheat	Maize	Rice	O Grains	Total
East Asia	-15.1	-4.6	0.5	-1.8	-21.0	-30.6	-27.6	-2.0	-4.7	-64.9
PRC	-8.8	7.8	0.9	-0.9	-0.9	-22.1	-9.2	-0.7	-3.4	-35.3
Republic of Korea	-4.6	-6.2	-0.1	-0.4	-11.2	-6.3	-9.1	-0.4	-0.5	-16.3
Other East Asia	-1.7	-6.3	-0.4	-0.5	-8.9	-2.3	-9.3	-0.9	-0.8	-13.3
South Asia	-4.9	-0.0	1.6	0.1	-3.3	-16.6	-1.1	0.7	-1.9	-18.9
India	0.4	0.0	0.7	0.1	1.3	-3.3	-0.7	0.2	-1.3	-5.2
Pakistan	-3.1	-0.0	1.3	-0.0	-1.8	-9.1	-0.1	1.4	-0.3	-8.1
Bangladesh	-1.3	-0.0	-0.1	-0.0	-1.5	-2.3	-0.0	-0.2	-0.0	-2.6
Other South Asia	-0.9	-0.0	-0.3	-0.0	-1.3	-1.8	-0.3	-0.7	-0.2	-3.0
Southeast Asia	-6.7	-2.6	6.5	-0.5	-3.2	-10.0	-6.0	11.5	-0.8	-5.2
Indonesia	-2.8	-0.5	-0.4	-0.0	-3.7	-4.1	-1.4	-1.5	-0.1	-7.1
Thailand	-0.6	0.0	5.5	-0.1	4.9	-0.7	-0.9	7.6	-0.2	5.8
Malaysia	-1.0	-2.0	-0.6	-0.1	-3.7	-1.5	-3.3	-0.9	-0.1	-5.9
Philippines	-2.0	-0.2	-0.2	-0.2	-2.6	-3.0	-0.6	0.4	-0.3	-3.5
Viet Nam	-0.3	0.1	1.9	-0.0	1.6	-0.5	0.1	3.3	-0.0	2.8
Myanmar	-0.0	0.0	0.4	-0.0	0.4	-0.1	0.0	2.6	-0.0	2.5
Other Southeast Asia	-0.0	0.0	-0.1	-0.0	-0.2	-0.0	-0.0	0.2	-0.0	0.2
Latin America	-8.6	-1.4	-1.7	-4.4	-16.0	-6.6	-1.7	-0.1	-7.2	-12.2
WANA	-22.1	-6.7	-3.3	-5.6	-37.7	-35.0	-10.1	-4.6	-7.4	-57.0
Sub-Saharan Africa	-5.2	-2.5	-3.4	-0.5	-11.6	-9.1	-2.4	-5.5	-1.0	-18.0
Developing	-62.8	-17.8	0.0	-12.7	-93.9	-108.3	-45.5	5.0	-23.0	-177.2
Developed	62.8	17.8	0.0	12.7	93.9	108.3	45.5	-5.0	23.0	177.2
Japan	-5.8	-16.6	-0.4	-6.3	-29.1	-6.6	-16.8	-1.4	-6.7	-31.5

Source: IFPRI IMPACT.

WANA will be the primary importer of livestock products by 2010, at 2.8 million mt. Some Latin American countries, on the other hand, will improve their exporter positions: Brazil, for example, will increase its net meat exports from 0.8 million mt to 1.2 million mt by 2010. Although Asia has long been considered to be a primary region for livestock imports, this situation has dramatically changed as a consequence of the Asian economic crisis. A combination of declining income growth and increased domestic prices has led to a more mixed picture, with some countries, especially in Southeast Asia, projected to be significant net exporters by 2010. Again, the developed countries will satisfy most of the increased importing demands, led by Australia, at 2.0 million mt, followed by the United States, at 1.5 million mt. Japan, on the other hand, is expected to remain the largest net livestock importing country, at 2.4 million mt in 2010 (Table A.17).

PROJECTIONS OF MALNUTRITION AMONG PRE-SCHOOL CHILDREN

Table A.18 shows that the projected number of malnourished pre-school children (less than 5 years of age) under the baseline scenario is expected to decline in developing countries by 20 million, from 185 million to 165 million children—a drop of only 11 percent. Despite the availability of food on a global level to meet effective demand at slowly declining world prices, food security in one region, Sub-Saharan Africa, will actually worsen, with the absolute number of malnourished children expected to grow from 27 million in 1993 to 36 million by 2010. Another region, WANA, shows no improvement in numbers of malnourished children (8 million in both 1993 and 2010). Within Asia, numbers of malnourished children are expected to decline in Southeast Asia (by 3 million in the 1993–2010 period), and the PRC (by 7 million in the same period). Within South Asia, although India shows significant progress in reducing malnutrition, it is projected

Table A.17: Net Trade for Livestock Products, 1993 and Projected 2010 (thousand metric tons)

	1993					2010				
	Beef	Pork	Sheep & Goat	Poultry	All Meats	Beef	Pork	Sheep & Goat	Poultry	All Meats
East Asia	-313	663	-20	-204	126	-587	667	-100	-162	-182
PRC	57	676	6	111	850	-54	391	-50	234	521
Republic of Korea	-217	-4	-11	-21	-253	-326	32	-16	-31	-341
Other East Asia	-153	-9	-15	-294	-471	-208	244	-35	-364	-363
South Asia	101	2	10	-4	109	149	-87	-23	-126	-87
India	99	1	9	1	110	269	-84	-53	-5	127
Pakistan	1	0	0	-5	-4	32	0	42	-77	-3
Bangladesh	0	0	0	0	0	-62	0	1	-36	-97
Other South Asia	1	1	1	0	3	-89	-3	-13	-9	-114
Southeast Asia	-151	195	-24	120	140	-350	567	-20	603	800
Indonesia	-47	69	-14	-111	-103	-29	181	11	-51	112
Thailand	0	0	0	178	178	-33	63	0	449	479
Malaysia	-67	116	-9	54	94	-143	187	-13	172	203
Philippines	-44	0	-1	-2	-47	-97	134	-14	67	90
Viet Nam	0	9	0	1	10	-47	-2	-1	-15	-65
Myanmar	0	0	0	0	0	-7	9	-3	-10	-11
Other Southeast Asia	7	1	0	0	8	6	-5	0	-9	-8
Latin America	690	-152	-36	112	614	453	-216	-43	-167	27
WANA	-669	-4	-336	-388	-1,397	-1,233	-15	-603	-942	-2,793
Sub-Saharan Africa	-14	-29	36	-97	-104	-241	-53	-45	-181	-520
Developing	-382	672	-423	-480	-613	-1,877	820	-934	-1,022	-3,013
Developed	382	-672	423	480	613	1,877	-820	934	1,022	3,013
Japan	-637	-628	-57	-456	-1,778	-772	-768	-59	-754	-2,353

Source: IFPRI IMPACT.

Table A.18: Number of Malnourished Children Under 5 Years of Age, 1993 and Projected 2010

	1993	2010
	(million children)	
Latin America	10	8
Sub-Saharan Africa	27	36
WANA	8	8
India	76	59
Other South Asia	24	24
Southeast Asia	16	13
PRC	24	17
Developing	185	165

Source: IFPRI IMPACT.

to remain the home of 59 million malnourished children by 2010, 40 percent of the total in developing regions. At the same time, moreover, child malnutrition figures for the rest of South Asia are projected to hold steady at 24 million. In short, under current likely trends, the ability to meet food needs on a global scale will continue to fail to translate into meeting food needs for each individual, and those living in certain parts of the developing world will continue to suffer.

Author Index

Abrol, I.P., 305
ADB, 233, 236, 237, 241, 247, 267, 269, 271, 272, 276, 277, 278
Adelman, I., 26
Ahmed, R., 113
Ahuja, V., 31, 38, 39, 44, 46, 230
Alexandratos, N., 294
Ali, M., 153
Alston, J.M., 151, 159
Ammar Siamwalla, 65, 206
Amsden, Alice, 10
Anderson, D., 82
Anderson, J. R., 195, 328
Anderson, K., 193, 196, 205, 207, 208
Angeles-Reyes, E., 24, 96
Antle, J. M., 154
APO, 142
Arens, P., 302, 328
Arunatilake, N., 211
Atinc, T.M., 271
Azam, Q.T., 151, 153

Baanante, C.A., 298, 308, 309
Bach, C.F., 204
Balisacan, A.M., 36,195
Ballabh, V., 337
Banik, A., 211
Bardhan, K., 35
Barker, R., 74
Bautista, R.M., 192, 196
Becker, G., 370
Bell, C., 21, 38, 39, 44, 46, 230
Benjavan, Rerkasem, 182, 185, 186, 187, 302
Berry, P., 195

Besley, T., 47
Bhagwati, J., 212
Bhalla, G.S., 100, 109, 202
Bhalla, S., 100, 109, 111
Bhatia, R., 298
Binswanger, H.P., 44, 73, 160, 336
Bird, R.M., 20
Blackgrun, H., 185
Blarel, B. 337
Blomquist, W., 174
Bloom, E. A., 151, 153
Blyn, G., 74
Bongaarts, J., 371
Bonny, S., 299
Boone, P., 235
Borenzstein, E., 9
Boserup, E., 110
Bosworth, B., 1, 7, 8, 9, 10, 11, 12, 16
Bouis, H., 162
Bourzat, D., 66
Boyce, J., 46
Braithwaite, J., 48, 247
van Breemen, N., 304, 305
Brooks, D. H., 183
Brooks, K., 233
Brown, L. R., 292
Bumb, B.L., 298, 308, 309
Burgess, R., 47
Buringh, P., 291
Byerlee, D., 24, 106, 153, 177

Cai, X., 227
Cao, Y.Z., 240
Canonero, G., 211
Carney, D., 181
Cassman, K.G., 154, 302

Chadha, G.K., 91,116
Chagnon, J., 36
Chambers, R., 305
Chen, E.K.Y., 6, 8
Chen, K., 8
Chen, S., 37
Choe, C., 230
Choi, J. S., 91
Chou, N. T., 301
Clark, J., 279, 280
Cohen, J.I., 314
Collins, S., 1, 7, 8, 9, 10, 11, 12, 16
Commandeur, P., 315
Corsetti, G., 257, 260, 262
Cotts Watkins, S., 371
Courbois, C., 167, 181, 182, 185, 186
Craig, B.J., 140, 151
Crook, F., 230
Crosson, P., 301, 302, 327, 328

Darwin, R., 318
Datt, G., 14, 32, 35, 37, 38
David, C.C., 77, 196, 197, 329
De Broeck, M., 233, 234
De Datta, S.K., 303
De Haan, A., 32, 33, 35, 36, 37
De Haan, C., 185
De Melo, N., 224
De Menil, G., 242, 243, 244, 245
Deb, N., 106
Deininger, K., 37, 44, 73
Delgado, C.L., 65, 167, 181, 182, 185, 186
DeRosa, D.A., 210
Desai, B.M., 153, 154
Desai, G.M., 174
Descalsota, J., 303
Dev, S.M., 154, 202
Devendra, C., 187
Dey, M., 149
Dimaranan, B., 188
Dogra, B., 305

Dollar, D., 195
Dorosh, P., 200, 201
Downing, T.E., 316, 317
Dregne, H.E., 301
Drysdale, P., 7
Dudal, R., 291

Ehui, S., 335, 341
Esty, D., 249
Evans, H.E., 106
Evans, P., 16
Evenson, R.E., 6, 149, 150, 151, 152, 153, 154, 311, 312, 314

Fabella, A., 228
Fan, G., 240
Fan, S., 44, 45, 78, 139, 140, 141, 143, 144, 147, 148, 149, 150, 153, 155, 156, 158, 159, 200, 324, 325, 327, 330, 332, 333, 341
Fane, G., 263, 281
FAO, 23, 164, 167, 168, 179, 216, 301, 309
FAOSTAT, 62, 63, 124, 126, 128, 129, 131, 132, 133, 135, 136, 137, 138
Farmer, B.H., 74
Feder, G., 44, 73, 230
Fforde, A., 228
Finance and Development, 269, 270
Fischer, S., 245
Fishlow, A., 10, 13
Francois, J. F., 207
Frankel, F. R., 74
Frankel, J. A., 210
Frederick, K.D., 225

Gaiha, R., 42
Garnaut, R., 257, 259, 264, 281, 283
Garrett, J.L., 324
GATT, 206
Gazmuri Schleyer, R., 73, 455
Gerpacio, R. V., 171, 302

Gibb, A., 21, 82, 108, 110
Goletti, F., 195
Gollin, D., 311, 312
Golubev, G. N., 227
Gomez, K. A., 303
Gonzales, L. A., 197
Graham, C., 246, 248
Green, D. J., 222, 223, 225, 228, 229, 240, 246
Griffin, K., 74
Grootaert, C., 247
Grossman, G. M., 8, 208

Haggblade, S., 21, 84, 87, 89, 94, 103, 105, 107, 108, 109, 110, 111, 117
Hakkeling, T. A., 300
Hamid, N., 231, 232
Hammer, J., 107
Han, K., 66
Hanstad, T., 230
Haque, T., 45
Harrington, L. W., 305
Hayami, Y., 118, 193, 196
Hazell, P., 21, 24, 42, 44, 45, 74, 75, 78, 84, 87, 89, 91, 94, 101, 102, 104, 105, 106, 107, 108, 109, 110, 111, 324, 327, 328, 329, 330, 334, 336, 340, 341, 344
Helpman, E., 8, 208
Henry, R., 183, 184
Herdt, R., 74
Hertel, T., 207
Ho, S. P. S., 87, 91, 111, 116
Hobbs, P., 155, 171, 302, 305
Hongladarom, C., 73
Hossain, M., 101, 106, 113, 171, 302, 326, 328
Hossain, S. I., 42
Huang, J., 155, 156, 162, 303, 304
Huang, Y., 7
Hyman, E.L., 117

IFAD, 35, 36, 44, 324
IFPRI, 355, 357, 360, 363, 367, 456, 459, 460, 461, 462, 463, 464, 469, 471, 472, 473, 476, 477, 479, 480, 482, 484, 486, 487
ILO, 74, 267, 270
IMF, 272, 286
Intal, J.P., 265
IPPC, 185
Islam, N., 67, 195, 216

Jalan, J., 48
Jamison, D., 42
Jayaraman, R., 113
Jha, D., 151
Jin, Z., 318
Johnston, B. F., 19, 24, 26, 110, 112
Jomo, K. S., 15
Joshi, V., 203
Josling, T.E., 209, 214, 215

Kada, R., 116
Kalirajan, K.P., 153, 154
Kane, Hal, 292
Karim, Z., 318
Kasryno, F., 146, 188, 197
Kathen, A., 313
Ke, B., 167
Keefer, P., 17
Kendall, H.W., 291, 298, 316, 317
Kennedy, E., 76
Kerr, J., 337
Khan, S.M., 211
Kilby, P., 22, 26, 110, 112
Kim, J., 5, 7
King, R.P., 24, 106
Kirsten, J., 336
Knack, S., 17
Knight, J., 78
Knowles, J.C., 267, 271
Kochlar, K., 285
Koo, W.W., 156

Korea Statistical Yearbook, 96, 97
Kostial, K., 233, 234
Krueger, A., 10, 12, 15, 212
Krugman, P., 5, 7
Kumar, P., 154, 155
Kumar, Vinod, 64, 65

Lanjouw, P., 113
Lau, L., 5, 7, 42
Lee, J.W., 14
Lee, Teng-hui, 20
Leiserson, M., 82
Leisinger, K.M., 313, 314, 315
Lele, U., 108
Lerman, Z., 233
Leuck, D., 310
Lewis, H., 210, 370
Lewis, J. D., 210, 370
Liedholm, C., 91
Lin, J. Y., 155, 231
Lindert, P. H., 196, 294, 303
Linnemann, H., 306
Lipton, M., 32, 33, 35, 36, 37, 40, 44, 46, 48, 74, 160, 176
Li Pun, H., 187
Liu, S.B., 187
Livernash, R., 313, 314, 315
Loungani, P., 285
Lucas, R.E., 8

Magrath, W., 302, 328
Malik, R. P. S., 298
Mao, W., 156
Marcoux, A., 249
Martin, W., 58, 204, 273
Mathai, G., 109
Matthews, R.B., 318
McBride, T., 311, 312
McDonald, B., 207, 298
McIntire, B. J., 66
McKibbin, W., 273

McLeod, R. H., 257, 265
McMillan, J., 155
Mcneely, J.A., 311
Mead, D. C., 103, 117
Mearns, R., 47, 336, 337
Meinzen-Dick, R., 295, 338
Mellor, J.W., 18, 19, 22, 24, 108, 110
Melmed,-Sanjak, J., 46
Meyer, Richard L., 69, 70, 71, 72
Michael, W., 271
Mingsarn Kaosa-ard, 181, 182, 185, 186, 187, 302
Minot, N., 195
Mishra, P.K., 340
Monke, E., 174
Moormann, F.R., 304, 305
Morris, M., 155, 171, 302
Mruthyunjaya, 155
Müller, K., 197
Mustafa, U., 304

Nagarajan, G., 69, 70, 71, 72, 101
Nagy, J.G., 151
Namboodiri, N.V., 153, 154
Neiss, H., 284
Nellor, D.C.L., 284
Noland, M., 273
Nordstrom, H., 207

Oldeman, L.R., 300, 301
Oram, P., 176
Osaka, H., 9
Oshima, H.T., 116
Ostrom, E., 338
Ostry, J.D., 9
Otsuka, K., 20, 111, 115, 118, 329, 337

Paarlberg, R., 294
Panagariya, A., 210
Pardey, P.G., 139, 140, 141, 143, 144, 147, 148, 149, 150, 151, 153, 156, 158, 159

Park, A., 39
Park, F.K., 116
Parker, S., 219, 222, 223, 239
Parry, M.L., 318
Pearse, A., 74
Pearson, S., 174
Pender, J., 337, 341
Penning de Vries, E. W. T., 306, 328
Perez, N.D., 146
Pernia, E.M., 267, 271
Persley, G.J., 313
Pesenti, P., 257, 260, 262
Pimentel, D., 291, 298, 316, 317
Ping, L., 230
Pingali, P.L., 65, 66, 154, 170, 171, 172, 175, 228, 302, 303, 308, 330
Pinstrup-Andersen, P., 74, 75, 329
Place, F., 341
Plucknett, D.L., 291, 307
Pomareda, C., 340
Poppele, J., 268
Postel, S., 295, 304
Pray, C., 151, 152, 154, 157
Pritchett, L., 268
Prosterman, R. L., 230

Quibria, M.G., 34, 36

Radelet, S., 257, 258, 259, 261, 283
Rae, A.N., 188, 189
Rahman, M., 211
Rahman, S. H., 202
Rajagopalan, V., 101, 102, 104, 107, 110
Rajapakse, P., 211
Ramasamy, C., 21, 74, 101, 102, 104, 107, 110
Rana, P.B., 222, 224, 229, 240
Ranade, A., 154, 202
Ranis, G., 24, 96, 267, 270
Rangarajan, C., 110
Rasmussen, L.N., 338

Ravallion, M., 32, 37, 38, 118
Reardon, T., 20, 91, 334
Reilly, J., 316
Ringler, C., 274
Robinson, S., 210
Rodrik, D., 10, 11, 13, 15, 16, 17
Roe, T.L., 25
Roell, A., 24
Roger, P.A., 175
Rohwer, J., 1
Rola, A.C., 175
von Roozendaal, G., 315
Roseboom, J., 139, 140, 149
Rosegrant, M.W., 6, 65, 66, 73, 146, 149, 151, 152, 153, 154, 155, 156, 157, 170, 171, 172, 173, 174, 175, 188, 274, 295, 296, 302, 303, 304, 308, 314, 330, 455
Rosenzweig, C., 318
Rothwell, G., 183, 184
Roubini, N., 257, 260, 262
Roumasset, J.A., 195
Rozelle, S., 39, 155, 156, 303, 304
Ryan, J., 91

Sachs, J.D., 10, 12, 257, 258, 259, 261, 283
Sahay, R., 245
Sahota, G. S., 7
Saith, A., 117
Samad, M., 304
Samarappuli, I. N., 302
San, N.N., 197
Sarris, A.H., 216
Scherr, S.J., 300, 301, 328, 336, 344
Schimmelpfennig, D., 317
Schultz, T.P., 370
Schwartz, S.J., 183
Seckler, D., 158, 296
Sen, Abhijit, 114
Sengupta, N., 211
Setboonsarng, S., 150

Severino, J.M., 255, 267, 268, 271, 272, 286
Shand, R.T., 153, 154
Sharma, M., 326
Shetty, S., 296
Singh, G., 202
Singh, I., 116, 337
Singh, R., 42, 100
Sinha, S., 40, 337
Slade, R., 21, 101, 104, 105, 107
Smale, M., 311, 312
Smalley, E., 187
Soesastro, H., 278, 282
Sombroek, W. G., 300
Song, L., 78
Squire, L., 37
Srinivasan, T. N., 34, 36, 203, 205, 206, 207, 211
Steinfeld, H., 185
Stevens, J., 204
Stewart, F., 24, 96, 267, 270
Stiglitz, J. E., 260, 282
Stoeckel, A., 273
Stone, B., 199
Stone, M. R., 285
Subbarao, K., 48
Sumarto, S., 268
Suyanto, S., 337
Svendsen, M., 170, 295

Tabatabai, H., 34
TAC, 324
Tangerman, S., 209, 214, 215
Taube, G., 224, 245
Teng, P. S., 328
Thapa, G. B., 149
Thiesenhusen, W., 46
Thimm, H. V., 177
Thomas, S., 67, 195, 216
Thomas, V., 253
Thorat, S., 44, 99, 333
Timmer, C. P., 19, 195, 215

Tomich, T. P., 19, 22, 26, 337
Tritt, G., 219, 222, 223, 239
Tuan, F., 200
Tyagi, D.S., 100
Tyers, R., 208

UN, 350, 352, 458
UNDP, 33
UNICEF, 268, 273
Unnevehr, L., 188
Uphoff, N., 338
USAID, 292

Valdés, A., 200, 201, 341
Van de Valle, D., 247
Van Lynden, G. W. J., 301
Van Zyl, J., 336
Vokes, R.W.A., 222, 223, 225, 228, 229, 232, 240, 246
Von Braun, J., 76, 77
Von Roozendaal, G., 315
Vyas, V. S., 109
de Vylder, S., 228

Wade, R., 257, 258
Walker, T., 91, 337
Wang, S., 39
Wang, Y., 253
Wanmali, S., 113
Warley, T.K., 209, 214, 215
Warner, A., 10, 12
Warr, P.G., 58, 258, 262, 265
WDI, 1, 2, 3, 23, 61, 221
Wei, S.J., 210
Wen, G.J., 230
Whalley, J., 155
Wiemer, 199
Wilkes, G., 311
Wodon, Q.T., 32, 42, 118
Wolfe, D., 316
Wong, C.P.W., 227
Woo, W.T., 219, 222, 223, 239, 240

Wood, S., 341
World Bank, 1, 7, 10, 12, 14, 16, 34, 37, 38, 41, 42, 43, 44, 54, 55, 140, 222, 240, 243, 244, 245, 249, 250, 266, 267, 268, 269, 272, 273, 276, 277, 279, 280, 281, 282, 285, 296, 299
World Bank/FAO/UNIDO/Industry Fertilizer Working Group, 309
WRI, 292
Wright, B.D., 311
WTO, 205
Wu, G., 39

Xuan, V. T., 228

Yadav, S., 73, 328, 455
Yao, Y., 230
Yeldan, E., 25
Young, A., 5, 16
Yudelman, 315

Zeigler, R., 328
Zettelmeyer, J., 224, 245
Zhu, L., 155

SUBJECT INDEX

A

AEZs. *See* agro-ecological zones
Africa, Sub-Saharan **291, 460-464, 466, 469, 471-473, 475-477, 479, 480, 485, 487**
Africa, West and North (WANA) **460-466, 469, 471-473, 475-477, 479, 480, 483-485, 487**
AFTA. *See* ASEAN Free Trade Area
Agreement on Technical Barriers to Trade **205**
Agreement on the Application of Sanitary and Phytosanitary Measures **205**
agriculture **18-21, 255, 272-275, 289, 290, 293-297, 315-319**
  CTEs, ETEs **223-237, 251, 252**
  decline in sector **58, 59**
  diversification **383**
  government expenditures **144**
  growth **21, 106, 123**
  protection of **192-197, 200, 201**
  reform, CTEs & ETEs **229-237**
  subsistence **34**
agro-ecological zones (AEZs) **73, 326, 327**
agroclimatic zones
  irrigated areas **324-326**
  rainfed areas **324-333**
agroindustrial sector **20**
AIDS (acquired immune dificiency syndrome) **43**
alkalinity **304**
America, Central **306**
America, North **306**

America, South **306**
Amu Darya (river) **226**
animal husbandry **161, 182**
animal raising **228**
APEC. *See* Asia Pacific Economic Cooperation
aquaculture
  inland **294**
  products **65**
Aral Sea **225-227, 249, 398**
arid lands **324**
arid zones **304**
ASEAN. *See* Association of South East Asian Nations
ASEAN Free Trade Area (AFTA) **209, 210**
Asia, Central **32, 33, 46, 48, 50, 51, 219-254, 379, 387, 388**
Asia, East
  food demand in **163-169, 179, 361, 457-462, 467-471, 474-478**
  food supply in **361, 457-462, 467-471, 474-478**
  growth, agricultural, in **60, 65, 137**
  growth, economic, in **5, 7, 8, 10-17, 378, 379**
  land reform in **46**
  policy, macroeconomic, in **190, 197**
  policy, trade, in **196, 197**
  poverty reduction in **29, 33**
  rural nonfarm sector in **94, 97**
  trade, agricultural **482, 484**
Asia Pacific Economic Cooperation (APEC) **209, 210, 278**

Asia, South
  agricultural expansion/loss in 293
  agriculture, structural changes in 59-63
  degradation, environmental, in 300
  fertilizer use in 308
  food demand in 163, 164, 166-168, 170, 360-362, 457-462, 467-471, 474-478
  food supply in 168, 170, 355-362, 367, 457-462, 467-471, 474-478
  gender, women, in 54
  growth, agricultural, in 59-63, 65, 130, 137, 139, 147-151
  growth, economic, in 3-5, 7, 21, 25
  investment, agricultural, in 158
  investment, public, in 158
  land reform in 46, 47
  malnutrition, child, in 366-371, 373, 483-485
  policy, macroeconomic, in 200-202
  policy, trade, in 200-202
  poverty in 29-33
  rural nonfarm sector in 84, 89
  trade, agricultural, in 482, 484
Asia, Southeast
  crisis, financial and economic, in 255-288
  degradation, environmental, in 300, 328
  fertilizer use in 308
  food demand in 164-168, 170, 360-362, 457-462, 467-471, 474-478, 482, 484
  food supply in 168, 170, 178, 179, 355-362, 367, 457-462, 467, 471, 474-478, 482, 484
  growth, agricultural, in 65, 130
  growth, economic, in 9, 11
  land, agricultural, expansion/loss in 293
  land reform in, 46

malnutirion, child, in 366-371, 373, 483-485
policy, macroeconomic in 196-200
policy, trade in 196-198
poverty reduction in 29-31, 34, 36
price stabilization in 194, 195
rural nonfarm sector 82, 84, 89, 94
trade, agricultural 482, 484
Association of South East Asian Nations (ASEAN) 188, 209, 210, 278

B

BAAC. *See* Bank for Agriculture and Agricultural Cooperatives
baht (Thailand) 259, 286
Bangladesh
  agriculture, importance of, in 21, 23
  agriculture, structural changes in, 59-63
  food demand in 459-464, 467-471, 474-478
  food supply in 459-464, 467-471, 474-478
  growth, agricultural, in 124-145, 158
  growth, economic, in 2-4, 12
  land reform in 46
  malnutrition, child, in 372, 373, 482, 484
  markets, rural financial, in 369-371
  policy, trade, in 201
  poverty reduction in, 33, 34, 36
  price stabilization in 194, 195
  research, agricultural, in, 148-150
  rural nonfarm sector in 83, 84, 108
  trade, agricultural 484, 484
bank closures 283
Bank for Agriculture and Agricultural Cooperatives 72
Bank Rakyat Indonesia unit desas

Indexes 495

(BRI-UD) **71**
bankruptcy-implementation procedures **389**
banks **282**
   agricultural development **69**
   closures **283**
barley **169**
basmati rice **201**
beef **165, 166, 178, 179, 188, 189, 363, 478, 482**
beef (production) **478**
BIMAS project, (Indonesia) **69, 70, 71**
biodiversity **323**
biophysical limits **290**
biosafety regulations **315**
biotechnology **187, 290, 312-315, 321, 404-406, 457**
   research **405**
   revolution **404**
breeding stock **188**
BRI-UD. *See* Bank Rakyat Indonesia
Brunei **210**
BULOG (Indonesia) **195**
bureaucracy
   efficiency **15**
   government **15**

C

Cambodia **328, 396**
canal irrigation **332**
CAP. *See* Common Agricultural Policy
capital **18, 19, 101, 245, 255, 260, 261, 281, 282**
   accumulation **5, 58**
   controls **281-283**
   human **12, 15, 41, 393**
   inputs **6**
   investment **69**
   markets **100, 260**

   nonequity inflows **283**
   private **1**
   reserves **262**
   short-term international **257**
capital-adequacy standards **283**
capital-asset ratios **262**
carbon dioxide (CO_2) **316-319**
cash crops **380**
cassava **302, 363**
cement **20**
Central America **307, 317**
Central Asia **47, 49, 51**
Central Asian Transition Economies (CTEs) **219-250**
cereals **65, 161-190, 294, 347, 458, 466-474, 481, 483**
   area **354, 470**
   consumption **365, 458**
   demand **162, 163, 274, 464**
   feed demand **165, 462**
   food demand **162-164, 460**
   growth rates **468**
   imports **356, 483**
   markets **347**
   net trade **485**
   prices **368**
   production **169, 347, 348, 350-366, 466-474, 481, 483**
   projections **458**
   trade **352, 356, 357, 363, 364, 482**
   yield **354, 472**
CGE. *See* computable general equilibrium model
CGIAR. *See* Consultative Group on International Agricultural Research
chaebols **261**
child labor **271**
Chile **282-283**
China. *See* PRC (People's Republic of China)
Chuan Leekpai, (Thailand) **265**
cities **293**

climate, changes in 315
greenhouse gases 316
CMEA. *See* Council for Mutual Economic Assistance
coconut 406
collectivization agricultural 224-229, 231, 232
commercial logging 250
commercialization agricultural 64-80, 383-385
Common Agricultural Policy (CAP) 214
computable general equilibrium (CGE) 26, 273-275
Consultative Group on International Agricultural Reseach (CGIAR) 326
contract procurement system 230
cooperatives 233
financial 66-72
copper 185
corn 302
corporate governance 277, 281
corrupt lending practices 263
corruption 402, 253, 278
cotton 201, 227, 245, 303
Council for Mutual Economic Assistance (CMEA) 223, 234
covariant risk 247
CPI. *See* Consumer Price Index
credit 382, 13, 14, 39, 188, 197
allocation 70, 263
contraction 285
guarantees 392, 69
institutions 25
markets 339
policies 237
programs 52
rural 342
services 44
subsidized 393, 70
crisis, Asian financial and economic 40, 72, 255-288, 379, 380, 401, 402
cropland 250, 292
expansion/loss 290-294
crops 65, 146, 291
area 290-294, 347
improvement 80
insurance 340
yields 130
CTEs See Central Asian Transition Economies
Cultural Revolution of 1966–76 (PRC) 223
currency
depreciation 272-273

D

dairy products 162, 188, 189, 458
food demand 162
production 359-361, 364, 365
trade 361, 362, 364
debt foreign 257
decentralization 403, 177, 279, 297
decentralization of gov't services 279
deficits 257, 259
budget 11
fiscal 67
deforestation 396, 304, 323
degradation
environmental 171, 172, 189, 289, 299-305, 309, 310, 319, 324, 395
land 299-305, 319-323, 327, 328, 352, 395
water quality 226, 227
dekhan (private farms) (Uzbekistan) 236
demand
linkages 385
deposit insurance schemes 263
depreciations 272
development

economic 57, 323, 324, 328-333
disease
  sexually transmitted 43
diversification 57, 64-80, 263
  agricultural 64, 74-78, 383-385
  economic 20, 334, 343
  effects on poverty 74-78
Doi Moi (Renovation Policy) (Viet Nam) 223
duties
  import 191

E

East and Southeast Asian Transitional Economies (ETEs) 219, 220, 223, 224, 226, 228, 229, 232, 237, 238, 246, 247, 249, 251, 253, 387
East and Southeast Asian countries 365
East Asia 348
East Asian NIEs 12, 15
Eastern Europe 223, 233, 247, 248, 308, 465
economic reform 349
economies of scale 335
economy
  rural nonfarm 20, 21, 81-121, 386-390
ecosystems 295
education 1, 16, 36, 41-43, 45, 79, 80, 266, 270-272, 277, 331, 341, 378, 383, 393, 402, 457
  farmer 177
  instruction 16
  investment in 15
  reforms 277
efficiency 15
effluent or pollution charges 297
eggs 457
electrification 385
  rural 45, 112, 331, 332, 385, 395

employment 267-277
and unemployment 267, 268
  rural 20, 83
  self 34
  shares 82, 85, 86
  women's 87-90, 107, 120
energy 241, 269
  demand & consumption 297-299
  prices 299
Engel's Law 58
environment 289-321
  degradation 156, 395
  externalities 399
  problems 154
  protection of 249
environmental degradation 156
equitability 35, 39, 246, 336-338
erosion 156
ETEs 387. See East and Southeast Asian Transitional Economies
ethnic tensions 265
Europe, Eastern 223, 233, 247, 248, 310, 460-465, 469, 471-473, 476, 477, 479, 480, 483-485
European Commission 214
European Union 214
eutrophication 310
evapotranspiration 304, 317
exchange foreign 67
exchange rates 10, 11, 191, 192, 198
  appreciation of 198, 201, 256, 258, 259, 285-287, 389
  controls 389
  depreciation 272-275, 288
  foreign
    appreciation of 389
    depreciation of 389
  regime 219
  misalignment 256
export-promotion policies 12
exports
  declines 258, 259

extensification 250
extension
  agricultural 45, 176, 177, 384, 385
  agricultural, decentralization, of 177
  agricultural, privatization, of 177

F

FAIR. *See* Federal Agriculture Improvement and Reform Act
family planning 34
FAO. *See* Food and Agriculture Organization
FDI. *See* foreign direct investment
Federal Agriculture Improvement and Reform (FAIR) Act (US) 214
feed demand 466, 467
feedgrains 214
feeding programs 43
fertility 371
  rates 366
fertilizer 20, 24, 66, 130, 132, 134, 144-147, 174, 192, 197, 298, 302, 303, 307-310, 319
  demand 308, 309
fiscal
  deficits 244, 389
  discipline 379, 18
  stabilization 244
fisheries 231
flooding 396, 339
food
  availability 367, 370
  demand 64, 345, 347, 348, 353-366, 370, 457, 458, 465-468
  processing 380, 252
  production 289, 321
  security 345, 347, 353, 371, 374, 375, 380
  self-sufficiency 18
  subsidies 49

  supply 18, 24, 345-371, 374, 375, 409, 457, 458
  trade 481-485
Food and Agriculture Organization (FAO) 324
food demand 162, 345, 466
  changes in 465
  prices 162
foreign direct investment (FDI) 14
forest 291
  degradation 250
fragile soils 334
France 299
fruits 231, 236
fuelwood consumption 250

G

gas-induced 318
GATT. *See* General Agreement on Tariffs and Trade
GDP. *See* gross domestic product
GDP per capita 4, 5
gender 34, 35
  differences 35
  women 34, 35, 41, 42, 87-90, 107, 120, 272, 276
gene banks 311
gene pool 315
General Agreement on Tariffs and Trade (GATT) 203-209
genetic engineering 313
  manipulation 307
  mapping techniques 313
  resources 406, 407
  ex situ 311
  in situ 311
geographic targeting of regions 52
germplasm 311
global warming 290, 316-319
globalization 40, 298
  economic 407-409

goat **165, 167, 178, 179, 478, 482**
gold **245**
governance **15, 255, 278-281, 287, 288, 378, 401-403**
  corporate **388**
grains See also barley, millet, oats etc. **460, 462**
  coarse **164, 168, 169**
  millet **169**
  oats **169**
  prices **231**
  production **231**
Grameen Bank, (Bangladesh) **71**
Great Depression (US) **223**
"green box" policies **206**
green revolution **28, 39, 74, 104, 134, 298, 299, 378, 382**
gross domestic product (GDP) **2, 3, 5, 11, 37, 49, 51, 58, 59, 61, 123, 140-143, 202-203, 207, 219, 220, 224, 254, 255**
  decline in, **255**
grounduts **302**
groundwater **174, 297**
growth **36, 127**
  agricultural **18-28, 36-39, 57, 106, 112, 123-160, 220, 289, 328, 329, 334, 335, 378-380**
  animal numbers **352**
  economic **21, 36, 37, 39, 81, 191, 289, 377-383**
  rates **2, 3**
  rural nonfarm **81-121**
  sources of **104-118**

H

Havana conference, 1947-1948 **203**
health care **34, 36, 42, 43, 76, 90, 80, 270-272, 275-277, 383, 393, 402**
  budgets **276**
  centers **276**

improvements **32**
  insurance **270**
  programs **276**
  schemes **271**
  services **43**
  status **266, 288**
Heavy Industries Corporation of Malaysia (HICOM) **14**
HICOM. See Heavy Industries Corporation of Malaysia
high-potential areas **40, 329, 330, 344**
high-risk zones **320**
high-yield varieties (HYVs) **34, 110, 331, 332**
holistic approach **336**
Hong Kong, China **2, 4, 6, 11-14, 18, 22, 27**
"hot spot" **328**
household plots **236**
households
  female headed **35**
human capital **381**
  formation **393, 271**
hybridization **457**
hyperinflation **220**
HYVs. See high-yield varieties

I

IARCs. See International Agricultural Research Centers
IFPRI. See International Food Policy Research Institute
ILO. See International Labor Organization
IMF. See International Monetary Fund
immunization **51**
IMPACT. See International Model for Policy Analysis of Agricultural Commodities and Trade
imports

duties on 191
quotas 191
income 21, 37-40, 58, 85, 87, 93, 104, 105, 110, 265, 268
   data 85
   distribution 37
   growth 349
   shares 85, 86
   sources of 20
income-based identification 53
income-generation programs 52
India
   agriculture, importance of, in 21, 23, 25
   agriculture, structural changes in 61-64
   degradation, environmental, in 304, 305
   food demand in 457-462, 467-471, 474-477
   food supply in 457-462, 467-471, 474-477
   growth, agricultural, in 124-145
   growth, economic, in 2-4, 107-110, 379
   land, agricultural, expansion/loss in 293
   land reform in 46, 47
   less favored areas (LFAs) in 324-327, 329-333
   malnutrition, child, in 369, 372, 373, 484, 485
   markets, rural financial, in 371
   policy, development, in 337
   policy, economic, in 115, 116
   policy, trade, in 12, 202, 203
   poverty, policies for 43, 45-47, 49, 51, 54
   poverty reduction in 30-35, 35, 37
   poverty, rural 35, 36
   productivity, labor 137-139
   productivity, land 133-136
   research, agricultural, public, in 148-150
   research, agricultural, private, in 157, 158
   rural nonfarm sector in 82, 85, 88, 91, 94, 98, 99, 105, 107, 110
   trade, international, in 479-484
Indonesia
   agriculture, structural changes in 59-63
   crisis, economic and financial, in 255, 256, 259-261, 265, 267-273, 284, 285
   degradation, environmental, in 302, 328
   food demand in 457-462, 467-471, 474-477
   food supply in 457-462, 467-471, 474-477
   governance in 17
   growth, agricultural in 124-145, 308
   growth, economic, in 2, 4, 7, 21-23
   livestock production in 188
   malnutrition, child, in 372, 396
   policy, macroeconomic, in 11, 197
   poverty, reduction in 31, 33, 35, 37, 43, 46, 53
   research, agricultural, private, in 158
   research, agricultural, public in 149, 150
   rural financial markets in 69-71, 108
   rural nonfarm sector in 83-85, 93-95, 98, 99, 114
   trade, agricultural, in 482, 484
industrial estates 115
industrial sector 19
industrialization 20, 21, 24, 180, 186, 222
   rural 115

rural policies **115**
inequality **246**
infant mortality **381**
inflation **11, 285, 286**
infrastructure **1, 12, 24, 34, 38, 44-46, 64, 66, 72, 110, 112, 113, 170, 323, 332, 333, 378, 379, 385-386, 457**
  rural **24, 44, 66, 72, 112, 382, 383**
inheritance rights **231**
institutions
  public, reform of **342**
insurance
  agricultural **340**
Integrated Child Development Services (India) **49**
Integrated pest management (IPM) **175**
Integrated Rural Development Program (India) **71**
intellectual property rights **314**
intensification
  agricultural **64, 77, 110, 170, 334, 344, 395, 396, 400, 401**
International Agricultural Research Centers (IARCs) **314, 315, 405**
international commodity markets **321**
international commodity prices **353, 363**
  cereals **353**
  livestock products (plus milk) **353**
  roots and tubers **353**
International Convention on Biodiversity **407**
International Crops Research Institute for the Semi-Arid Tropics **405**
International Food Policy Research Institute (IFPRI) **274, 325, 344, 455–484**
International Labor Organization (ILO) **267**

International Model for Policy Analysis of Agricultural Commodities and Trade (IMPACT) **274, 345, 346, 353-371, 456-487**
International Monetary Fund (IMF) **283-287**
International Rice Research Institute (IRRI) **303, 405**
International Trade Organization **203**
inventions **157**
investment **10, 25, 26, 324**
  agricultural **144, 145, 154, 160, 320, 348-354, 458, 459**
  decline in **123, 154**
  foreign **184, 407, 408**
  international **256-260, 262-266**
  less favored areas **324, 328-333, 340-341**
  private **156-159, 384**
  public **25, 66, 141-144, 146, 320, 330, 350, 381-394, 457-459**
  social **350-352, 366, 370, 373**
  technological **110**
  water **351**
IPM. *See* Integrated pest management
IRRI. *See* International Rice Research Institute
irrigation **45, 197, 295, 301, 319, 331, 332, 350, 396, 457**
  system efficiency **297**
tubewell **156-158**

J

Jakarta **268**
Japan
  food demand in **457-462, 467-471, 474-478**
  food supply in **457-462, 467-471, 474-478**

governance in 17
growth, agricultural, in 25
policy, industrial, in 13-15
rural nonfarm sector in 111, 115, 116
tariffs in 206
trade, agricultural 482, 484
Java 302, 308
Jin Tanakan Mai (New Economic Policy) (Lao PDR) 223
jute 20

K

Kazakhstan 219
poverty reduction in 33
reform, economic, in 219-222, 233, 234, 241, 244
kilocalories 366
kolkhoz (collective farm) 225, 236
Korea, Republic of 382, 59, 255, 286
agriculture, importance of 21, 23, 25, 27
agriculture, structural changes in 59-63
crisis, economic and financial in 255, 256, 258-261, 263, 25-267, 270, 272, 284
education in 15
fertilizer use in 308
food demand in 457-462, 267-271, 474-478
food supply in 457-462, 267-271, 474-478
growth, agricultural, in 124-145, 148-150
growth, economic, in 2-4, 6-9
land reform in 46
policy, industrial, in 13, 14
policy, macroeconomic, in 11, 114, 193

policy, trade, in 12, 13
rural nonfarm sector in 87, 91, 93-97, 114, 116
tariffs in 206
trade, agricultural 482, 484
Kyrgyz Republic 204, 219
poverty reduction in 33
reform, economic, in 219-222, 224, 232, 241, 244
safety nets in 51

L

labor 18, 19
agricultural 132
force 60, 62, 63
inputs 6
opportunity cost of 66, 79
productivity growth 134
"pulled out" of farming 97
"pushed out" of agriculture 100
rural 26, 27, 39, 65, 66
land 75, 290, 321
agricultural 323-344
distribution 336, 337
distribution equitability 39
expansion/loss 290-294
irrigated 34
landlessness 336
ownership 388, 34
property rights in 386-388, 396
rainfed 301
redistribution 382
reform 25, 46, 47, 335-336
resources 290
rights 387
land reform 46-48, 336
land to labor ratio 128
land-tenure insecurity 230
Lao PDR. *See* Lao People's Democratic Republic (PDR)

Lao People's Democratic Republic (PDR)
poverty reduction in 36
reform, economic, in 219-223, 228, 229, 231, 238, 240, 242, 244
state-owned enterprises in 238, 240
Latin America 291, 459-466, 469, 471-473, 476, 477, 479, 480, 484, 485, 487
law, rule of 15
lending requirements 69
less favored areas (LFAs) 323-344
investment in 386, 395-398, 400, 401, 405
LFAs. *See* less favored areas
liberalization 383, 66, 67, 219
agricultural 390
economic 40, 67, 389-391
trade 341, 342, 389, 408, 409
licensing 120
life expectancy 248
linkages
demand 24
economic 380, 119
liquidity injections 284
literacy 15, 32, 152, 153, 248, 277
livestock 65, 161-190, 347, 359-362, 474-483
cattle 478
chicken 478
consumption 161-190
demand 476
food demand 161, 162, 180
markets 364
policies 188
pork 364
production 179, 180, 187, 347, 396, 397, 478
products 125, 236, 480, 483, 487
supply 161
trade 359-361
yield 351

loans 272
interest rates 69
nonperforming 262

M

maize 164, 167-169, 198, 317, 318, 363, 460, 462, 466, 467, 470, 474, 481, 483
malaria 43
Malaysia
agriculture, importance of, in 21, 23
agriculture, structural changes in 59-64
crisis, financial and economic, in 255-261, 272, 274, 280
degradation, environmental, in 328
food demand in 457, 462, 467-471, 474-477
food supply in 457, 462, 467-471, 474-477
growth, agricultural, in 124, 125
growth, economic, in 2-4, 7, 9
livestock production in 184, 185, 188
malnutrition, child, in 373
policy, industrial, in 14
policy, macroeconomic, in 198
policy, trade, in 198
poverty reduction in 30, 31, 33, 34, 36, 53
productivity, agricultural, in 148-150
rural nonfarm sector in 101, 105, 107
trade, agricultural 479, 480, 484
malnutrition 412, 266, 269, 353, 373
child 326, 348, 366-375, 410, 485, 487
manufacturing 20, 21, 23
manufacturing sectors 267

markets **382, 321, 332, 457**
  capital **97**
  cereal **347, 348, 353-366**
  credit **39**
  financial **255, 257**
  labor **97**
  output **39**
  product **39**
  rural **97**
  rural financial **66, 68-72, 80**
Masagana 99 (Philippines) **69**
meals **457**
meat **166-178, 347, 348, 359-362, 364, 365, 457, 458**
  consumption **347, 359, 365, 474-477**
  demand **165, 474-477**
  production **359, 478**
mechanization **382**
  agricultural **133, 134**
Mekong River **228**
Mekong River Basin **36**
Mercado Comun del Sul, (MERCOSUR) **209**
MERCOSUR. *See* Southern Common Market
MFOs. *See* microfinance organizations
microfinance **117, 119, 120**
  organizations (MFOs) **71**
micronutrient deficiency **303**
migration **20, 41, 269**
  out-migration **20, 323, 334, 343, 400**
milk **457**
  consumption **362, 365**
  production **361**
millet **169, 317**
mining **241, 294**
monetary policies **283**
money supply **283**
  growth **244**
Mongolia **219, 234–35**

poverty reduction in **33**
reform, economic, in **219, 224, 229, 232, 234, 235, 246, 248**
monoculture systems **302**
"moral hazard" **260**
morbidity **394**
Myanmar **59, 210, 292, 308**
  food demand in **457-462, 467-470, 474-478**
  food supply in **457-462, 467-470, 474-478**
  growth, agricultural, in **124, 125, 148-150**
  growth, economic, in **2-4, 59, 61-63**
  trade, international in **479-484**

N

NAFTA. *See* North American Free Trade Agreement
NARs. *See* National Agricultural Research Institute
National Agricultural Research Institutes (NARs) **314**
National Food Authority (Philippines) **195**
natural resources **346**
neem **406**
Nepal
  agriculture, structural changes in **61-63**
  degradation, environmental **305**
  fertilizer use in **158**
  growth, agricultural in **124-145, 148, 150**
  growth, economic, in **2-4**
New Economic Mechanism (Lao PDR) **223**
newly industrializing economies (NIEs) **11**
NGOs. *See* nongovernment

organizations
NIEs. *See* newly industrializing economies
nitrogen
 availability 305
non-tariff barriers 205
nonagricultural sector 19, 81-121, 319
nonfarm
 activities 103
 economy 81-121
 employment 85, 90
 goods 379
 growth rates 94
 income shares 87
 sector 81-121
 services 379
nongovernment organizations (NGOs) 53, 118, 279, 280, 339, 344, 401-403
noninterventionist policies 13
North America 292, 310
North American Free Trade Agreement (NAFTA) 209
North China Plain 294, 304
nutrients 300, 309
nutrition 43, 76, 80, 383

O

oats 169
OECD. *See* Organization for Economic Cooperation and Development
Open Economic Zones 243
"open" economies 13
open trade policies 378, 379
Organization for Economic Cooperation and Development (OECD) 27
"orphan" commodities 160
output price policies 396

"overindustrialization" 224
oversight
 financial institutions 72, 200, 260-263, 280, 281
ownership rights 339

P

paddy rice 169
Pakistan
 agriculture, importance of, in 21-23
 degradation, environmental, in 304-305
 food demand in 457-462, 467-471, 474-478
 food supply in 457-462, 467-471, 474-478
 growth, agricultural, in 59-63, 124, 125, 152, 153
 growth, economic, in 2-4, 7
 land, agricultural expansion/loss in 293
 malnutrition, child, in 372, 373
 policy, macroeconomic, in 196, 201
 policy, trade, in 12
 poverty reduction in 32, 33, 46, 54
 research, agricultural in 148-152
 rural nonfarm sector in 83, 84, 88
 trade, agricultural 482, 484
parastatals 402
patents 406
People's Democratic Republic (Lao PDR) 219, 223
peso (Philippines) 286
pest control 175, 338
pesticides 24, 174, 175, 192, 298
Philippines
 agriculture, importance of, in 17, 23, 25
 agriculture, structural changes in 59-63

crisis, economic and financial, in 255, 256, 259-261, 265, 272, 274, 285
fertilizer use in 308
food demand in 457-462, 467-471, 474-477
food supply in 457-462, 467-471, 474-477
growth, agricultural, in 124, 125, 148-150
growth, economic, in 2-4, 7, 379
land, agricultural, expansion/loss in 293
policy, macroeconomic, in 196, 198, 199
poverty reduction in 30, 31, 34, 36
price stabilization in 194, 195
research, agricultural, private, in 158
rural financial markets in 69
rural nonfarm sector in 83, 88, 93-96, 98, 99, 108, 117
trade, agricultural 484, 484
policies
agricultural 348-354
crop and livestock 186-188
financial 262, 280, 281
fiscal 10, 283
industrial 10, 13-15, 117, 262
irrigation 172
macroeconomic 19, 40, 67, 113, 191-218, 242-245, 255, 263, 378, 389, 390
macroeconomic reforms 113
market-oriented 11, 120
market 11, 12, 19, 25
monetary 283-287
price 191-218
protectionist 13
trade 1, 10-12, 19, 40, 113, 120, 191-218, 242-245
Ponggol Pigwaste Plant (Singapore) 185

population 248, 349
childhood 366
density 386, 110
growth 350
pressures 289
reduction 370
rural 98
pork 165-167, 178, 179, 189, 475, 478, 482
potatoes 232, 363, 364
poultry 65, 165-167, 178-180, 183-186, 363, 475, 478, 482
production 478
technology 385
poverty 29-55, 81, 91, 103, 104, 268-272, 287, 412
absolute 34, 40
in transition economies 246
level 268
reduction 29, 30, 32, 36-39, 324, 325, 328-333, 381
reduction, pace of 30-32
rural 33-40, 54, 324
"severe poverty" 36
transient 48
transitional 246, 248
urban 34, 38, 0
PPPs. *See* purchasing power parities (PPPs)
PRC (People's Republic of China)
agriculture, structural changes in 59-63
degradation, environmental, in 303-305
food demand in 167, 183-186, 457-462, 467-471, 474, 478
food production in 355, 357, 358, 360
food supply in 183-186, 457-462, 467-471, 474, 478
growth, agricultural, in 21, 23, 25, 124, 125, 148-150

Indexes 507

growth, economic, in 2-4
health care in 43
land, agricultural, expansion/loss in 293-295
land reform in 46
less favored areas (LFAs) in 325, 328
livestock demand in 167
livestock supply in 183-186
malnutrition, child, in 359, 367, 483, 485
markets, rural financial, in 69
policy, macroeconomic 199
poverty reduction in 30, 31, 33
price stabilization in 194, 195, 242
research, agricultural, in 148-150
reform, ecopnomic in 219-223, 226, 227, 229-231, 238-240, 250, 252
rural industrialization in 115
rural nonfarm sector 115
trade, agricultural, in 482, 484
trade, international, in 204, 205
prices
  controls 244
  distortions 240
  food 20
  liberalization 235
policies 191, 196
  reforms 232, 321
  stabilization 194, 195, 242-245
private plots 227
private-sector
  research 157
privatization 177, 219, 297
product markets 382
productivity
  agricultural 39, 43, 123, 289, 300-303, 305-319, 330, 351
  cereal-crop 169-171
  labor 6-9, 27,43, 134-139, 327
  land 134-136
property rights 46-48, 66, 72, 73, 76, 80, 159, 219, 230, 237, 253, 303, 315, 321, 383, 386, 387, 406
  communal 388
  intellectual 406
  usufructuary 231
protection
  agricultural 192, 194-197, 200, 201
  industry 192
prudential regulation 281
public
  expenditures 140, 144
  intervention 28
  sanitation 350, 351
Public Distribution System (India) 49
public research 383
purchasing power parities (PPPs) 139-141

Q

quotas
import 191

R

R&D. *See* research and development
R&E. *See* research and extention
rainfed areas 301, 327
rationing 49, 244
reform
  corporate 280, 281
  economic 378
  education 277
  financial 264, 265, 280, 281, 407
  institutional 277-279, 288, 402, 403
  macroeconomic 383
  markets 341, 342
  policy 48, 348
regional disparities 395, 77
regional trade agreements (RTAs) 209
regulation 159

financial institutions 72, 260-263, 280, 281
religious institutions 403, 279
research
  agricultural 45, 66, 147, 151-155, 176, 187, 307, 339, 340, 383-385, 403-406
  livestock 187
  research and development (R&D) 157
research and extention (R&E) 45
resources
  agricultural 335, 336
  conservation 311
  plant genetic 290, 310-315
revolution
  agricultural rural 377
  rice 161, 164, 168-171, 201, 214, 295, 318, 363, 364, 460, 462, 466, 467, 474, 481
Rice Biotechnology Network 405
risk management 339
roads 45, 112, 225,, 331, 332, 385, 386
  rural 112
Rockefeller Foundation 405
roots and tubers 358, 363, 456, 458
  production 354, 355
  trade 358, 359, 364, 365
RTA. *See* regional trade agreements
rubber 273, 302
rupee (Pakistan) 331
rupiah (Indonesia) 286
rural
  towns 111
rural financial markets 383, 391
rural nonfarm
  enterprises 231
  sector 81-121
Russian Federation 204
rye 169

S

SAARC. *See* South Asian Association for Regional Cooperation
safety-net programs 48-50, 270-272, 275-277
SAFTA. *See* South Asian Free Trade Area
salinization 156, 172, 227, 296, 301, 302, 304, 305
SAMs. *See* Social Accounting Matrices (SAMs)
sanitation 80, 276, 457
SAPTA. *See* South Asian Preferential Trading Arrangement
SARC. *See* South Asian Regional Cooperation
savings 1, 19
semiarid zones 304
service sector 19, 20
SEZs. *See* Special Economic Zones
sheep 165-167, 178, 179, 363, 478, 482
Singapore 6, 7, 9, 11-13, 15, 17, 18, 22, 27, 185
Social Accounting Matrices (SAMs) 102
SOEs. *See* state-owned enterprises
soil
  chemistry parameters 303
  depth 302
  erosion 300
  fertility 323
  fertility decline 302
  moisture 336
  nitrogen 303
salinity 153
salinization 300
  toxins 305
Soil Degradation in South and Southeast Asia (ASSOD) 300
sorghum 169, 317

South America 307
South Asian Association for Regional Cooperation (SAARC) 209, 211
South Asian Free Trade Area (SAFTA) 211
South Asian Preferential Trading Arrangement (SAPTA) 209, 211
Southeast Asia 255, 295, 317
Southern Common Market (MERCOSUR) 209
Soviet Union 223, 225-228, 233, 235, 246-248, 251, 304, 308, 379, 483 *See also* Russian Federation
sovkhoz (state farm) 225, 236
soybeans 317, 458
Special Economic Zones (SEZs) 239
Sri Lanka
  agriculture, structural changes in 61, 63
  degradation environmental 302
  growth, agricultural, importance of 21, 23
  growth, agricultural in 124-145, 148-150
  growth, economic in 2-4
  health and nutrition 43
  policy, macroeconomic 196
  policy, trade in 12
  poverty, reduction in 49, 54
  rural nonfarm sector in 83, 84, 88
  water supply 295
state-owned enterprises (SOEs)
  reform of 237-242, 252, 388
Sub-Saharan Africa 291, 354, 466
subsidies 49, 115, 116, 120, 146, 188, 224, 309, 342, 382
  agricultural input 197-203
  explicit 238
  implicit 238
  negative effects 145
subsistence agriculture 270
sugar cane 201, 317

support services 80
surpluses
  agricultural 19
Syr Darya (river) 226, 227

T

Taipei,China 382, 46
  growth, agricultural, importance of 20, 25, 27
  growth, economic, in 2, 6, 9, 11, 12-15, 19
  land reform in 46
  rural nonfarm sector in 91, 111, 116
  trade, international 204
Tamil Nadu Nutrition Program (India) 49
tariffs 203, 283
  agricultural 203
  "dirty tariffication" 203, 206-208
taxation 198, 203-208
  agriculture 193, 206, 207
  benefits 391, 116
  breaks 115
  deductions 277
  incentives 282
  laws 280
  of agriculture 193
  pasture 189
  subsidies 13
technologies
  agricultural 177, 335, 336, 384, 385
  change 74
  package 39
  scale-neutral 75
  transfer 184
telecommunications 386, 112, 241, 341
temperature
  increases in 316-319
terms of trade
  agricultural 19

textiles 20
TFP. *See* total factor productivity
Thailand
  agriculture, structural changes in 59-63
  crisis, economic and financial, in 267, 269-273, 284
  currency crisis in 263
  education in 15
  fertilizer use in 308
  food demand in 457-462, 467-471, 474-478
  food supply in 457-462, 467-471, 474-478
  growth, agricultural, in 124-145, 148-150
  growth, economic, in 2-4, 9
  land, agricultural expansion/loss in 293
  land reform in 46
  malnutrition, child, in 373
  markets, rural financial, in 71
  policy, macroeconomic, in 11, 198
  policy, trade, in 13
  poverty reduction in 30, 31, 33, 34, 36, 43, 46, 53
  productivity, economic, in 7
  rural nonfarm sector in 83, 84, 87, 88, 91, 108
trade, agricultural 482, 484
tiger economies 11
total factor productivity (TFP) 5-9, 45, 149-155, 299
  source of growth 151-155
towns
  rural 20, 81, 395
"township-village enterprises" (TVEs) 115
toxic wastes 226
tractors 132
tradable property rights 297
trade 68, 89, 191, 196, 333, 455

agricultural 214, 215
agricultural commodities 363
controls
  quantitative 243
  liberalization 11, 244, 318, 398
  restrictions 389, 191
trade agreements
  regional 209-213
trade-related intellectual property (TRIPs) 205
trade-related investment measures (TRIMs) 205
training 120
transformation 400, 412
  economic 1, 9, 10, 19, 21, 28, 81, 377
  rural nonfarm 81
transport
  subsidies 240
transport and communications 385, 386
transportation 80
  prices 269
tree crop products
  cocoa 125
  palm oil 125
  rubber 125
trickle-down benefits 329
TRIMs. *See* trade-related investment measures
TRIPs. *See* trade-related intellectual property
tubewells 156-158
Turkmenistan 33
turmeric 406
TVE. *See* "township-village enterprises"

U

UN. *See* **United Nations**
underemployment 267
unemployment 288

benefits 50
insurance programs 277
urban 248
UNICEF. *See* United Nation's Children's Fund
United Nations (UN) 350, 352
United Nations Children's Fund 268
United States 130, 134, 173, 174, 206, 213, 215, 304, 459, 475, 483
  Great Depression 214, 223
UR. *See* Uruguay Round
urbanization 162, 290, 291, 292
Uruguay round (UR) 203-207, 209, 216, 217, 390
user fees 412
Uzbekistan 219, 236
  poverty in 33
  reform, agricultural, in 232, 235, 236, 241, 245
  reform, ecomonic, in 219, 221, 222, 224, 232

V

vaccinations 271, 276
value added
  agricultural 107, 108
vegetables 65, 232, 236, 303
Viet Nam
  agriculture, structural changes in 59-63
  degradation, environmental, in 328, 396
  fertilizer use in 308
  food demand in 457-462, 467-471, 474-478
  food supply in 457-462, 467-471, 474-478
  growth, agricultural, in 124-145
  growth, economic, in 2-4
  land reform in 46
  poverty reduction in 246, 247

price stabilization in 194-195
reform, economic, in 219-223, 228, 229, 231, 238, 240, 242, 244
trade, agricultural 482, 484
Virgin Lands (Soviet Union) 235

W

wages 267
  farm 100
  nonfarm 100
WANA. *See* **West Asia and North Africa**
waste 185
waste-dilution 227
waste-disposal 396
water
  agricultural 294-297, 319, 320
  allocation 172
  availability 172, 173, 457, 458
  demand 173, 294-297, 319, 320
  endowments 332
  ground 296, 297
  nontraditional sources 294
  pricing 172, 173
  quality problems 396
  rights of users 173
  supply 294-297, 319, 350-352, 398, 399, 402, 458, 459
  surface 296, 297
  systems 297
waterlogging 153, 173, 300-305
watershed regions 396
watershed-development projects 333
welfare benefits 252
welfare effects 270
West Asia and North Africa (WANA) 348, 465
Western Arabia 317
Western Europe 310
wetland ecosystems 250
wheat 164, 167-169, 201, 214,

295, 317, 318, 348, 460, 462, 465, 467, 474, 481
wheat varieties 384
"win-win" strategy 45
wind erosion 300
women 87, 89, 103, 120, 272
  access to services
  basic education 55
  family planning 55
  health services 55
  nutrition 55
  sanitation 55
water 55
  farmers 335
  female-headed household 248
  property rights 48
  status 55
woodland 291
World Trade Organization (WTO) **203-209**, 217, 390
WTO. *See* World Trade Organization

Y

yield insurance 340
yields **404**, 326

Z

zamindari (India, Pakistan) **47**
zinc **185**